AF388901

Enological Repercussions of Non-*Saccharomyces* Species 2.0

Enological Repercussions of Non-*Saccharomyces* Species 2.0

Editor

Antonio Morata

MDPI • Basel • Beijing • Wuhan • Barcelona • Belgrade • Manchester • Tokyo • Cluj • Tianjin

Editor
Antonio Morata
Department of Food Science and Technology,
Universidad Politécnica de Madrid (UPM)
Spain

Editorial Office
MDPI
St. Alban-Anlage 66
4052 Basel, Switzerland

This is a reprint of articles from the Special Issue published online in the open access journal *Fermentation* (ISSN 2311-5637) (available at: https://www.mdpi.com/journal/fermentation/special_issues/non_saccharomyces_2).

For citation purposes, cite each article independently as indicated on the article page online and as indicated below:

LastName, A.A.; LastName, B.B.; LastName, C.C. Article Title. *Journal Name* **Year**, *Volume Number*, Page Range.

ISBN 978-3-0365-0150-5 (Hbk)
ISBN 978-3-0365-0151-2 (PDF)

Cover image courtesy of Prof. María Antonia Bañuelos.

Contents

About the Editor

Antonio Morata Professor of Food Science and Technology at the Universidad Politécnica de Madrid (UPM), Spain, specialized in wine technology. Coordinator of Master in Food Engineer at UPM. Professor of Enology and Wine Technology in the European Master of Viticulture and Enology, Euromaster Vinifera-Erasmus+. Spanish delegate at the group of Experts in Wine Microbiology and Wine Technology of the International Organizations of Vine and Wine (OIV). Author of more than 75 research articles, 3 books, 4 edited books, 6 Special Issues and 16 book chapters. Editorial Board Member of Fermentation and Beverages (MDPI).

 *fermentation*

Editorial

Enological Repercussions of Non-*Saccharomyces* Species 2.0

Antonio Morata

enotecUPM, Chemistry and Food Technology Department, Escuela Técnica Superior de Ingeniería Agronómica, Alimentaria y de Biosistemas, Universidad Politécnica de Madrid, Avenida Complutense S/N, 28040 Madrid, Spain; antonio.morata@upm.es

Received: 9 November 2020; Accepted: 13 November 2020; Published: 17 November 2020

Abstract: Non-*Saccharomyces* yeast species are currently a biotechnology trend in enology and broadly used to improve the sensory profile of wines because they affect aroma, color, and mouthfeel. They have become a powerful biotool to modulate the influence of global warming on grape varieties, helping to maintain the acidity, decrease the alcoholic degree, stabilize wine color, and increase freshness. In cool climates, some non-*Saccharomyces* can promote demalication or color stability by the formation of stable derived pigments. Additionally, non-*Saccharomyces* yeasts open new possibilities in biocontrol for removing spoilage yeast and bacteria or molds that can produce and release mycotoxins, and therefore, can help in reducing SO_2 levels. The promising species *Hanseniaspora vineae* is analyzed in depth in this Special Issue in two articles, one concerning the glycolytic and fermentative metabolisms and its positive role and sensory impact by the production of aromatic esters and lysis products during fermentation are also assessed.

Keywords: non-*Saccharomyces*; sensory improvement; dealcoholization; SO_2; grape variety; *Hanseniaspora vineae*; *Brettanomyces bruxellensis*; *Pichia guilliermondii*; *Metschnikowia pulcherrima*; *Schizosaccharomyces pombe*; *Lachancea thermotolerans*

Non-*Saccharomyces* Species in Wine Biotechnology

Some non-*Saccharomyces* yeast species have a powerful impact on wine aroma [1–3] by the release of fermentative aromatic compounds (*Torulaspora delbrueckii*, *Candida stellate*, *Starmerella bacillaris*, *Wickerhamomyces anomalus*, *Hanseniaspora vineae*, *Schizosaccharomyces pombe*) [4–11], the production of varietal aromas such as thiols 3 MH and 3 MHA (*Pichia kluivery*) [3,12], or the expression of exocellular enzymatic activities (*Metschnikowia pulcherrima*) [13]. The selection of the optimal strains of these species [14] according to the specific production of the previously described aromatic compounds can even improve the effect on the sensory profile of wines during fermentation. Additionally, the weak implantation of non-*Saccharomyces* species during must fermentation and the low competitiveness with *Saccharomyces cerevisiae* make necessary the elimination of wild yeasts from grapes to ensure a suitable impact [15]. To reach this goal, emerging non-thermal technologies open new possibilities in the effective implantation at an industrial scale of non-*Saccharomyces* starters.

One of the key non-*Sacchaccaromyces* yeasts is currently *Hanseniaspora vineae*, as it has high effectiveness in modulating the aromatic profile of neutral varieties by the production of acetate esters, especially 2-phenylethyl acetate [3,10] and benzyl acetate [3,16], that are impact compounds of the floral aroma of rose petals and jasmine flowers [3]. Additionally, *H. vineae* releases during fermentation large amounts of cell wall polysaccharides that make it interesting in the fermentation and ageing of lees of white neutral varieties. Moreover, the better adaptation of *H. vineae* to the fermentative process than other *Hanseniaspora* fruit clade species has been highlighted, in terms of fermentative performance: growth, fermentation kinetics, and alcohol tolerance [17]. The use of suitable levels of

SO_2 in grape must has been observed to have a positive effect in the selection of some *Hanseniaspora* spp. favoring the production of acetate esters, especially significant amounts of 2-phenylethyl acetate [18]. SO_2 management can be an interesting tool to modulate wild non-*Saccharomyces* populations for improving the aroma in uninoculated wines.

Another hot topic in the use of non-*Saccharomyces* yeasts is the adaptation to winemaking of specific grape varieties in global warming-affected climatic regions [19]. In warm areas, the winey and flat profile show even aromatic varieties can be improved by using non-*Saccharomyces* yeasts in mixed and sequential fermentations with *Saccharomyces cerevisiae*. In such conditions, some species behave as powerful biotools to improve wine freshness [3]. Among them, *Torulaspora delbrueckii*, *Lachancea thermotolerans*, and *Metschnikowia pulcherrima* [19,20], together with some apiculate yeasts [10,18,20], are key species. *Torulaspora delbrueckii* was the first species produced, and broadly used at an industrial level, because of the effect on wine aroma and mouthfeel [3,6]. *Lachancea thermotolerans* applications are increasing due its role in modulating wine acidity by the formation of lactic acid from sugars with a clear repercussion in pH control in warm areas [21,22]. The use of both species together in mixed fermentation helps to improve the sensory profile and freshness of wines from warm areas [19].

Another concomitant problem in warm areas is the excessive alcoholic degree, and several technologies are being developed to manage the high alcoholic degree. Among them, the use of non-*Saccharomyces* yeasts in which the metabolization of some sugars is derived to alternative products to ethanol is currently being studied [23,24]. The formation of glycerol, lactic acid, or yeast biomass [21,23,24] is a natural way to derive sugars used for the production of ethanol towards molecules or structures with repercussions in the sensory profile.

Biocontrol and bioprotection are research fields that are being strongly developed in wine biotechnology. Some non-*Saccharomyces* open interesting possibilities to exclude or eliminate undesired yeasts during fermentation because of their spoilage role in the production of defective compounds, such as ethylphenols by *Brettanomyces* or volatile acidity/ethyl acetate by some apiculate yeasts [25]. Some non-*Saccharomyces* with hydroxycinnamate decarboxylase activity can promote the formation of vinylphenolic pyranoanthocyanins during fermentation, therefore favoring the immobilization of ethylphenol precursors in stable pigments [26]. Additionally, the control and elimination of microorganisms that produce toxic molecules for human health, like biogenic amines or fungal toxins, have been studied [27,28].

Lastly, non-*Saccharomyces* species can be considered a new source of bioproducts or bioadditives with improved features that open new possibilities in wine biotechnology [28]. The use of non-*Saccharomyces* as a source of antimicrobial peptides can control toxin-producing or spoilage molds or undesired yeast or bacteria. Production in cocultures or the addition of molecules from non-*Saccharomyces* can promote the development of starters for alcoholic or malolactic fermentation. The application of non-*Saccharomyces* or their derivatives as oxygen consumers or reducers can control oxidation during fermentation and stabilization to reduce SO_2 levels. Many other alternative emerging uses of derived products from non-*Saccharomyces* species will be available for industrial applications soon.

Conflicts of Interest: The author declares no conflict of interest

References

1. Ciani, M.; Maccarelli, F. Oenological properties of non-*Saccharomyces* yeasts associated with wine-making. *World J. Microbiol. Biotechnol.* **1997**, *14*, 199–203. [CrossRef]
2. Jolly, N.P.; Augustyn, O.P.H.; Pretorius, I.S. The role and use of non-*Saccharomyces* yeasts in wine production. *S. Afr. J. Enol. Vitic.* **2006**, *27*, 15–39. [CrossRef]
3. Morata, A.; Escott, C.; Bañuelos, M.A.; Loira, I.; del Fresno, J.M.; González, C.; Suárez-Lepe, J.A. Contribution of Non-*Saccharomyces* Yeasts to Wine Freshness. A Review. *Biomolecules* **2020**, *10*, 34. [CrossRef] [PubMed]

4. Loira, I.; Vejarano, R.; Bañuelos, M.A.; Morata, A.; Tesfaye, W.; Uthurry, C.; Villa, A.; Cintora, I.; Suárez-Lepe, J.A. Influence of sequential fermentation with *Torulaspora delbrueckii* and *Saccharomyces cerevisiae* on wine quality. *LWT Food Sci. Technol.* **2014**, *59*, 915–922. [CrossRef]

5. Loira, I.; Morata, A.; Comuzzo, P.; Callejo, M.J.; González, C.; Calderón, F.; Suárez-Lepe, J.A. Use of *Schizosaccharomyces pombe* and *Torulaspora delbrueckii* strains in mixed and sequential fermentations to improve red wine sensory quality. *Food Res. Int.* **2015**, *76*, 325–333. [CrossRef]

6. Ramírez, M.; Velázquez, R. The Yeast *Torulaspora delbrueckii*: An Interesting but Difficult-To-Use Tool for Winemaking. *Fermentation* **2018**, *4*, 94. [CrossRef]

7. García, M.; Esteve-Zarzoso, B.; Cabellos, J.M.; Arroyo, T. Advances in the Study of *Candida stellata*. *Fermentation* **2018**, *4*, 74. [CrossRef]

8. Padilla, B.; Gil, J.V.; Manzanares, P. Challenges of the Non-Conventional Yeast *Wickerhamomyces anomalus* in Winemaking. *Fermentation* **2018**, *4*, 68. [CrossRef]

9. Martin, V.; Valera, M.J.; Medina, K.; Boido, E.; Carrau, F. Oenological Impact of the *Hanseniaspora/Kloeckera* Yeast Genus on Wines—A Review. *Fermentation* **2018**, *4*, 76. [CrossRef]

10. Del Fresno, J.M.; Escott, C.; Loira, I.; Herbert-Pucheta, J.E.; Schneider, R.; Carrau, F.; Cuerda, R.; Morata, A. Impact of *Hanseniaspora vineae* in Alcoholic Fermentation and Ageing on Lees of High-Quality White Wine. *Fermentation* **2020**, *6*, 66. [CrossRef]

11. Loira, I.; Morata, A.; Palomero, F.; González, C.; Suárez-Lepe, J.A. *Schizosaccharomyces pombe*: A Promising Biotechnology for Modulating Wine Composition. *Fermentation* **2018**, *4*, 70. [CrossRef]

12. Anfang, N.; Brajkovich, M.; Goddard, M.R. Co-fermentation with *Pichia kluyveri* increases varietal thiol concentrations in Sauvignon Blanc. *Aust. J. Grape Wine Res.* **2009**, *15*, 1–8. [CrossRef]

13. Morata, A.; Loira, I.; Escott, C.; del Fresno, J.M.; Bañuelos, M.A.; Suárez-Lepe, J.A. Applications of *Metschnikowia pulcherrima* in Wine Biotechnology. *Fermentation* **2019**, *5*, 63. [CrossRef]

14. Suárez-Lepe, J.A.; Morata, A. New trends in yeast selection for winemaking. *Trends Food Sci. Technol.* **2012**, *23*, 39–50. [CrossRef]

15. Morata, A.; Loira, I.; Vejarano, R.; González, C.; Callejo, M.J.; Suárez-Lepe, J.A. Emerging preservation technologies in grapes for winemaking. *Trends Food Sci. Technol.* **2017**, *67*, 36–43. [CrossRef]

16. Martin, V.; Giorello, F.; Fariña, L.; Minteguiaga, M.; Salzman, V.; Boido, E.; Aguilar, P.S.; Gaggero, C.; Dellacassa, E.; Mas, A.; et al. De novo synthesis of benzenoid compounds by the yeast *Hanseniaspora vineae* increases the flavor diversity of wines. *J. Agric. Food Chem.* **2016**, *64*, 4574–4583. [CrossRef] [PubMed]

17. Valera, M.J.; Boido, E.; Dellacassa, E.; Carrau, F. Comparison of the Glycolytic and Alcoholic Fermentation Pathways of *Hanseniaspora vineae* with *Saccharomyces cerevisiae* Wine Yeasts. *Fermentation* **2020**, *6*, 78. [CrossRef]

18. Cuijvers, K.; Van Den Heuvel, S.; Varela, C.; Rullo, M.; Solomon, M.; Schmidt, S.; Borneman, A. Alterations in Yeast Species Composition of Uninoculated Wine Ferments by the Addition of Sulphur Dioxide. *Fermentation* **2020**, *6*, 62. [CrossRef]

19. Escribano-Viana, R.; Garijo, P.; López-Alfaro, I.; López, R.; Santamaría, P.; Gutiérrez, A.R.; González-Arenzana, L. Do Non-*Saccharomyces* Yeasts Work Equally with Three Different Red Grape Varieties? *Fermentation* **2020**, *6*, 3. [CrossRef]

20. Romani, C.; Lencioni, L.; Biondi Bartolini, A.; Ciani, M.; Mannazzu, I.; Domizio, P. Pilot Scale Fermentations of Sangiovese: An Overview on the Impact of *Saccharomyces* and Non-*Saccharomyces* Wine Yeasts. *Fermentation* **2020**, *6*, 63. [CrossRef]

21. Morata, A.; Loira, I.; Tesfaye, W.; Bañuelos, M.A.; González, C.; Suárez Lepe, J.A. *Lachancea thermotolerans* Applications in Wine Technology. *Fermentation* **2018**, *4*, 53. [CrossRef]

22. Vaquero, C.; Loira, I.; Bañuelos, M.A.; Heras, J.M.; Cuerda, R.; Morata, A. Industrial Performance of Several *Lachancea thermotolerans* Strains for pH Control in White Wines from Warm Areas. *Microorganisms* **2020**, *8*, 830. [CrossRef] [PubMed]

23. García, M.; Esteve-Zarzoso, B.; Cabellos, J.M.; Arroyo, T. Sequential Non-*Saccharomyces* and *Saccharomyces cerevisiae* Fermentations to Reduce the Alcohol Content in Wine. *Fermentation* **2020**, *6*, 60. [CrossRef]

24. Ivit, N.N.; Longo, R.; Kemp, B. The Effect of Non-*Saccharomyces* and *Saccharomyces* Non-*Cerevisiae* Yeasts on Ethanol and Glycerol Levels in Wine. *Fermentation* **2020**, *6*, 77. [CrossRef]

25. Peña, R.; Vílches, J.; Poblete, C.; Ganga, M.A. Effect of *Candida intermedia* LAMAP1790 Antimicrobial Peptides against Wine-Spoilage Yeasts *Brettanomyces bruxellensis* and *Pichia guilliermondii*. *Fermentation* **2020**, *6*, 65. [CrossRef]

26. Morata, A.; Escott, C.; Loira, I.; Del Fresno, J.M.; González, C.; Suárez-Lepe, J.A. Influence of *Saccharomyces* and non-*Saccharomyces* Yeasts in the Formation of Pyranoanthocyanins and Polymeric Pigments during Red Wine Making. *Molecules* **2019**, *24*, 4490. [CrossRef]

27. Vilela, A. Non-*Saccharomyces* Yeasts and Organic Wines Fermentation: Implications on Human Health. *Fermentation* **2020**, *6*, 54. [CrossRef]

28. Vejarano, R. Non-*Saccharomyces* in Winemaking: Source of Mannoproteins, Nitrogen, Enzymes, and Antimicrobial Compounds. *Fermentation* **2020**, *6*, 76. [CrossRef]

Publisher's Note: MDPI stays neutral with regard to jurisdictional claims in published maps and institutional affiliations.

Article

Impact of *Hanseniaspora Vineae* in Alcoholic Fermentation and Ageing on Lees of High-Quality White Wine

Juan Manuel Del Fresno [1], **Carlos Escott** [1], **Iris Loira** [1], **José Enrique Herbert-Pucheta** [2], **Rémi Schneider** [3], **Francisco Carrau** [4], **Rafael Cuerda** [5] and **Antonio Morata** [1,*]

[1] enotecUPM, Chemistry and Food Technology Department, Escuela Técnica Superior de Ingeniería Agronómica, Alimentaria y de Biosistemas, Universidad Politécnica de Madrid, Avenida Complutense S/N, 28040 Madrid, Spain; juanmanuel.delfresno@upm.es (J.M.D.F.); carlos.escott@gmail.com (C.E.); iris.loira@upm.es (I.L.)

[2] Consejo Nacional de Ciencia y Tecnología-Laboratorio Nacional de Investigación y Servicio Agroalimentario y Forestal, Universidad Autónoma Chapingo, Carretera México-Texcoco Km 38.5, Chapingo C.P., Estado de México 56230, Mexico; jeherbert@conacyt.mx

[3] Oenoborands SAS Parc Agropolis II-Bât 5 2196 Bd de la Lironde-CS 34603, CEDEX 05, 34397 Montpellier, France; Remi.SCHNEIDER@oenobrands.com

[4] Área Enología y Biotecnología de Fermentaciones, Facultad de Química, Universidad de la Republica, Gral. Flores 2124, Montevideo 11800, Uruguay; fcarrau@fq.edu.uy

[5] Comenge Bodegas y Viñedos SA, Curiel de Duero, 47316 Valladolid, Spain; cuerda@comenge.com

* Correspondence: antonio.morata@upm.es

Received: 10 June 2020; Accepted: 28 June 2020; Published: 1 July 2020

Abstract: *Hanseniaspora vineae* is an apiculate yeast that plays a significant role at the beginning of fermentation, and it has been studied for its application in the improvement of the aromatic profile of commercial wines. This work evaluates the use of *H. vineae* in alcoholic fermentation compared to *Saccharomyces cerevisiae* and in ageing on the lees process (AOL) compared to *Saccharomyces* and non-*Saccharomyce*s yeasts. The results indicated that there were not significant differences in basic oenological parameters. *H. vineae* completed the fermentation until 11.9% *v/v* of ethanol and with a residual sugars content of less than 2 g/L. Different aroma profiles were obtained in the wines, with esters concentration around 90 mg/L in *H. vineae* wines. Regarding the AOL assay, the hydroalcoholic solutions aged with *H. vineae* lees showed significantly higher absorbance values at 260 (nucleic acids) and 280 nm (proteins) compared to the other strains. However, non-significant differences were found in the polysaccharide content at the end of the ageing process were found compared to the other yeast species, with the exception of *Schizosaccharomyces pombe* that released around 23.5 mg/L of polysaccharides in hydroalcoholic solution. The use of *H. vineae* by the wineries may be a viable method in fermentation and AOL to improve the quality of white wines.

Keywords: *Hanseniaspora vineae*; alcoholic fermentation; non-*Saccharomyces*; ageing on lees; polysaccharides; white wines

1. Introduction

The inoculation of commercial *S. cerevisiae* yeast strains is the most common practice in the industrial elaboration of commercial wines. However, nowadays, winemakers are trying to obtain quality wines with different organoleptic characteristics. In this regard, the use of different species of yeast could be interesting. Many studies have been done with respect to obtaining differentiated quality products and the use of non-*Saccharomyces* yeasts for this purpose [1–3]. The use of *H. vineae* in wineries could be a good alternative to the traditional *Saccharomyces* fermentations. This yeast and others apiculate yeast of the genus *Hanseniaspora/Kloeckera* are the main species present on mature grapes and play a significant role at the beginning of fermentation, producing enzymes and aroma compounds that expand the diversity of wine colour and flavor [4]. Normally, *H. vineae* appears in the first stages of the fermentation but it is quickly dominated by *S. cerevisiae* [5]. The main interest in this yeast is due to the aromatic profile of the wines obtained [6]. This yeast produces a fruity and floral aroma due to the increased amounts of acetate esters, primarily 2-phenylethyl acetate [7] and benzyl acetate. Other authors [8,9] investigated the potential of to the genus *Hanseniaspora* to produce acetate esters. In the same way, the modulation of the aeration during the growing stage of these yeasts can increase the aromatic diversity and quality of the wine obtained [10]. In addition, the *H. vineae* species can be used in pure culture because this yeast might reach about 10% of the alcohol by volume of fermentative capacity under winemaking conditions [4]. In this respect, we conducted a semi-industrial assay in this study using *H. vineae* in pure culture compared to *S. cerevisiae* in the control.

Additionally, in this study, the use of *H. vineae* in aging on lees (AOL) has been assayed in comparison with other yeast species. The AOL technique consists of a long contact of the yeast lees with the wine. During this contact, the yeast autolysis is produced with the breakdown of cell membranes, the release of intracellular constituents, the liberation of hydrolytic enzymes and the hydrolysis of intracellular biopolymers into low molecular weight products [11]. Among these compounds, the polysaccharides have an effect on the physico-chemical properties of the wine, as well as on the sensory properties [12]. The AOL improves the aromatic and gustatory complexity of wine, mainly by improving its body and reducing its astringency [13]. The main problem of this technique is that the AOL is a slow process, many studies have been done trying to accelerate the cell lysis like the use of emerging physical technologies such as high hydrostatic pressures and ultrasounds [14]. Another technique to reduce the ageing time is the use of yeast species that have a high capacity to release polysaccharides into the wine. In previous studies, [15] certain wine spoilage yeasts like *Saccharomycodes ludwigii*, *Zygosaccharomyces bailii*, and *Brettanomyces bruxellensis* were shown to produce a greater quantity of polysaccharides compared to *S. cerevisiae* strains. In the same way, these authors classified the released polysaccharides according to their composition. Therefore, the AOL may depend on the yeast used and its cell wall polysaccharide composition.

The main objective of this work is to obtain information about the use of *H. vineae* in alcoholic fermentation as well as in the AOL technique.

2. Materials and Methods

2.1. Yeast Species Used in Alcoholic Fermentation

The *H. vineae* yeast strain used in this study was isolated by Professor Francisco Carrau (Facultad de Química, Universidad de la República, Montevideo, Uruguay) and it is currently under evaluation by "Oenobrands SAS, France".

The yeast strain Fermivin 3C (*S. cerevisiae*) used as control in this study is a selected yeast marketed by "Oenobrands SAS, France".

2.2. Alcoholic Fermentation Conditions

The Albillo grape variety (*Vitis vinifera* L.) was fermented at "Comenge Bodegas y Viñedos SA" (Curiel de Duero, Spain). The white must was fermented in triplicate in 120 L stainless steel barrels. The fermentation process was monitored following the daily variation of density and temperature. The samples were taken once at the end of the fermentation.

2.3. Yeast Species Used in Ageing on Lees

Two strains of *S. cerevisiae* were used as controls in the AOL assay, the strains 7VA and G37 (SC7VA, SCG37), both yeasts were isolated by the Chemistry and Food Technology Department of ETSI Agronómica, Alimentaria y de Biosistemas, Universidad Politécnica de Madrid.

Three species of non-*Saccharomyces* yeasts were used, the same *H. vineae* strain that had been previously used in the alcoholic fermentation trial, as well as *Lachancea thermotolerans* L31 strain (L31) isolated and selected by enotecUPM (Food Technology Department, ETSIAAB, Universidad Politécnica de Madrid) and *Schizosaccharomyces pombe* 938 (SP938, IFI, CSIC, Madrid, Spain).

The yeast lees biomass used for the AOL assay was obtained by growing in 2 L of YEPD medium enriched with 100 g/L of glucose. The growth was carried out at 25 °C for three days. Then, the biomass was washed three times with deionized water, discarding the supernatant after each centrifugation, at 1200 rcf, for 3 min.

2.4. Ageing on Lees Conditions

The AOL was done in hydroalcoholic solution (13.5% *v/v*) sulphited to 60 mg/L with $K_2S_2O_5$ and the pH was adjusted to 3.5 with phosphoric acid. The samples were prepared in triplicates, using ISO flasks of 0.5 L. The dosage of yeast lees was 6 g/L and the ageing process was done at 16 °C in a dark room for 156 days. The samples were mixed once a week to simulate a *bâtonnage* process.

2.5. Basic Oenological Parameters Analysis

The values of ethanol (% *v/v*), pH, total acidity (g/L) expressed as tartaric acid, volatile acidity (g/L) expressed as acetic acid, malic acid (g/L), lactic acid (g/L) and glucose/fructose content (g/L) were obtained by Fourier transform infrared spectroscopy (FTIR), using an OenoFoss™ instrument (FOSS Iberia, Barcelona, Spain).

2.6. NMR Spectroscopy

NMR spectra of a triplicate set of Albillo white wines fermented with *H. vineae* and *S. cerevisiae* yeast strains, were carried out on a Bruker 600 Avance III HD spectrometer, equipped with a 5-mm 1H/D TXI probehead equipped with a z-gradient at 298 ± 0.1 K of temperature. The following set of NMR experiments were conducted:

(a) Standard 1H-one-dimensional NMR experiment was carried out as step for calibration of the water-to-ethanol multi-presaturation module: with 4 transients of 32,768 complex points, having recycling delays of 5 s and with acquisition times of 1700 milliseconds, produced an experimental time of 26 s. No apodization function was applied during Fourier Transform.

(b) {1Hwater_presat NMR}: 1D single pulse NOESY experiment with a homemade shaped-pulse water-to-ethanol presaturation during both the relaxation delay (5 s) and mixing times (100 milliseconds), with a 8.18×10^{-4} W power irradiation level for the solvent signals' elimination, centering the transmitter frequency at 4.7 ppm and shifting the decoupler frequency between 3.55 ppm (CH2-ethanol) and 1.08 ppm (CH3-ethanol) for accurate multi-presaturation of all signals [16,17] were acquired for each sample as follows: a total of 128 transients were collected into 32,768 complex data points, with a spectral width of 9615.4 Hz and acquisition times of 1700 ms, produce experimental times of 10'58''.

(c) NMR post-processing was carried out as follows: ppm calibration and manual phase corrections were conducted with the use of Bruker TopSpin 4.0.8 software. Global and soft baseline corrections, least-squares NMR alignments, variable size bucketing and data matrix normalization were carried out with NMRProcFlow [18]. Scaling and statistical analysis workflow for obtaining the Principal Component Analysis to determine relationships between *H. vineae* and *S. cerevisiae* wine samples, from the constant sum normalized NMR data matrix, were developed with the BioStatFlow 2.9.2 software. Identified metabolites were quantified (Table 1) through qNMR methods [19,20] routinely used in oenology [21,22].

Table 1. Targeted metabolites concentration (mg/L) of *Saccharomyces cerevisiae* and *Hanseniaspora vineae* wine samples obtained with the PULCON-NMR method [21].

mg/L	Fermentation with *Saccharomyces cerevisiae*	Fermentation with *Hanseniaspora vineae*
Furfural	1.47 ± 1.14 [a]	3.29 ± 2.82 [a]
Formiate	2.44 ± 0.66 [a]	3.05 ± 0.84 [a]
Shikimic	1.54 ± 0.21 [a]	1.85 ± 0.19 [a]
Fumaric	0.58 ± 0.44 [a]	0.53 ± 0.17 [a]
Sorbic	1.63 ± 1.44 [a]	2.73 ± 3.33 [a]
β-Glucose	500.02 ± 58.39 [b]	365.60 ± 37.23 [a]
Fructose	695.11 ± 146.39 [a]	803.69 ± 238.53 [a]
Citrate	244.05 ± 25.82 [a]	255.38 ± 7.52 [a]
Succinate	291.47 ± 28.40 [a]	233.36 ± 25.83 [a]
Glutamine	54.03 ± 10.14 [a]	59.20 ± 5.41 [a]
Acetate	289.70 ± 18.64 [a]	274.73 ± 22.25 [a]
Proline	34.17 ± 7.66 [a]	42.29 ± 6.35 [a]
γ-Aminobutyric acid	67.68 ± 5.11 [a]	73.61 ± 7.32 [a]
Arginine	28.43 ± 11.10 [a]	44.00 ± 26.46 [a]
Alanine	119.82 ± 42.98 [a]	150.20 ± 78.16 [a]
Lactic	156.56 ± 31.04 [a]	174.25 ± 44.30 [a]
Threonine	188.38 ± 70.77 [a]	230.49 ± 78.09 [a]
Valine	52.72 ± 18.84 [a]	37.63 ± 17.85 [a]
Isoleucine	29.33 ± 9.08 [a]	37.18 ± 7.66 [a]

[a] Means with the same letter are not significantly different ($p < 0.05$).

2.7. Volatile Compounds from the Alcoholic Fermentation Analysis

The volatile compounds of the wines obtained in fermentation assay were measured using an Agilent Technologies 6850 gas chromatograph, equipped with an integrated flame ionization detector (GC-FID) and DB-624 column (60 m × 250 μm × 1.40 μm). Analyses were performed according to the method described by [23]. The injector temperature was 250 °C, and the detector temperature was 300 °C. The column temperature was 40 °C for the first 5 min, rising linearly by a 10 °C/min until reaching 250 °C; this temperature was maintained for 5 min. Hydrogen was used as the carrier gas. The flow rate was 22.1 L/min. The injection split ratio was 1:10. The detection limit was 0.1 mg/L.

Calibration was performed using the following external standards: acetaldehyde, metanol, 1-propanol, diacetyl, ethyl acetate, 2-butanol, isobutanol, 1-butanol, acetoin, 2-methyl-1-butanol, 3-methyl-1-butanol, isobutyl acetate, ethyl butyrate, ethyl lactate, 2.3-butanediol, 3-ethoxy-1-propanol, isoamyl acetate, hexanol, 2-phenyl ethanol and 2-phenylethyl acetate.

2.8. Proteins and Nucleic Acids Estimation by Absorbance at 260 and 280 nm

The absorbance measurements were done through the ageing after centrifugation (1200 rcf for 3 min) using a 1-cm path-length quartz cuvette. All spectrometric measurements were obtained using an 8453 spectrophotometer from Agilent Technologies™ (Palo Alto, CA, USA).

2.9. Polysaccharides Analysis (HPLC-RI)

The polysaccharides content was measured after 156 days of ageing in the AOL assay, using an HPLC-RI technique. An 1100 HPLC chromatograph (Agilent Technologies, Palo Alto, CA, USA) equipped with a refractive index detector with Ultrahydrogel 250 molecular exclusion column (Waters) was used, according to the method described by [24]. The eluent was 0.1 M $NaNO_3$ in deionized water (MilliQ). A calibration curve constructed from the following pullulan standards (polymaltotriose) (Shodex, Showa Denko K.K, Japan) were used to determine the concentration of polysaccharides in the samples: P-800 (788 kDa), P-400 (404 kDa), P-200 (212 kDa), P-100 (112 kDa), P-50 (47.3 kDa), P-20 (22.8 kDa), P-10 (11.8 kDa) and P-5 (5.9 kDa).

2.10. Statistical Analysis

Statgraphics v.5 software (Graphics Software Systems, Rockville, MD, USA) was used to calculate means, standard deviations, analysis of variance (ANOVA), least-significant difference (LSD) test and principal component analysis (PCA). The LSD test was used to detect significant differences between means. Significance was set at $p < 0.05$.

3. Results and Discussion

3.1. Basic Oenological Parameters

In general, no significant differences were found in the wines fermented by *H. vineae* compared to conventional wines fermented by *S. cerevisiae*, with the exception of the total acidity parameter. The *S. cerevisiae* wines showed 0.5 g/L more total acidity than the *H. vineae* wines (Table 2). However, no differences in lactic acid, malic acid and volatile acidity content were found, therefore, the decrease of total acidity may be due to the precipitation of tartaric acid during the alcoholic fermentation. It is important to mention that these differences were not reflected in the pH values, since the pH was similar in all the wines studied.

Table 2. Ethanol content (% *v/v*), pH, total acidity (g/L) as tartaric acid, volatile acidity as acetic acid (g/L), malic acid (g/L), lactic acid (g/L) and glucose and fructose (g/L) after fermentation process. Mean ± SD for three replicates.

	Fermentation with *Saccharomyces cerevisiae*	Fermentation with *Hanseniaspora vineae*
Ethanol (% *v/v*)	11.93 ± 0.15 [a]	11.90 ± 0.00 [a]
pH	3.17 ± 0.03 [a]	3.21 ± 0.02 [a]
Total Acidity (g/L)	6.30 ± 0.10 [b]	5.80 ± 0.17 [a]
Volatile Acidity (g/L)	0.45 ± 0.07 [a]	0.36 ± 0.02 [a]
Malic Acid (g/L)	2.00 ± 0.10 [a]	1.87 ± 0.06 [a]
Lactic Acid (g/L)	0.10 ± 0.10 [a]	0.00 ± 0.00 [a]
Gluc and Fruc (g/L)	1.67 ± 0.60 [a]	1.07 ± 0.38 [a]

[a] Means with the same letter are not significantly different ($p < 0.05$).

Regarding the residual sugar content, both yeasts have been able to ferment all the sugar, with final concentrations in the wine below 2 g of residual sugar per litre. These results are in line with those obtained by other authors that compared both yeast species in Macabeo and Merlot grape wines [25]; nevertheless, [26] found 0.5 g/L of glucose and frutose more in *H. vineae* wines than in *S. cerevisiae* wines before the malolactic fermentation. This fact is linked to the glycolytic power—all wines showed similar ethanol contents around 11.9% *v/v*. These results indicate that both yeast species may produce wines with similar basic oenological parameters.

Targeted NMR analysis allowed the identification and quantification (Table 1) of typical wine metabolites in both *H. vineae* and *S. cerevisiae* samples: furfural (9.64 ppm), formiate (8.41 ppm), shikimic acid (6.87 ppm), fumaric acid (6.4 ppm), β-glucose (4.55 ppm), fructose (4.04 ppm), citrate (2.84 ppm), succinate (2.66 ppm), glutamine (2.25 ppm), acetate (2.01 ppm), proline (2.05 ppm), γ-aminobutyric acid (1.96 ppm), arginine (1.70 ppm), alanine (1.55 ppm), threonine (1.28 ppm), valine (1.1 ppm) and isoleucine (0.91 ppm). With the results obtained by NMR, a principal component analysis (PCA) was performed. Using the 2D-projections (PC1 = 43.1%, PC2 = 24.2%), slight overlaps were observed amongst groups (Figure 1A). The distribution was better explained with the first three components (PC1 = 43.1%, PC2 = 24.23% and PC3 = 13.59%). Even though the results were not statistically significant between the two yeasts studied (Table 1), the PCA made it possible to differentiate the wines studied into two independent clusters corresponding with the two target yeasts (Figure 1). Chemical shift loading plots (Figure 1B) show a set of relevant resonances that permits the discrimination between yeasts by PCA: formiate (8.4123 ppm, PC1 [+], PC2 [+]); shikimic (6.8740 ppm, PC1 [−], PC2 [−]); β-glucose (4.5395 ppm, PC1 [−], PC2 [−]); fructose (4.0375 ppm, PC1 [+], PC2 [−]); citrate (2.8415 ppm, PC1 [−], PC2 [−]); succinate (2.6655 ppm, PC1 [+], PC2 [−]); all amino acids present positive PCA 2 (glutamine 2.2465 ppm, PC1 [+], PC2 [+]; alanine 1.551 ppm, PC1 [+], PC2 [+], valine 1.0595 ppm, PC1 [+], PC2 [+] and isoleucine 0.9140 ppm, PC1 [−], PC2 [+]) and acetate (2.0925 ppm, PC1 [−], PC2 [−]). These results allow us to differentiate the metabolism of both yeasts, even though these differences were not quantitatively observed. It is noted that we identified the same separation between the must fermented by *H. vineae* and *S. cerevisiae* when the PCA was done on fermentative volatile compounds (Figure 2).

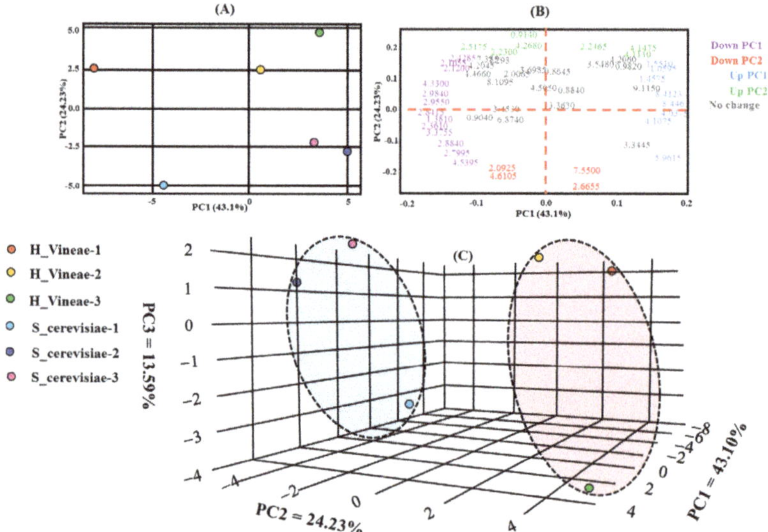

Figure 1. Principal component analysis (PCA) score plots comprising the 67.33% variance (**A**) and 80.92% variance (**C**) and chemical shift loading plots (**B**) obtained by a variable NMR bucketing procedure) of the data-reduced NMR fingerprints of Albillo white wines fermented at two different conditions. Red and blue ovals (89% confidence intervals) represent respectively *H. vineae* and *S. cerevisiae* fermentation groups, each analyzed in triplicate.

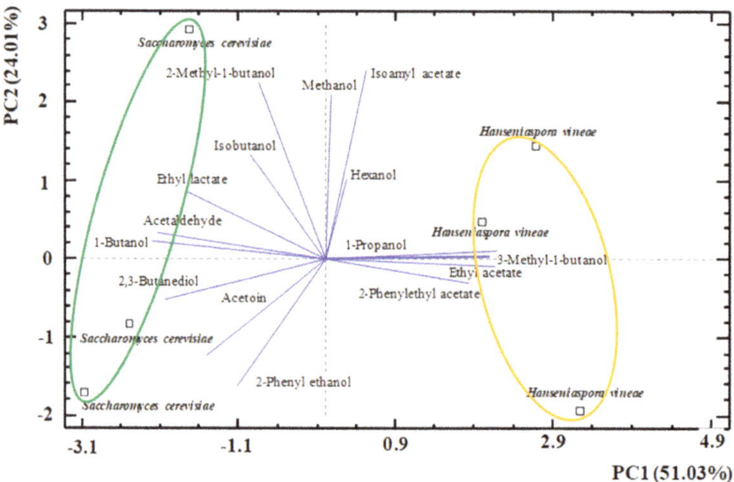

Figure 2. Principal component analysis (PCA) of the fermentative volatile compounds.

3.2. Volatile Compounds from the Alcoholic Fermentation

Considering the total volatile compounds identified, *S. cerevisiae* produced a larger amount of volatile compounds (Table 3) with around 1200 mg/L. In this regard the concentration of acetaldehyde and 2,3-butanediol have a special importance. The amount of these compounds was significantly higher in the wines from *S. cerevisiae*. Similar results were obtained after the fermentation of artificial red must [27].

Both yeast species did not show significant differences in the sum of higher alcohols. It interesting to point out that other authors reported a decrease in higher alcohols after the fermentation of the Chardonnay grapes must have with *H. vineae* compared with that of *S. cerevisiae* [5].

The fermentation with *H. vineae* resulted in increases in acetate esters and some ethyl esters, like ethyl acetate with concentrations around 79 mg/L. These results are similar to the results obtained by [5].

2-Phenylethyl acetate is an ester with strong aromatic power and its perception threshold reported is 250 µg/L [28]. This compound is associated with fruity, floral and honey aromas [25]. The 2-phenylethyl acetate concentration was significantly higher in *H. vineae* wines than in *S. cerevisiae* wines (Table 3). This fact has been reported by several authors [25,29] who identified up to 50 times more abundance of this compound in wines fermented by *H. vineae*. However, no significant differences in 2-phenylethanol content were found. This can be due to the fact that there are significant differences between these two yeast species in the acetylation step due to an increase in the copy number of the acetyl transferases genes in *H. vineae* [29].

In addition, the "odour activity values" (OAV) were calculated (see Table 3). It allows us to estimate the contribution of a specific compound to the aroma of the wine [30]. Among the compounds that have been identified, only ethyl acetate, 2-methyl-1-butanol, 2.3-butanediol, isoamyl acetate, hexanol and 2-phenylethyl acetate have obtained an OAV greater than one. It must again be emphasized the importance of the 2-phenylethyl acetate. This compound had 31.84 OAV and statistically higher concentrations in *H. vineae* than in *S. cerevisiae* wines. In this regard, the concentration identified as 2-phenylethyl acetate had an important organoleptic repercussion in the wines obtained by the fermentation of *H. vineae*, providing fruity, floral and honey aromas to these wines.

A principal component analysis (PCA) was done for the 15 volatile compounds identified after the fermentation process (Figure 2) and it allowed to differentiate the aromatic profile between the yeasts studied. The distribution was explained with the first two components. The compounds 2-phenylethyl acetate, ethyl acetate, 3-methyl-1-butanol, 1-propanol, hexanol, isoamyl acetate and methanol are associated positively with the PC1. A cluster including the wines fermented by *H. vineae* was found in the positive values of the PC1 with the highest concentration of these volatiles. It is noteworthy the contribution of the 2-phenylethyl acetate produced by the metabolism of this yeast species; on the contrary, on the negative values of the principal component PC1, a cluster composed of the wines fermented by *S. cerevisiae* was identified, including the contribution of 2-phenyl ethanol and indicating the difference between the two yeast species in the acetylation of this compound.

Table 3. Concentration of volatile compounds produced by fermentation (mg/L), measured by GC-FID. Mean ± standard deviation of three replicates. Different letters indicate values with statistical significant differences ($p < 0.05$).

	Fermentation with *Saccharomyces cerevisiae*	Fermentation with *Hanseniaspora vineae*	OAV *Saccharomyces cerevisiae*	OAV *Hanseniaspora vineae*	Odor Character	Perception Threshold (mg/L)
Acetaldehyde	26.34 ± 2.19 [b]	19.86 ± 2.29 [a]	0.263	0.198	pungent, fruity, suffocating, fresh, green [4]	100–125 [31]
Methanol	43.03 ± 2.76 [a]	42.42 ± 1.56 [a]	0.059	0.063	pungent odor [6]	668 [32]
1-Propanol	20.49 ± 1.72 [a]	28.68 ± 0.52 [b]	0.041	0.057	alcohol, ripe fruit [7]	500 [1] [33]
Diacetyl	nd	nd	-	-	pleasant, buttery [4]	0.2 [33]
Ethyl acetate	45.99 ± 2.72 [a]	79.26 ± 3.31 [b]	3.832	6.605	fruity, sweet, fingernail polish, etherous [4]	12 [2] [34]
2-Butanol	nd	nd	-	-	medicinal, wine-like [7]	150 [2] [33]
Isobutanol	23.80 ± 0.57 [a]	22.64 ± 3.49 [a]	0.595	0.566	Coca [5]	40 [2] [33]
1-Butanol	3.97 ± 0.10 [b]	0.00 ± 0.00 [a]	0.026	0	Medicinal [7]	150 [2] [33]
Acetoin	5.66 ± 0.21 [a]	5.50 ± 0.09 [a]	0.037	0.036	from buttery to plastic [6]	150 [1] [33]
2-Methyl-1-butanol	22.96 ± 0.90 [a]	25.29 ± 0.15 [b]	0.574	0.632	pungent odor [6]	40 [1] [35]
3-Methyl-1-butanol	112.83 ± 17.57 [a]	101.43 ± 17.53 [a]	2.821	5.536	pungent odor [6]	40 [1] [35]
isobutyl acetate	nd	nd	-	-	sweet, ester, medicinal [4]	1.6 [3] [33]
Ethyl butyrate	nd	nd	-	-	strawberry, apple, banana [7]	0.02 [1] [35]
Ethyl lactate	7.97 ± 0.13 [a]	7.15 ± 0.81 [a]	0.569	0.511	Floral [5]	14 [1] [35]
2.3-Butanediol	892.38 ± 97.06 [b]	643.99 ± 28.48 [a]	5.94	4.293	creamy, buttery [8]	150 [1] [36]
3-ethoxy-1-propanol	nd	nd	-	-	-	0.1 [1] [37]
Isoamyl acetate	3.63 ± 0.44 [a]	3.71 ± 0.22 [a]	121	123.66	banana, fresh [4]	0.03 [2] [33]
Hexanol	4.64 ± 0.05 [a]	4.75 ± 0.32 [a]	1.16	1.187	green [5]	4 [1] [33]
2-Phenyl ethanol	9.67 ± 0.33 [a]	9.52 ± 0.30 [a]	0.976	0.976	rose talc, honey [8]	10 [32]
2-Phenylethyl acetate	6.33 ± 0.45 [a]	7.96 ± 0.34 [b]	25.32	31.84	Flowery [7]	0.25 [1] [33]
Higher Alcohols	189.76 ± 18.63 [a]	187.55 ± 21.03 [a]				
Esters	55.96 ± 2.73 [a]	90.93 ± 3.13 [b]				
Total Volatiles	1.229.71 ± 77.42 [b]	1.002.14 ± 33.29 [a]				

Different letters indicate values with statistical significant differences ($p < 0.05$). [1] In wines; [2] In hydroalcoholic solution 10% *v/v*; [3] In beer; [4] from [38]; [5] [39]; [6] [34]; [7] [34]; [8] [32].

3.3. Intracellular Components and Polysaccharides Content Measured in the Ageing on Lees

The relative measurement of the intracellular components release has been done by the UV absorbance at 260 and 280 nm [40,41]. These measurements correspond to the relative amount of nucleic acids and proteins, respectively [42].

Regarding the monitoring at 260 nm, the samples with HV yeast lees showed the highest values during the entire ageing period. However, the SCG37 samples showed the lowest absorbance values without significant differences with SP938 through the AOL stage. It is also interesting to note the difference between the two *Saccharomyces* strains studied, the SC7VA samples showed absorbance values around 0.4–1 AU, while the lees of the yeast SCG37 resulted in lower values, around 0.1–0.2 AU. These results may indicate that the same yeast species can show different capacities for releasing cellular compounds depending on the strain used.

Similar results were obtained in the monitoring of 280 nm absorbance, but in this case no significant differences were obtained between the HV and SC7VA samples during the 91 days of ageing. These values indicated that both yeasts could be used to accelerate the release of cellular compounds. Therefore, the use of HV and SCVA yeast strains could be indicated to perform an AOL process.

The polysaccharides released after the action of glucanases are a good indicator of the autolysis process, being the parietal mannoproteins the majority of these polysaccharides [12]. After 156 days of ageing, the samples on SP938 lees have shown the highest content of polysaccharides with values around 23.5 mg/L. This quick releasing of compounds from the *Schizosaccharomyces* cell wall has already been observed by other authors [12]. It is interesting to stress the fact that the SP938 samples did not show the greatest absorbance values at 260 and 280 nm (Figure 3). This is possibly due to the fact that the high molecular weight polysaccharides do not have absorbance at these wavelengths as nucleic acids and proteins.

The HV samples showed a polysaccharides content of around 11 mg/L; this concentration was not statistically significant with respect to samples aged on the lees of the two *Saccharomyces* yeast strains (Figure 4). In the same way, it was not significantly different from the results obtained in L31 samples. The results obtained in the hydroalcoholic solution of these three yeast species were similar to the result of other assays with *Saccharomyces* previously done [13]. In other words, the yeast *H. vineae* could be an alternative to replace *S. cerevisiae* yeast in an AOL process after the alcoholic fermentation.

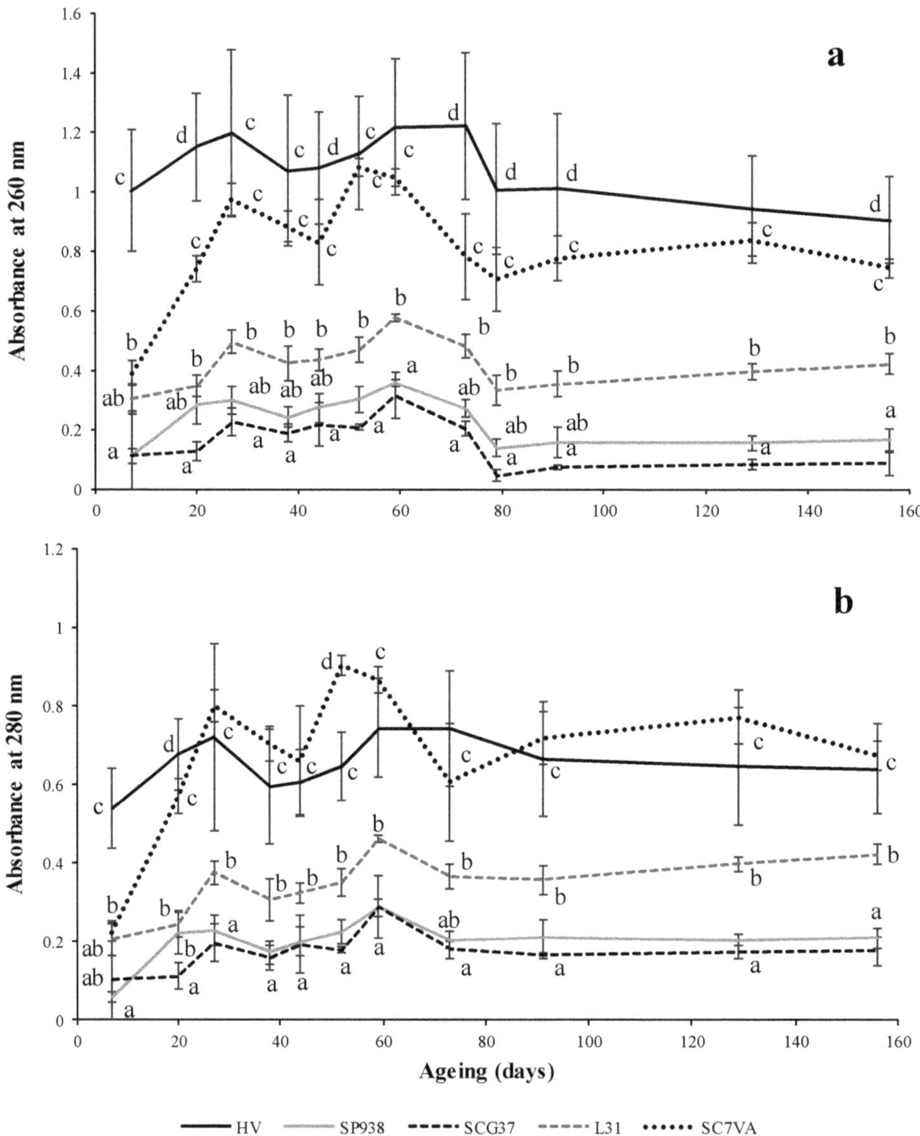

Figure 3. Evolution of the absorbance at 260 nm (**a**) and at 280 nm (**b**) in hydroalcoholic solutions, throughout 156 days of ageing on lees. HV (*Hanseniaspora vineae*); SP938 (*Schizosaccharomyces pombe* strain 938); SCG37 (*Saccharomyces cerevisiae* strain G37); L31 (*Lachancea thermotolerans* strain L31); SC7VA (*Saccharomyces cerevisiae* strain 7VA). Mean ± standard deviation of three replicates. Different letters in the same day indicate values with statistically significant differences ($p < 0.05$).

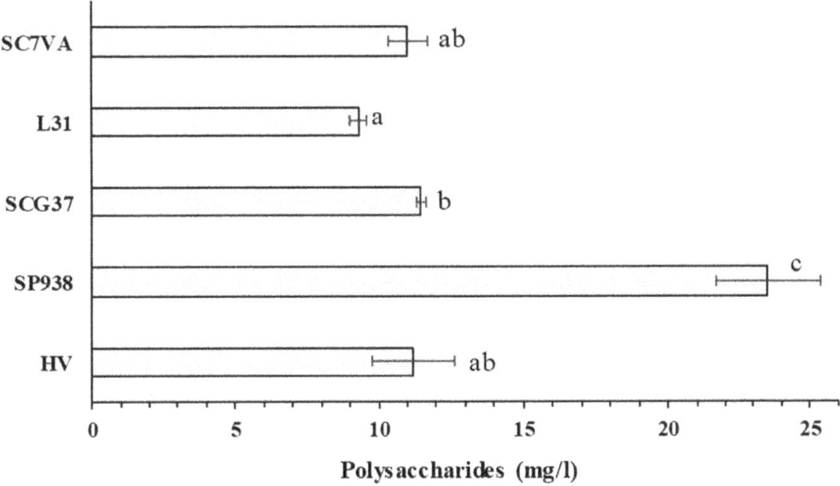

Figure 4. Polysaccharides content (mg/L) after 156 days of ageing on lees in hydroalcoholic solution. HV (*Hanseniaspora vineae*); SP938 (*Schizosaccharomyces pombe* 938 strain); SCG37 (*Saccharomyces cerevisiae* G37 strain); L31 (*Lachancea thermotolerans* L31 strain); SC7VA (*Saccharomyces cerevisiae* 7VA strain). Mean ± standard deviation of three replicates. Different letters indicate values with statistically significant differences ($p < 0.05$).

4. Conclusions

The use of *H. vineae* yeast in alcoholic fermentation resulted in wines with similar basic oenological parameters like the wines obtained by the *S. cerevisiae* fermentation. However, different aromatic profiles were identified by the PCA. Two clusters were shown with more production of acetate esters and ethyl esters by *H. vineae*. This yeast stands out for its higher production of 2-phenylethyl acetate, thus enhancing the fruity character of the wines.

The monitoring of the absorbance at 260 and 280 nm allowed to obtain a relative amount of nucleic acids and proteins released during the AOL process. In this context, the *H. vineae* yeast lees resulted in higher values of absorbance at these wavelengths throughout the ageing process. Nevertheless, the measurement of polysaccharides concentration by HPLC-RI after 156 days of ageing showed that there were no significant differences between the use of *H. vineae* yeast lees and the rest of yeast species studied, with the exception of the *S. pombe* samples.

H. vineae is an interesting yeast species to be used in alcoholic fermentation that can provide wines with more esters. In the same way, this yeast could be used in AOL processes because it is apparently quick to transfer certain cellular compounds. Nevertheless, further studies are necessary to obtain information on the cell wall polysaccharides released by this yeast and their sensory repercussion on aged wines.

Author Contributions: J.M.D.F. performed the analysis of the ageing on lees trial and drafted the manuscript; C.E. performed the analysis of the fermentation trial; I.L. revised and corrected the manuscript; J.E.H.-P. performed the analysis by NMR spectroscopy; R.S. performed the experimental design; F.C. revised and corrected the manuscript; R.C. performed the fermentations assays in the winery; and A.M. undertook the study's conceptualization, coordinated the investigation and revised the manuscript. All authors have read and agreed to the published version of the manuscript.

Funding: This research was funded by Projects FPA190000TEC2407-Oenological evaluation of one selected strain of *Hanseniaspora vineae* & ANII, ALI_2_2019_1_155314 *H. vineae* Project FQ—Lage y Cia-Uruguay.

Acknowledgments: J.E.H.-P. acknowledges the Mexican Ministry of Science and Technology (Consejo Nacional de Ciencia y Tecnología, CONACyT) junior research funding program No. 682 "Cátedras, CONACyT, Laboratorio Nacional de Investigación y Servicio Agroalimentario y Forestal" as well as the CONACyT grant program

Fermentation **2020**, *6*, 66

No. INFRA 269012 and Instituto Politécnico Nacional (FIDEICOMISO) for providing the 600 MHz NMR spectrometer resources.

Conflicts of Interest: The authors declare no conflicts of interest.

References

1. Canonico, L.; Solomon, M.; Comitini, F.; Ciani, M.; Varela, C. Volatile profile of reduced alcohol wines fermented with selected non-*Saccharomyces* yeasts under different aeration conditions. *Food Microbiol.* **2019**, *84*, 103247. [CrossRef] [PubMed]
2. Hranilovic, A.; Li, S.; Boss, P.K.; Bindon, K.; Ristic, R.; Grbin, P.R.; Van der Westhuizen, T.; Jiranek, V. Chemical and sensory profiling of Shiraz wines co-fermented with commercial non-Saccharomyces inocula. *Aust. J. Grape Wine Res.* **2018**, *24*, 166–180. [CrossRef]
3. Del Fresno, J.M.; Morata, A.; Loira, I.; Bañuelos, M.A.; Escott, C.; Benito, S.; González Chamorro, C.; Suárez-Lepe, J.A. Use of non-*Saccharomyces* in single-culture, mixed and sequential fermentation to improve red wine quality. *Eur. Food Res. Technol.* **2017**, *243*, 2175–2185. [CrossRef]
4. Martin, V.; Valera, M.; Medina, K.; Boido, E.; Carrau, F. Oenological Impact of the *Hanseniaspora/Kloeckera* Yeast Genus on Wines—A Review. *Fermentation* **2018**, *4*, 76. [CrossRef]
5. Medina, K.; Boido, E.; Fariña, L.; Gioia, O.; Gomez, M.E.; Barquet, M.; Gaggero, C.; Dellacassa, E.; Carrau, F. Increased flavour diversity of Chardonnay wines by spontaneous fermentation and co-fermentation with *Hanseniaspora vineae*. *Food Chem.* **2013**, *141*, 2513–2521. [CrossRef]
6. Lleixà, J.; Martín, V.; Giorello, F.; Portillo, M.C.; Carrau, F.; Beltran, G.; Mas, A. Analysis of the NCR Mechanisms in *Hanseniaspora vineae* and *Saccharomyces cerevisiae* During Winemaking. *Front. Genet.* **2019**, *9*, 747. [CrossRef]
7. Martin, V.; Giorello, F.; Fariña, L.; Minteguiaga, M.; Salzman, V.; Boido, E.; Aguilar, P.; Gaggero, C.; Dellacassa, E.S.; Mas, A.; et al. De novo Synthesis of Benzenoid Compounds by the yeast *Hanseniaspora vineae* Increases Flavor Diversity of Wines. *J. Agric. Food Chem.* **2016**, *64*, 4574–4583. [CrossRef]
8. Morata, A.; Escott, C.; Bañuelos, M.A.; Loira, I.; del Fresno, J.M.; González, C.; Suárez-Lepe, J.A. Contribution of Non-*Saccharomyces* Yeasts to Wine Freshness. A Review. *Biomolecules* **2019**, *10*, 34. [CrossRef]
9. Viana, F.; Gil, J.V.; Genovés, S.; Vallés, S.; Manzanares, P. Rational selection of non-*Saccharomyces* wine yeasts for mixed starters based on ester formation and enological traits. *Food Microbiol.* **2008**, *25*, 778–785. [CrossRef]
10. Yan, G.; Zhang, B.; Joseph, L.; Waterhouse, A.L. Effects of initial oxygenation on chemical and aromatic composition of wine in mixed starters of *Hanseniaspora vineae* and *Saccharomyces cerevisiae*. *Food Microbiol.* **2020**, *90*, 103460. [CrossRef]
11. Pérez-Serradilla, J.A.; de Castro, M.D.L. Role of lees in wine production: A review. *Food Chem.* **2008**, *111*, 447–456. [CrossRef]
12. Palomero, F.; Morata, A.; Benito, S.; Calderón, F.; Suárez-Lepe, J.A. New genera of yeasts for over-lees aging of red wine. *Food Chem.* **2009**, *112*, 432–441. [CrossRef]
13. Del Fresno, J.; Morata, A.; Escott, C.; Loira, I.; Cuerda, R.; Suárez-Lepe, J. Sonication of Yeast Biomasses to Improve the Ageing on Lees Technique in Red Wines. *Molecules* **2019**, *24*, 635. [CrossRef] [PubMed]
14. Morata, A.; Palomero, F.; Loira, I.; Suárez-Lepe, J.A. New Trends in Aging on Lees. In *Red Wine Technology*; Elsevier: New York, NY, USA, 2019; pp. 163–176. ISBN 9780128144008.
15. Giovani, G.; Rosi, I.; Bertuccioli, M. Quantification and characterization of cell wall polysaccharides released by non-*Saccharomyces* yeast strains during alcoholic fermentation. *Int. J. Food Microbiol.* **2012**, *160*, 113–118. [CrossRef] [PubMed]
16. Herbert-Pucheta, J.E.; Mejía-Fonseca, I.; Zepeda-Vallejo, L.G.; Milmo-Brittingham, D.; Maya, G.P. The "Wine-T1" NMR experiment for novel wine-metabolome fingerprinting with nuclear-spin relaxation. *BIO Web Conf.* **2019**, *12*, 02029. [CrossRef]
17. Herbert-Pucheta, J.E.; Padilla-Maya, G.; Milmo-Brittinham, D.; Lojero, D.; Gilmore, A.M.; Raventós-Llopart, L.; Hernández-Pulido, K.E.; Zepeda-Vallejo, L.G. Multivariate spectroscopy for targeting phenolic choreography in wine with A-TEEMTM and NMR crosscheck non-targeted metabolomics. *BIO Web Conf.* **2019**, *15*, 02006. [CrossRef]

18. Jacob, D.; Deborde, C.; Lefevbre, M.; Maucourt, M.; Moing, A. NMRProcFlow: A graphical and interactive tool dedicated to 1D spectra processing for NMR-based metabolomics. *Metabolomics* **2017**, *13*, 36. [CrossRef] [PubMed]

19. Cerceau, C.I.; Barbosa, L.C.A.; Filomeno, C.A.; Alvarenga, E.S.; Demuner, A.J.; Fidencio, P.H. An optimized and validated 1H NMR method for the quantification of α-pinene in essential soils. *Talanta* **2016**, *150*, 97–103. [CrossRef]

20. Liu, Y.; Fan, G.; Zhang, J.; Zhang, Y.; Li, J.; Xiong, C.; Zhang, Q.; Li, X.; Lai, X. Metabolic discrimination of sea buckthorn from different *Hippophae* species by 1H NMR based metabolomics. *Sci. Rep.* **2017**, *7*, 1–10. [CrossRef]

21. Wider, G.; Dreier, L. Measuring protein concentrations by NMR spectroscopy. *J. Am. Chem. Soc.* **2006**, *128*, 2571–2576. [CrossRef]

22. Godelman, R.; Fang, F.; Humpfer, E.; Schutz, B.; Bansbach, M.; Schafer, H.; Spraul, M. Targeted and nontargeted wine analysis by 1H NMR spectroscopy combined with multivariate statistical analysis. Differentiation of important parameters: Grape variety, geographical origin, year of vintage. *J. Agric. Food Chem.* **2013**, *61*, 5610–5619. [CrossRef] [PubMed]

23. Abalos, D.; Vejarano, R.; Morata, A.; González, C.; Suárez-Lepe, J.A. The use of furfural as a metabolic inhibitor for reducing the alcohol content of model wines. *Eur. Food Res. Technol.* **2011**, *232*, 663–669. [CrossRef]

24. Loira, I.; Vejarano, R.; Morata, A.; Ricardo-Da-Silva, J.M.; Laureano, O.; González, M.C.; Suárez-Lepe, J.A. Effect of *Saccharomyces* strains on the quality of red wines aged on lees. *Food Chem.* **2013**, *139*, 1044–1051. [CrossRef]

25. Lleixà, J.; Martín, V.; Portillo, M.d.C.; Carrau, F.; Beltran, G.; Mas, A. Comparison of Fermentation and Wines Produced by Inoculation of *Hanseniaspora vineae* and *Saccharomyces cerevisiae*. *Front. Microbiol.* **2016**, *7*, 338. [CrossRef]

26. Martin, V.; Fariña, L.; Medina, K.; Boido, E.; Dellacassa, E.; Mas, A.; Carrau, F. Oenological attributes of the yeast *Hanseniaspora vineae* and its application for white and red winemaking. *BIO Web Conf.* **2019**, *12*, 02010. [CrossRef]

27. Medina, K.; Boido, E.; Fariña, L.; Dellacassa, E.; Carrau, F. Non-*Saccharomyces* and *Saccharomyces* strains co-fermentation increases acetaldehyde accumulation: Effect on anthocyanin-derived pigments in Tannat red wines. *Yeast* **2016**, *33*, 339–343. [CrossRef] [PubMed]

28. Carrau, F.M.; Medina, K.; Farina, L.; Boido, E.; Henschke, P.A.; Dellacassa, E. Production of fermentation aroma compounds by *Saccharomyces cerevisiae* wine yeasts: Effects of yeast assimilable nitrogen on two model strains. *FEMS Yeast Res.* **2008**, *8*, 1196–1207. [CrossRef]

29. Giorello, F.; Valera, M.J.; Martin, V.; Parada, A.; Salzman, V.; Camesasca, L.; Fariña, L.; Boido, E.; Medina, K.; Dellacassa, E.; et al. Genomic and transcriptomic basis of *Hanseniaspora vineae*'s impact on flavor diversity and wine quality. *Appl. Environ. Microbiol.* **2019**, *85*. [CrossRef]

30. Gómez-Míguez, M.J.; Cacho, J.F.; Ferreira, V.; Vicario, I.M.; Heredia, F.J. Volatile components of Zalema white wines. *Food Chem.* **2007**, *100*, 1464–1473. [CrossRef]

31. Zea, L.; Serratosa, M.P.; Mérida, J.; Moyano, L. Acetaldehyde as Key Compound for the Authenticity of Sherry Wines: A Study Covering 5 Decades. *Compr. Rev. Food Sci. Food Saf.* **2015**, *14*, 681–693. [CrossRef]

32. Martínez-García, R.; García-Martínez, T.; Puig-Pujol, A.; Mauricio, J.C.; Moreno, J. Changes in sparkling wine aroma during the second fermentation under CO_2 pressure in sealed bottle. *Food Chem.* **2017**, *237*, 1030–1040. [CrossRef] [PubMed]

33. Bartowsky, E.J.; Pretorius, I.S. Microbial formation and modification of flavor and off-flavor compounds in wine. In *Biology of Microorganisms on Grapes, in Must and in Wine*; Springer: Berlin/Heidelberg, Germany, 2009; pp. 209–231. ISBN 9783540854623.

34. Peinado, R.A.; Moreno, J.; Bueno, J.E.; Moreno, J.A.; Mauricio, J.C. Comparative study of aromatic compounds in two young white wines subjected to pre-fermentative cryomaceration. *Food Chem.* **2004**, *84*, 585–590. [CrossRef]

35. Vilanova, M.; Martínez, C. First study of determination of aromatic compounds of red wine from *Vitis vinifera* cv. Castañal grown in Galicia (NW Spain). *Eur. Food Res. Technol.* **2007**, *224*, 431–436. [CrossRef]

36. Moreno-Arribas, M.V.; Polo, M.C. Volatile and Aroma Compounds. In *Wine Chemistry and Biochemistry*; Springer: New York, NY, USA, 2008; p. 249.

37. Peinado, R.A.; Moreno, J.; Medina, M.; Mauricio, J.C. Changes in volatile compounds and aromatic series in sherry wine with high gluconic acid levels subjected to aging by submerged flor yeast cultures. *Biotechnol. Lett.* **2004**, *26*, 757–762. [CrossRef]
38. Murnane, S.S.; Lehocky, A.H.; Owens, P.D. *Odor Thresholds for Chemicals with Established Occupational Health Standards*, 2nd ed.; American Industrial Hygiene Association: Falls Church, VA, USA, 2013; ISBN 9781935082385.
39. Sun, Q.; Gates, M.J.; Lavin, E.H.; Acree, T.E.; Sacks, G.L. Comparison of odor-active compounds in grapes and wines from *Vitis vinifera* and non-foxy American grape species. *J. Agric. Food Chem.* **2011**, *59*, 10657–10664. [CrossRef] [PubMed]
40. Aronsson, K.; Rönner, U.; Borch, E. Inactivation of Escherichia coli, *Listeria innocua* and *Saccharomyces cerevisiae* in relation to membrane permeabilization and subsequent leakage of intracellular compounds due to pulsed electric field processing. *Int. J. Food Microbiol.* **2005**, *99*, 19–32. [CrossRef] [PubMed]
41. Martínez, J.M.; Cebrián, G.; Álvarez, I.; Raso, J. Release of mannoproteins during *Saccharomyces cerevisiae* autolysis induced by pulsed electric field. *Front. Microbiol.* **2016**, *7*. [CrossRef]
42. Bunduki, M.M.C.; Flanders, K.J.; Donnelly, C.W. Metabolic and structural sites of damage in heat- and sanitizer-lnjured populations of *Listeria monocytogenes*. *J. Food Prot.* **1995**, *58*, 410–415. [CrossRef]

 *fermentation*

Article

Comparison of the Glycolytic and Alcoholic Fermentation Pathways of *Hanseniaspora vineae* with *Saccharomyces cerevisiae* Wine Yeasts

María José Valera [1], Eduardo Boido [1], Eduardo Dellacassa [2] and Francisco Carrau [1,*

[1] Área de Enología y Biotecnología de Fermentaciones, Facultad de Química, Universidad de la República, 11800 Montevideo, Uruguay; mariajose_valera_martinez@hotmail.com (M.J.V.); eboido@fq.edu.uy (E.B.)
[2] Laboratorio de Biotecnología de Aromas, Facultad de Química, Universidad de la República, 11800 Montevideo, Uruguay; edellac@fq.edu.uy
* Correspondence: fcarrau@fq.edu.uy

Received: 11 July 2020; Accepted: 29 July 2020; Published: 3 August 2020

Abstract: *Hanseniaspora* species can be isolated from grapes and grape musts, but after the initiation of spontaneous fermentation, they are displaced by *Saccharomyces cerevisiae*. *Hanseniaspora vineae* is particularly valuable since this species improves the flavour of wines and has an increased capacity to ferment relative to other apiculate yeasts. Genomic, transcriptomic, and metabolomic studies in *H. vineae* have enhanced our understanding of its potential utility within the wine industry. Here, we compared gene sequences of 12 glycolytic and fermentation pathway enzymes from five sequenced *Hanseniaspora* species and *S. cerevisiae* with the corresponding enzymes encoded within the two sequenced *H. vineae* genomes. Increased levels of protein similarity were observed for enzymes of *H. vineae* and *S. cerevisiae*, relative to the remaining *Hanseniaspora* species. Key differences between *H. vineae* and *H. uvarum* pyruvate kinase enzymes might explain observed differences in fermentative capacity. Further, the presence of eight putative alcohol dehydrogenases, invertase activity, and sulfite tolerance are distinctive characteristics of *H. vineae*, compared to other *Hanseniaspora* species. The definition of two clear technological groups within the *Hanseniaspora* genus is discussed within the slow and fast evolution concept framework previously discovered in these apiculate yeasts.

Keywords: glycolysis; yeast; pyruvate kinase; non-*Saccharomyces*; fermentation evolution clade

1. Introduction

One of the main characteristics of yeast affecting their oenological use is their capacity to ferment sugars. Non-*Saccharomyces* yeasts have traditionally been considered bad fermenters. For that reason, selected strains of *S. cerevisiae* have been used in the oenological industry to ensure that complete fermentation occurs [1,2]. Recently, however, wineries have been encouraged to apply new, non-*Saccharomyces* species in winemaking processes to provide distinguishable flavours within wines [3–6]. Non-*Saccharomyces* species have been used to produce different aromas and flavours, compared with *Saccharomyces* strains [7,8]. Therefore, many efforts have been made to identify non-conventional yeast strains for oenological purposes [4,8,9].

The selection of oenological yeasts is commonly accomplished by identifying species from raw material. Spontaneously fermented grape musts are the niche that is most commonly used to identify novel strains that are both capable of fermenting sugars and confer desirable flavours to wines [10,11]. *Hanseniaspora* is the most abundant genus on grapes and grape juices. Studies have shown that up to 75% of the yeast population during the early stages of fermentation is made up of *Hanseniaspora* species [12,13]. After the first 48–72 h of spontaneous fermentation, the percentage of

Hanseniaspora species present decreases and the *S. cerevisiae* strains correspondingly increase. However, some *Hanseniaspora* species have been detected throughout the fermentative process [14].

Researchers have maintained that observed changes in yeast populations during fermentation occur, at least in part, because *Hanseniaspora* species are sensitive to ethanol [15]. *S. cerevisiae* is able to produce high quantities of ethanol rapidly. Therefore, this species dominates the fermentation process until sugar is completely depleted. Recent studies have shown the effects of antimicrobial peptides secreted by *S. cerevisiae* throughout the fermentation [16,17] inhibiting the growth of non-*Saccharomyces* yeasts. Therefore, the reputation of *Hanseniaspora* species as poor fermenters may be due to the presence of other inhibitors and not directly related to their reduced capacity to ferment sugars.

Not all the *Hanseniaspora* strains have the same properties. Species of the genus produce different secondary metabolites and exhibit different fermentative behaviours. In fact, differences could even be detected between strains belonging to the same species [18,19]. *H. vineae* is an epiphyte yeast that is not easily isolated from fruit, a feature it shares with all the *S. cerevisiae* strains [20]. *H. vineae* can be isolated from samples after one or two days of spontaneous fermentation in wines and other fruit beverages such as cider [19]. This highlights the distinct behaviour of the species, compared with the majority of *Hanseniaspora* apiculate yeasts, which are commonly isolated from the skin of grapes or grapevine soil. The ability of strains identified as *H. vineae* to complete grape juice fermentation has been demonstrated via single inoculation [7]. Moreover, selected strains of *H. vineae* contribute positively to wine aromas by providing floral and honey notes, even when sequential inoculation with *S. cerevisiae* strains was performed [21]. Our assessment of *H. vineae* showed that levels of phenylpropanoid flavour compounds synthetized from grape must were elevated compared with other yeasts. In *H. vineae*, the presence of metabolic pathways that actively transform aromatic amino acids explains the elevated phenyl acetate ester and benzenoid derived compounds synthesis compared to other yeasts and these flavour compounds provide fruity and flowery aromas [21–23].

Although several phenotypic studies have been carried out throughout wine fermentation using non-*Saccharomyces* species, there is a lack of information regarding the genetic basis of observed characteristics in non-*Saccharomyces* strains [8]. Due to the development of next generation sequencing, genomes of *Hanseniaspora* species from wine have been recently sequenced [24–27]. Further work will be needed to determine which genes are responsible for each function. In previous studies, the aromatic profile of *H. vineae* was correlated with genomics and transcriptomics data [22]. However, genes involved in glycolysis and fermentative behaviour in the species remain unknown. In *S. cerevisiae*, genes necessary for fermentation have been reported using mutant analysis. All of these, were grouped in a "fermentome" [28]. The genome of *H. guilliermondii* has been recently analysed [27] and the presence and absence of genes involved in the glycolytic and fermentative pathways compared with *S. cerevisiae* and other *Hanseniaspora* species were reported. Moreover, the *H. uvarum* glycolytic pathway has been assessed in a study that revealed the catalytic potentials of enzymes involved in the route [29]. The authors showed that the main glycolytic enzyme of *H. uvarum*, pyruvate kinase, had a 15-fold lower enzymatic activity than that of the *S. cerevisiae* enzyme.

The aim of this work is to establish the differences and similarities between *H. vineae*, other *Hanseniaspora* species, and *S. cerevisiae* regarding glycolytic and fermentative behaviour. In the present study, a comparative analysis of the fermentative capacity of *H. vineae* was performed using genetic and transcriptomic data. Characterization of the glycolytic and fermentative potential of *H. vineae* will enhance our understanding about the mechanisms and the regulation of the fermentative process in a non-*Saccharomyces* yeast. *Hanseniaspora* genus studies might help reveal new signs of *S. cerevisiae* domestication mechanisms for wine production.

2. Materials and Methods

2.1. Yeast Strains

Yeast strains used for this study are listed in Table 1.

Table 1. Yeast strains used in this study.

Species	Strain	Source	Use
H. vineae	T02/19AF	Fermenting Tannat grape must (Uruguay)	Genomic, transcriptomic, phenotypic analysis
H. vineae	T02/05AF	Fermenting Tannat grape must (Uruguay)	Genomic and phenotypic analysis
H. osmophila	AWRI3579	Fermenting Chardonnay grape must (Australia)	Genomic and phenotypic analysis
H. uvarum	AWRI1280	Fermenting Tannat grape must (Uruguay)	Phenotypic analysis
H. uvarum	AWRI3580	Fermenting Chardonnay grape (Australia)	Genomic analysis
H. opuntiae	AWRI3578	Fermenting Chardonnay grape (Australia)	Genomic analysis
H. valbyensis	NRRL Y-1626	Soil (Denmark)	Genomic analysis
H. guilliermondii	UTAD222	Grape must (Portugal)	Genomic analysis
S. cerevisiae	288Sc	Laboratory strain	Genomic analysis
S. cerevisiae	ALG804	Oenological yeast (Oenobrands®)	Phenotypic analysis

2.2. Fermentation in Natural Grape Must

Chardonnay grape must containing 300 mg N/L and 200 g/L of sugars at pH 3.5 was treated with 200 mg/L dimethyldicarbonate to prevent microorganism growth. Pre-cultures of *H. vineae* T02/19AF, *H. vineae* T02/05AF, *H. uvarum* AWRI1280, *H. osmophila* AWRI3579, and *S. cerevisiae* ALG804 were isolated from the Chardonnay grape must and incubated at 25 °C for 12 h in a rotary shaker at 150 rpm. Then, 125-mL Erlenmeyer flasks closed with cotton plugs used to simulate microaerobic conditions were inoculated with 75 mL of must containing 1×10^5 cells/mL. Static batch fermentations were conducted at 20 °C to simulate winemaking conditions.

2.3. Growth Kinetics in Different Types of Media

Six types of growth media were prepared using yeast nitrogen base (YNB) (Difco, Detroit, MI, USA) as a sole nitrogen source (6.7 g/L). Media were supplemented with the following carbon sources: Glucose, fructose, sucrose, xylose, glycerol, and maltose (2% w/w). YNB that lacked a carbon source was used as a negative control.

Chardonnay must used in the fermentation analysis was also used to measure the growth kinetics of yeast strains tested. Moreover, synthetic media that mimicked must fermentations at pH adjusted to 3.5 and ethanol concentrations of either 5% or 10% were used (20 g/L glucose, 4 g/L tartaric acid; 0.134 g/L sodium acetate; 5 g/L glycerol; and 1.7 g/L YNB) (v/v).

Pre-cultures of *H. vineae* T02/19AF, *H. vineae* T02/05AF, *H. uvarum* AWRI1280, *H. osmophila* AWRI3579, and *S. cerevisiae* ALG804 were prepared in yeast extract peptone dextrose (YPD) media (1% yeast extract and 2% peptone, 2% glucose) via incubation for 12 h in a rotary shaker at 150 rpm and 25 °C. These pre-cultures were used to inoculate fermentations carried out in microtitler plates at a final volume of 250 μL. Inoculates producing 1×10^5 cells/mL in media were used for all strains and treatments. All conditions were tested in triplicate. Absorbance at 620 nm was measured at 30-min intervals for 48 h at 25 °C using an automatic plate reader (Tecan, Männedorf, Switzerland) and data were acquired with the Magellan software for further statistical analyses.

2.4. Fermentation Ability in Different Carbon Sources

The carbohydrate fermentation capacity was tested using Durham tubes immersed in media to detect gas production. Each type of medium tested was inoculated to produce a final concentration of 1×10^6 cells/mL in a final volume of 8 mL performed in triplicate. Results were visually assessed after a 48 and 96 h static incubation period at 28 °C.

2.5. Genomic Analysis

Genomic DNA was obtained from *H. vineae* cultures grown in a YPD medium at 30 °C using the Wizard Genomic DNA Purification Kit (Promega, NY, USA), according to the manufacturer's instructions.

The Illumina Genome Analyzer Iix platform in paired end mode was used to perform genomic sequencing as described previously [22]. Gene prediction was carried out using Augustus [30] trained with *S. cerevisiae* gene models. Peptide predictions were then annotated using BLASTp (cutoff for e-value 1^{-10}) against *S. cerevisiae* proteins, obtained from the *Saccharomyces* Genome Database [31].

A dendrogram was constructed using the sequences of nine genes encoding components of pathways related to glycolysis and fermentation from the *Hanseniaspora* species and *S. cerevisiae*. The genes assessed were *CDC19*, *FBA1*, *PGI1*, *PFK1*, *PFK2*, *HXK2*, *ENO1*, *PGK1*, and *PDC1*. *Schyzosaccharomyces pombe* was used as an external group. Neighbour joining and Kimura 2-parameter methods were carried out using the MEGA version 4 software [32,33].

2.6. Transcriptomic Analysis

Fermentations were performed in triplicate using chemically defined grape (CDG) must with a composition similar to that of natural grape juice, but devoid of grape precursors. Components of CDG must were defined as described in Carrau et al. [34], with modifications. Briefly, glucose and fructose were added in equimolar concentrations until a total sugar concentration of 200 g/L was reached. Vitamins and salts were added as previously described [35]. Yeast available nitrogen (YAN) content was adjusted to 100 mg N/L. Of this total, 50 mg N/L corresponded to amino acids and 50 mg N/L corresponded to diammonium phosphate (DAP) supplementation, as described previously [35]. The pH of the media was adjusted to 3.5 using HCl and a final concentration of 10 mg/L ergosterol was the only lipid provided.

Pre-cultures of *H. vineae* T02/19AF were prepared in a CDG medium and incubated 12 h in a rotary shaker at 150 rpm and 25 °C. The pre-cultures were subsequently used to inoculate fermentation reactions carried out in 250 mL Erlenmeyer flasks that were closed with cotton plugs to simulate microaerobic conditions. For all strains, fermentations were performed using 125 mL CDG and an inoculum to produce 1×10^5 cells/mL in the final medium. Static batch fermentations were conducted at 20 °C to simulate winemaking conditions.

Wine samples for transcriptomic analyses were taken during the fermentation process at day 1 (exponential growth), day 4 (end of exponential phase), and day 10 (stationary phase of fermentation). For transcriptomic studies, total RNA obtained from *H. vineae* T02/19AF isolated from three replicates sampled from three different fermentation stages (days 1, 4, and 10) were analysed independently. The nine samples were paired-end sequenced using Illumina MySeq. Trinity software was used to assemble raw reads from transcriptomic analyses and further statistical analyses were performed as specified by Giorello et al. [22].

2.7. Statistical Analysis

All the treatments were performed in triplicate and the statistical error was calculated as the standard deviation of all data analysed. To compare growth and fermentation kinetics, variance comparison was performed by the ANOVA test carried out with STATISTICA 7.0 software. Differences in the mean absorbance or weight loss were evaluated using the Tukey test.

3. Results and Discussion

3.1. Fermentative Capacity of H. vineae in Different Media

Hanseniaspora species used a limited number of carbon sources, which may have been related to the reduced competitiveness of the species throughout fermentations [27]. Regarding growth in different carbon sources (Figure 1A), growth of all the *Hanseniaspora* strains tested on both glucose and fructose had kinetics similar to that of *S. cerevisiae* ALG804. The media supplemented with sucrose was fermented by *S. cerevisiae* in a similar manner as that of media containing simple hexose. *H. uvarum* AWRI1280, however, did not grow on media containing sucrose. *H. vineae* T02/05AF, *H. vineae* T02/19AF, and *H. osmophila* AWRI3579 were able to grow on and ferment sucrose to an extent. Invertase gene (*SUC2*) is present in the genome of *H. vineae*. *SUC2* is highly expressed on day 4 of fermentation reactions, but not day 1 or 10. However, other invertase homologs were not observed in the genomes of any other *Hanseniaspora* species except *H. osmophila* [25]. Recently, Steenwyk et al. [36] grouped *H. vineae* and *H. osmophila* within the slower-evolving linage of *Hanseniaspora*. In this branch, the *SUC2* gene is present. This is different in species of the fast evolving linage including *H. uvarum*, *H. opuntiae, H. valbyensis,* and *H. guilliermondii,* which might have lost the gene as a result of rapid mutation rates [36]. The same fact was detected with another key gene that show *Saccharomyces* wine yeast adaptations. Increased sulfite tolerance conferred by *SSU1* (Table 2) is present in *H. vineae* and *H. osmophila* and it is absent in the other *Hanseniaspora* species. The presence of *SUC2* and *SSU1* genes are indicators of adaptations to alcoholic fermentation in yeast [37].

Glycerol was not used as a unique carbon source for *H. vineae* in accordance with data reported by Albertin et al. [38]. However, Hv*GUT1* and Hv*GUT2* genes were present in the genomes of both *H. vineae* strains analysed. In addition, xylose was not used by the *H. vineae* strains as expected. A finding that was likely due to the lack of enzymes needed to carry out the xylose conversion. The group of genes were also determined to be absent in *H. guilliermondii, H. uvarum,* and *H. opuntiae* [27]. However, *H. vineae* T02/05AF and T02/19AF have the ability to grow weakly when maltose is provided as a sole carbon source, despite the fact that they were not able to ferment the sugar. The same behaviour was also observed for *H. jakobsenii* [36].

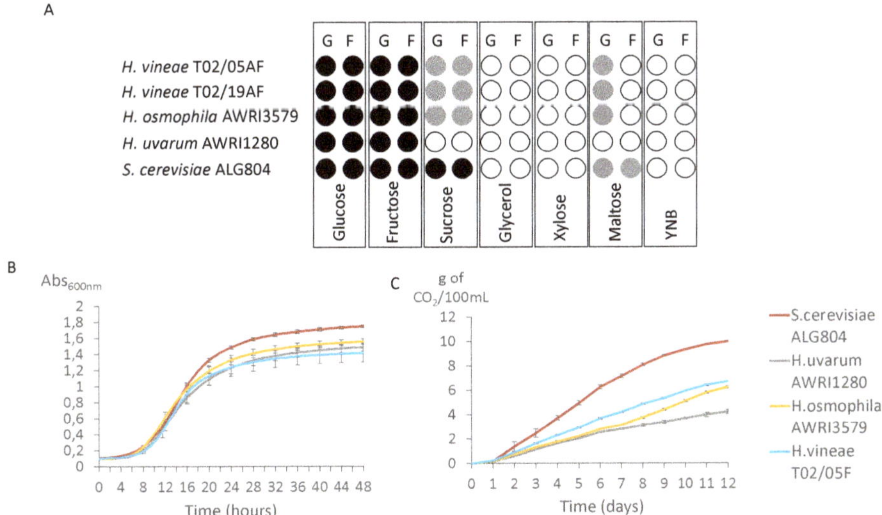

Figure 1. Capacity of *Hanseniaspora vineae* and *Saccharomyces cerevisiae* to grow and ferment under varied conditions. (**A**) Growth of *H. vineae* and *S. cerevisiae* (G) and the capacity of the species ferment (F) when six different carbon sources (2% w/w) were provided. Yeast nitrogen base (YNB) that lacked a carbohydrate was used as a negative control. Black filled circles indicate that full growth and fermentation were observed, grey circles indicate the moderate capacity of species to grow and ferment, and white circles indicate that the species was not able to grow or ferment. (**B**) Growth kinetics of *Hanseniaspora spp.* and *S. cerevisiae* on the Chardonnay grape juice measure as increased absorbance over a period of 48 h. (**C**) Fermentation kinetics of the three strains in the Chardonnay grape juice after 12 days are shown. Growth and fermentation experiments were performed using independent triplicate samples and error bars express standard deviation.

Table 2. Genes involved in sugar transport, glycolysis, and alcoholic fermentation from *S. cerevisiae* and *H. vineae*. Gene copy numbers are detailed in brackets.

	S. cerevisiae	*H. vineae*
Sugar transport and sensors	*HXT* (x17); *SNF3*; *RGT2*; *FPS1*; *GPR1*; *GUP1*; *GUP2*; *STL1*; *JEN1*; *ASC1*; *ASC2*; *GPA2*	*HXT* (x2); *SNF3*; *GPR1*; *GUP1*; *STL1* (x2); *JEN1*; *ASC1*; *GPA2*
Glycolysis	*HXK1*; *HXK2*; *PGI1*; *PFK1*; *PFK2*; *FPBA1*; *TPI1*; *TDH1*; *TDH2*; *TDH3*; *PGK1*; *GPM1*; *ENO1*; *ENO2*; *CDC19*; *PYK2*	*HXK2*; *PGI1*; *PFK1*; *PFK2*; *FPBA1*; *TPI1*; *TDH2*; *TDH3*; *PGK1*; *GPM1*; *ENO1*; *ENO2*; *CDC19*
Alcoholic fermentation	*PDC1*; *PDC2*; *PDC5*, *PDC6*; *ADH* (x8)	*PDC1*; *ADH* (x8)
Key genes of wine yeasts adaptations	*SSU1*; *CUP1* (x2); *SUC2*; *THI5*; *THI11*, *THI12*, *THI13*; *THI14*; *THI16*; *THI20*; *THI21*; *THI72*; *THI73*; *THI80*; *TPC1*	*SSU1*; *SUC2* *THI7*; *THI72*; *THI80*; *TPC1*

As expected, *Saccharomyces* was able to grow and ferment sugars faster than *Hanseniaspora* species and significant differences between the species occurred after 16 h. The growth kinetics of the three *Hanseniaspora* strains tested were similar on grape must (Figure 1B), however fermentation kinetics of *H. vineae* and *H. osmophila* revealed that these species consume sugars significantly faster than *H. uvarum* (Figure 1C).

3.2. Sugar Transport

The transport of sugars into the cytosol of cells is a key step of the glycolytic pathway. *S. cerevisiae* is able to detect extracellular nutrients and make metabolic adjustments that rapidly facilitate the use of extracellular compounds [39].

Of the multiple sensors described in *S. cerevisiae*, *H. vineae* possessed the following genes Hv*SNF3*, Hv*GPA2*, Hv*GPR1*, and Hv*ASC1*, which were determined to be associated with the hexose sensing capacity of both T02/19AF and T02/05AF strains (Table 2). Sc*SNF3* encodes a low glucose sensor present in the plasma membrane that is involved in the regulation of glucose transport and also has the capacity to sense fructose and mannose in *S. cerevisiae*. Expression of the gene in *H. vineae* increases throughout fermentation (Figure 1). Sc*GPA2*, Sc*GPR1*, and Sc*ASC1* are hexose sensors that have been reported to be necessary for fermentation and are part of the "fermentome" in *S. cerevisiae*. Deletion of the genes was previously reported to induce protracted fermentation [28]. Hv*GPA2* and Hv*GPR1* have similar expression patterns throughout the fermentation process. The genes are maximally expressed on day 4 and their expression levels decrease at day 10. On the other hand, Hv*ASC1* is most highly expressed on the first day of fermentation and levels were drastically reduced both on day 4 and 10 relative to day 1 (Figure 2).

S. cerevisiae possesses 20 sequences putatively associated with the hexose transport [40]. *H. guilliermondii* UTAD222 possess 22 sugar transporters, and based on their DNA sequences, ten were predicted to be associated with the hexose transport, all of them were most similar to *HXT2* [27]. A comparison of sugar transporters of both sequenced strains of *H. vineae* with *S. cerevisiae* revealed that T02/05AF had two copies of the Hv*HXT6* gene and one copy of Hv*HXT1*. Sequences homologous to Sc*HXT2* were not found in the species. However, strain T02/19AF was determined to have a single copy of Hv*HXT6*. Expression levels of the gene increased after day 10 of fermentation. No sequences homologous to *HXT1* were identified. Sc*HXT1* is a low affinity hexose and pentose transmembrane transporter and is paralogous to Sc*HXT6* [41]. The Sc*HXT6* gene encodes a high affinity hexose transmembrane transporter that transports glucose, fructose, and mannose [42]. Tondoni et al. [43] revealed that in *S. cerevisiae* and *Torulaspora delbrueckii*, *HXT6* is most highly expressed throughout late stages of fermentation (Figure 2). In addition, both T02/05AF and T02/19AF strains have one sequence that is homologous to the *S. cerevisiae* Sc*HXT7* gene. Sc*HXT7* is a high-affinity glucose transporter that is very similar to Sc*HXT6* [42,44]. This gene is maximally expressed at day 4 at the end of the exponential phase of fermentation in *H. vineae*. Both *H. vineae* strains sequenced lacked polyol transporters (such as Sc*HXT13*, Sc*HXT17*, or Sc*HXT16* of *S. cerevisiae*) needed for the uptake of sorbitol and mannitol.

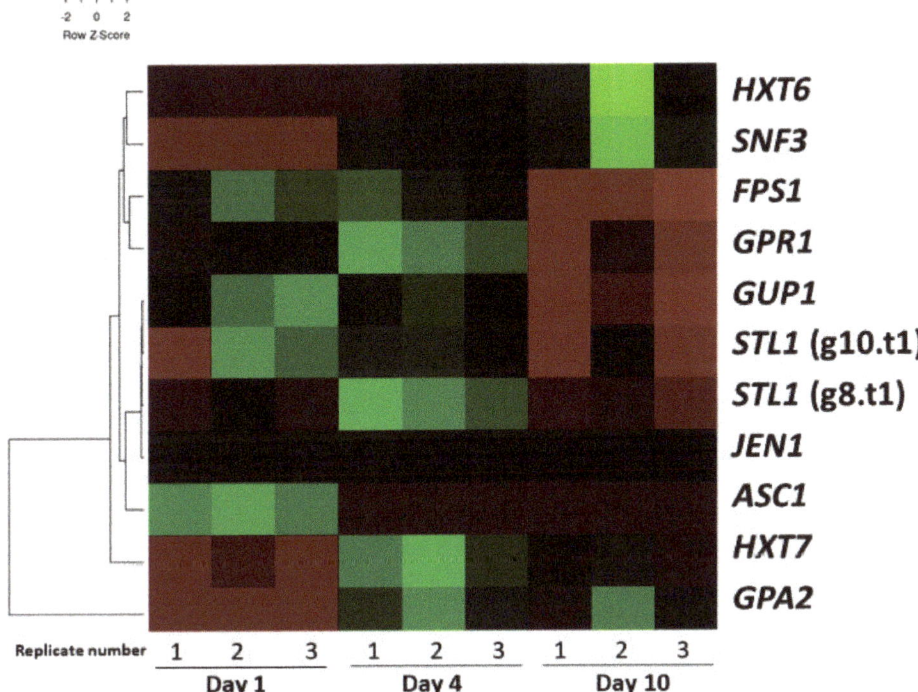

Figure 2. Heatmap depicting the expression levels of genes putatively involved in sugar detection and transport in *H. vineae* after 1, 4, and 10 d of fermentation. The green colour indicates elevated expression levels and the red colour indicates reduced expression levels. Data are shown in triplicate.

Other homologs of sugar transporters have been found in *H. vineae*. Three tandem copies of Hv*STL1* were identified in both T02/19AF and T02/05AF strains that shared homology with a glycerol proton symporter of the plasma membrane, which has been shown to be inactivated in response to glucose in *S. cerevisiae* [45]. Other *Hanseniaspora* species sequenced, such as *H. osmophila*, *H. opuntiae*, *H. guilliermondii*, *H. uvarum*, and *H. valbyensis* also possessed between two and four copies of the gene. According to the transcriptomic analyses, just one of the copies identified was differentially expressed throughout fermentation in *H. vineae* (Figure 2).

Hv*FPS1*, a putative plasma membrane channel involved in glycerol and xylitol movement, is present in the genome of both T02/19AF and T02/05AF. Expression of the gene is elevated near the beginning of fermentation reactions (days 1 and 4) and decreased at day 10 (Figure 1). One copy of Hv*GUP1* was present in each strain analyzed as well as Hv*JEN1*. Moreover, it was suggested that Sc*GUP1* participates in glycerol transport and Sc*JEN1* mediates the high-affinity uptake of lactate, pyruvate, and acetate so that they can be used as carbon sources in *S. cerevisiae* [46,47].

3.3. Glycolytic Pathway in H. vineae Strains

The first enzyme of the glycolysis pathway is a hexokinase (Figure 3). Sc*HXK2* phosphorylates glucose in the cytosol. In *S. cerevisiae*, this isoform is principally responsible for glucose activation, which is needed to initiate glycolysis when glucose is provided as a carbon source and inhibits Sc*HXK1* [41]. However, in *H. vineae*, Hv*HXK2* was the only enzyme identified with putative hexokinase activity, amino acid homology was higher compared to other species of this genus.

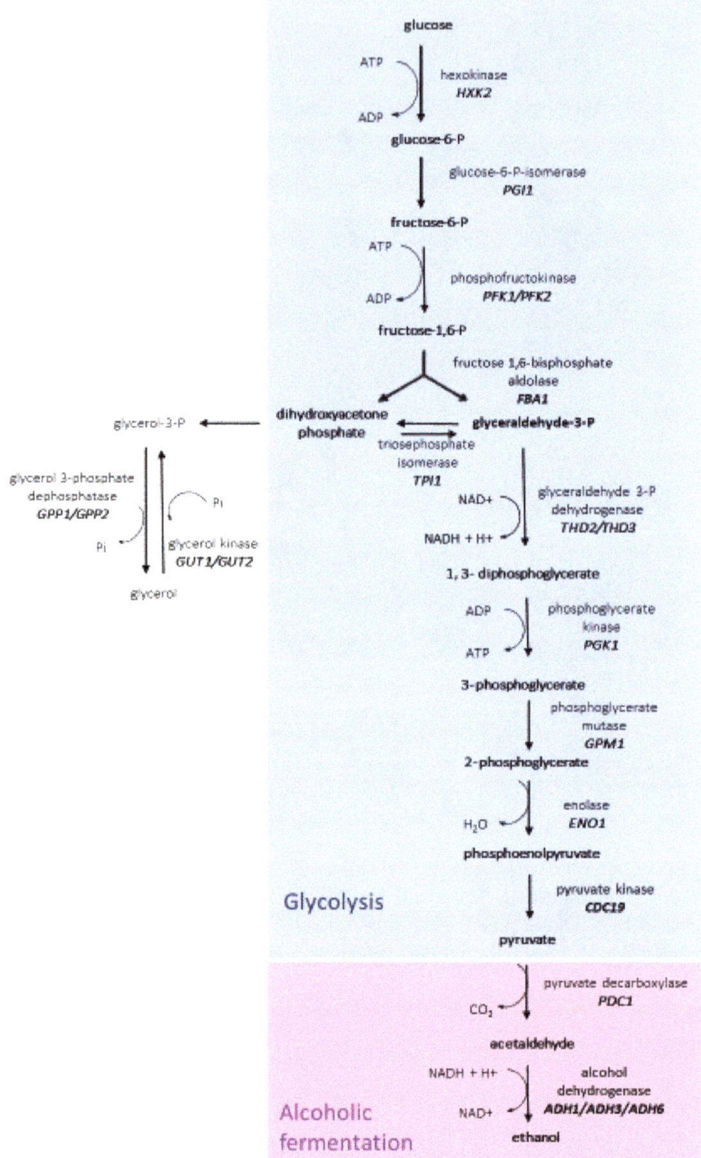

Figure 3. Glycolysis and alcoholic fermentation pathways in yeast. Genes putatively predicted to be involved in the catabolic pathway based on sequence data from genomic analyses of *Hanseniaspora vineae* strains are presented.

Phosphofructokinase activity was determined to be the second-most important glycolytic enzyme. The enzyme determines fermentation capacity and is indispensable for anaerobic growth. In *S. cerevisiae*, the enzyme is composed of two alpha and beta subunits that are encoded by Sc*PFK1* and Sc*PFK2*, respectively. *Hanseniaspora* strains possess sequences homologous to both Sc*PFK1* and Sc*PFK2* subunits and similar to *S. cerevisiae*, the subunits form a hetero-octameric complex [29]. Protein sequences of both Hv*PFK1* and Hv*PFK2* were most similar to *S. cerevisiae* (76.78% and 79.24%) and *H. osmophila* (76.46% and 77.50%) relative to the other *Hanseniaspora* species assessed (Figure 4A). Phosphofructokinase only works in the forward direction and is not involved in gluconeogenesis. In fact, three activities are required for gluconeogenesis: Pyruvate carboxylase, phosphoenolpyruvate carboxykinase, and fructose-1,6-bisphosphatase. No genes encoding the key gluconeogenic enzymes have been identified in *H. vineae*, *H. guilliermondii*, *H. uvarum*, *H. osmophila*, or *H. valbyensis* [27]. This explains why *Hanseniaspora* species are not able to grow when non-carbohydrate precursors such as pyruvate, amino acids, or glycerol are provided as energy sources. This is different than *S. cerevisiae*, which is able to grow on a variety of carbon sources including ethanol and lactate [48].

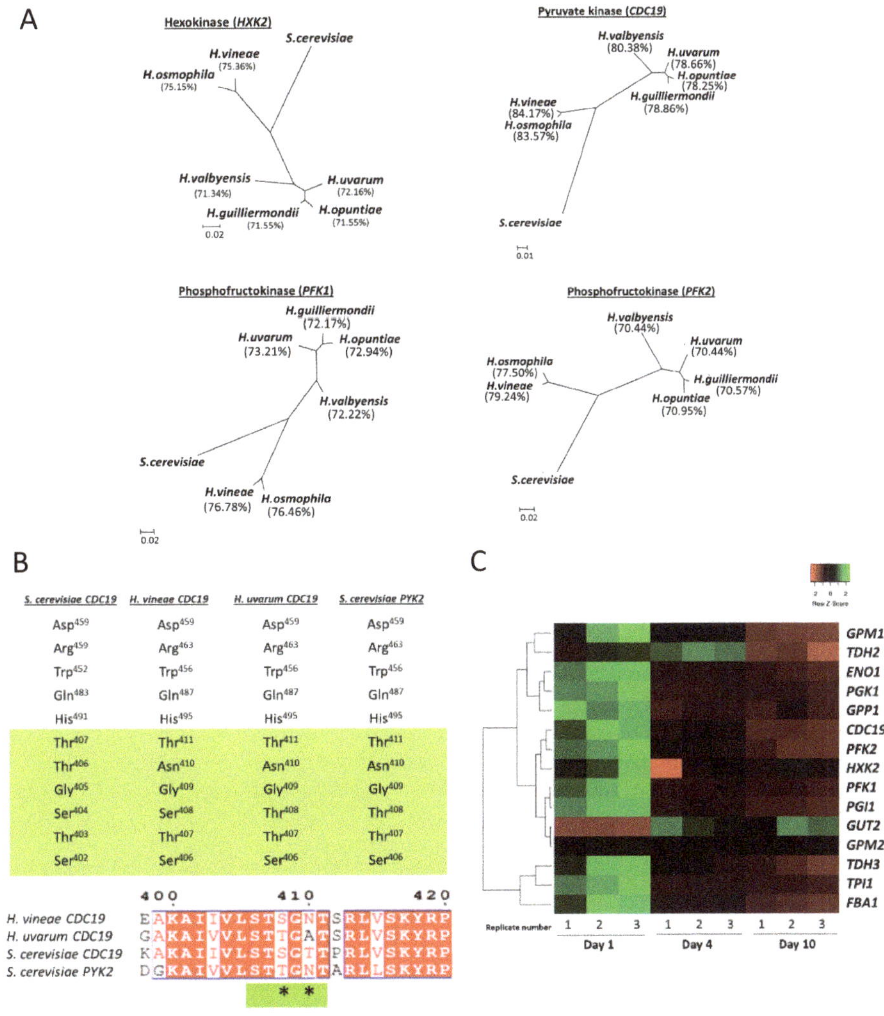

Figure 4. Genes involved in glycolysis. (**A**) Dendrograms showing the genetic distances between predicted amino acid sequences of enzymes involved in glycolysis from seven *Hanseniaspora* species and the *Saccharomyces cerevisiae* S288c strain. Amino acid homology was calculated for each *Hanseniaspora* strain against *S. cerevisiae*. (**B**) Amino acid sequences that correspond to the binding domains of fructose

1,6-bisphosphate inducer of pyruvate kinase in *S. cerevisiae*. Sc*CDC19* and Sc*PYK2* genes were compared with predicted sequences of *CDC19* from *H. uvarum* and *H. vineae*. Amino acids corresponding to the region that differ from *CDC19* and *PYK2* are highlighted in green and the position of residues are marked with an asterisk (*). (**C**) A heatmap describing the expression levels of genes putatively determined to be involved in glycolytic pathways of *H. vineae* 1, 4, and 10 days after the initiation of fermentation. Green and red colours indicate high and low levels of expression, respectively. Data are shown in triplicate.

Predicted amino acid sequences of phosphoglucose isomerase (*PGI1*) from *H. vineae* and *H. osmophila* were 86% similar to that of *S. cerevisiae*. Predicted *PGI1* amino acid sequences from *H. uvarum*, *H. valbyensis*, *H. guilliermondii*, and *H. opuntiae* were approximately 71% similar to *S. cerevisiae*. This tetrameric enzyme is involved in the interconversion of glucose-6-phosphate and fructose-6-phosphate. Phosphoglucose isomerase activity has also been associated with the regulation of the cell cycle and gluconeogenic events of sporulation in *S. cerevisiae* [49,50].

Two copies of the *S. cerevisiae* Sc*ENO1* gene that encodes an enolase were identified in *H. osmophila*, while only one copy was identified in other sequenced *Hanseniaspora* species. This enzyme catalyzes the conversion of 2-phosphoglycerate to phosphoenolpyruvate during glycolysis. Glyceraldehyde-3-phosphate dehydrogenase (GAPDH) is a tetramer that catalyzes the conversion of glyceraldehyde-3-phosphate to 1,3 bis-phosphoglycerate. Three unlinked genes, Sc*TDH1*, Sc*TDH2*, and Sc*TDH3*, encode related, but not identical, polypeptides that form catalytically active homotetramers with different specific glyceraldehyde 3-phosphate dehydrogenase activities in *S. cerevisiae* [51,52]. In *H. vineae*, only *TDH2* and *TDH3* homologues have been identified, and both were differentially expressed throughout fermentation (Figure 4C).

H. vineae strains T02/19AF and T02/05AF possess Hv*GUT1* and Hv*GUT2* genes. Both Sc*GUT1* and Sc*GUT2* are associated with glycerol kinase activities in the cytoplasm and mitochondria, respectively. Glycerol degradation is a two-step process that is mediated by *GUT1* and/or *GUT2*. Under aerobic conditions, *S. cerevisiae* is able to utilize glycerol as a sole carbon and energy source [53]. Both of the enzymes have homologs that have been identified in *H. vineae* and *H. osmophila*, other *Hanseniaspora* species such as *H. uvarum*, *H. guilliermondii*, and *H. opuntiae* lack homologous of these genes [27].

Several specific activities associated with glycolytic enzymes of *S. cerevisiae* and *H. uvarum* have high degrees of similarity, which highlights the general conservation of glycolytic pathways and the downstream reactions involved in ethanol production [29]. Pyruvate kinase is a key enzyme that catalyses an irreversible step of the glycolytic pathway. The position of the enzyme at the branchpoint between fermentation and respiration makes it a key determinant energy metabolism [54]. Recent work revealed that the pyruvate kinase activity enhanced the capacity of *S. cerevisiae* to ferment sugars versus *H. uvarum* [29]. The predicted proteins, Cdc19p, of *H. vineae* and *H. osmophila* are more homolog to the corresponding Cdc19p of *S. cerevisiae* than those of *H. uvarum* and other *Hanseniaspora* species (Figure 4A). When residues of the catalytic domain of ScCdc19p [55] are compared with those of *H. uvarum* and *H. vineae*, only one amino acid difference was identified; Asp^{265} was substituted with Gly^{269} in *H. uvarum* and *H. vineae* (Figure 4B). However, in the binding site of the allosteric activator, fructose 1,6-bisphosphate, two amino acid differences between *H. uvarum* and *S. cerevisiae* and one between *H. vineae* and *S. cerevisiae* were identified. The two differences identified between *H. uvarum* and *S. cerevisiae* are at the same positions (Figure 4B) as those identified in the *PYK2* gene of *S. cerevisiae*, a paralog of *CDC19* that is characterized by its low pyruvate kinase activity compared with the pyruvate kinase protein encoded by *CDC19* (formerly *PYK1*) [54].

Expression levels of 13 *H. vineae* genes involved in the glycolytic pathway mainly decreased from day 1 to day 4 of fermentation and were maintained throughout the stationary phase (Figure 4C). This finding is in agreement to previous observations in *S. cerevisiae* [43]. However, levels of Hv*TDH2* expression remained high at both days 1 and 4 and decreased expression levels were observed at day 10. Additionally, expression of Hv*GUT2* peaked at days 4 and 10, and increased expression levels of the gene were not detected at day 1. Finally, Hv*GPM2* was not expressed under the conditions assessed.

3.4. Alcoholic Fermentation in H. vineae Strains

The pyruvate decarboxylase activity plays a key role in the alcoholic fermentation pathway. Three different pyruvate decarboxylase isozymes have been identified in the genome of *S. cerevisiae*: Sc*PDC1*, Sc*PDC5,* and Sc*PDC6*. The function of pyruvate decarboxylase is the degradation of pyruvate into acetaldehyde and carbon dioxide. The enzyme is responsible for transferring the final product of glycolysis (pyruvate) to ethanol production [56]. In *H. vineae*, no sequences homologous to Sc*PDC5* and Sc*PDC6* were found and Hv*PDC1* was the only pyruvate decarboxylase isozyme identified in the species. In *S. cerevisiae*, Sc*PDC1* was strongly expressed in fermenting cells. The enzyme is conserved among yeast, bacteria, and plants. It is regulated by glucose and ethanol concentrations and also by itself [57]. The active enzyme has a homotetrameric structure and the enzyme has two known cofactors: Thiamin diphosphate (ThDP) and Mg^{2+} [58–60]. In *H. vineae*, genes involved in thiamine biosynthesis have not been identified and a similar finding was also reported in *H. guilliermondii* [27] and most other *Hanseniaspora* species [36]. It has been suggested that this may contribute to the low alcoholic fermentative capacity of *Hanseniaspora* species, the phenotype has been shown to be related to the weak pyruvate kinase activity of *H. uvarum* [29]. *S. cerevisiae* genes associated with thiamine production are upregulated in the stationary phase of growth. Oenological strains with improved expression levels of the genes have corresponding elevated rates of fermentation [61]. This phenomenon may result from vitamin depletion that occurs after the exponential phase.

Alcohol dehydrogenases, which catalyse the conversion of acetaldehyde to ethanol are key fermentative enzymes. Many alcohol dehydrogenases have been identified in *S. cerevisiae* including Sc*ADH1*, Sc*ADH2*, Sc*ADH3*, Sc*ADH4*, Sc*ADH6*, and Sc*ADH7*. Many homologues of *S. cerevisiae* alcohol dehydrogenases have been found in the *H. vineae* genome. *H. vineae* has the same number of copies of the genes as *S. cerevisiae*. Eight alcohol dehydrogenase genes are present in *H. vineae* species, compared to six in *H. osmophila*, and four in other sequenced species of *Hanseniaspora* such us *H. uvarum*, *H. guilliermondii*, *H. valbyensis*, and *H. opunti*ae. This may explain the improved adaptation of *H. vineae* to alcohol fermentation relative to other *Hanseniaspora*. It is noteworthy that of the eight Hv*ADH* sequences found in the genome of *H. vineae*, at least three Hv*ADH6* genes are encoded in tandem. Increased copies of the gene may be associated with increased fermentation capacity, indicating that the alcohol dehydrogenase activity might be a key feature of alcoholic fermentation adaptations [62]. *H. vineae* has an enhanced tolerance to ethanol (Figure 5B) versus *H. uvarum* and *H. osmophila*, which are unable to grow in media containing 10% ethanol.

H. vineae and *H. osmophila* genes encoding putative alcohol dehydrogenases were grouped in two main clusters that contained either *ADH1*, *ADH2* and *ADH3* or *ADH6* and *ADH7* (Figure 5A), this is in agreement with the two multigenic families reported by Giorello et al. [22]. The clusters were formed according to the clustal alignment of predicted protein sequences, however, regarding adscription by a single homology with *S. cerevisiae ADHs* in the databases [22] produced some discrepancies. Therefore, Hv*ADH6* homologs from *H. vineae* and *H. osmophila* were removed from the Hv*ADH6* and Hv*ADH7* cluster. Moreover, the Hv*ADH1* homologous sequence of *H. vineae* is grouped in the cluster of Sc*ADH6* and Sc*ADH7*.

Hv*ADH* genes display different expression patterns (Figure 5C). Two of four paralogous copies of Hv*ADH6* were not differentially expressed at the time points analysed. Expression of one copy of *ADH6* significantly declined between days 1 and 4 of fermentation. In addition, the expression of one copy of *ADH3* was elevated on day 4 relative to day 1 (Figure 5C). These behaviours are similar to those of aryl alcohol dehydrogenases that facilitate the production of increased levels of alcohol by *S. cerevisiae* [63]. Therefore, Hv*ADHs* may be important for reducing levels of fusel aldehydes by producing increased levels of alcohol in *H. vineae* [22].

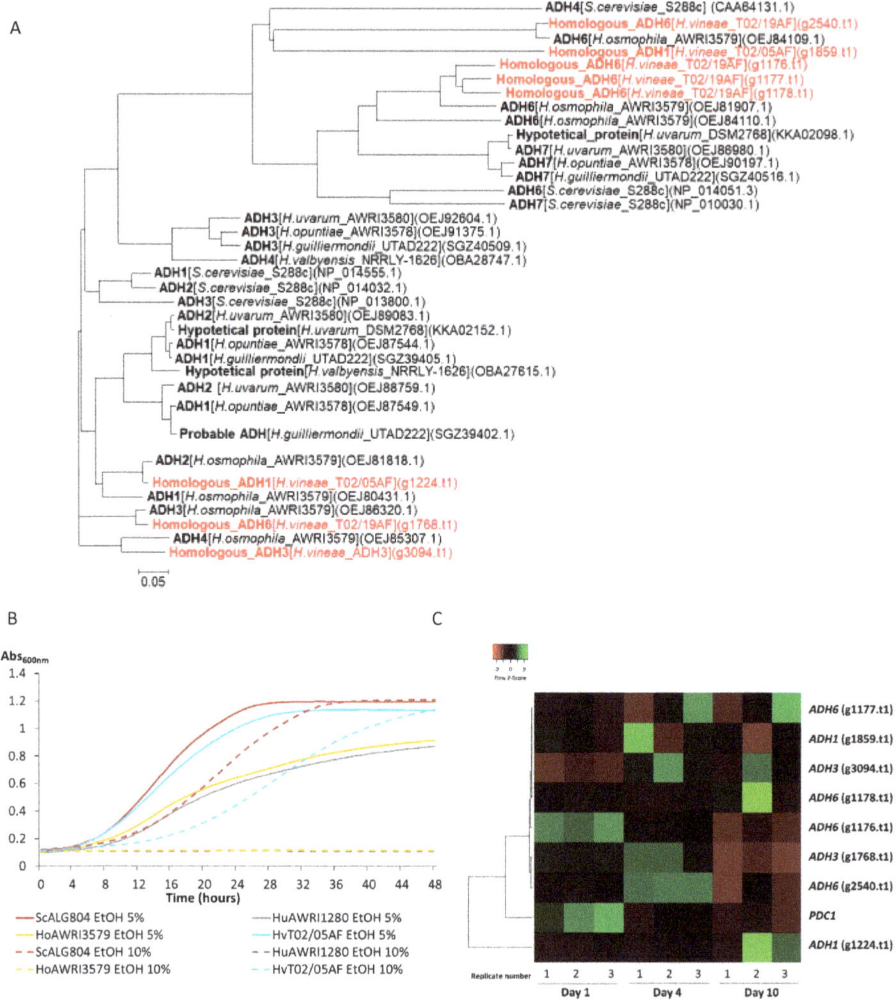

Figure 5. Characteristics that facilitate fermentation. (**A**) Dendrogram depicting relationships between the predicted amino acid sequences of several putative *ADH* genes. *Hanseniaspora vineae* sequences are indicated in red. (**B**) Growth of *Hanseniaspora* species and *Saccharomyces cerevisiae* in synthetic wine containing 5% of ethanol (solid line) and 10% ethanol (dotted line) for 48 h. Error bars are not shown to enhance clarity. SD < 0.05 for all samples. (**C**) A heatmap depicting expression levels of genes putatively involved in glycolytic pathways in *H. vineae* after 1, 4, and 10 days of fermentation. Green and red colours indicate high and low levels of expression, respectively. Data are shown in triplicate.

3.5. Hanseniaspora Genus as an Evolution Model for Alcoholic Fermentation Adaptations

The glycolytic potential of two strains of *H. vineae* were analysed using genetic, transcriptomic, and phenotypic data. Results explained the good performance of the species with respect to fermenting wine [7,21]. Findings also showed that the *H. vineae* behaviour was similar to traditional *S. cerevisiae* strains used in winemaking. Due to the outstanding capacity of *H. vineae* to produce aromatic metabolites, it was necessary to compare the capacities of the *H. vineae* strains to produce ethanol with

S. cerevisiae. The high degree of similarity between glycolytic and alcoholic fermentation enzymes of *H. vineae* and *H. osmophila* with *S. cerevisiae* showed that the two species should be classified as fermenters, while the remaining *Hanseniaspora* species assessed were adapted to the fruit niche and were correspondingly included in the fruit group. In our experience, *H. vineae* strains cannot be isolated from the fresh grape fruits [19]. A dendrogram of concatenated DNA sequences from seven glycolytic and fermentation genes (Figure 6) indicated the presence of two clades of *Hanseniaspora* species, similar to findings of Steenwyk et al. [36] determined using genes from the DNA repair processes present within the genus. Interestingly, the fruit and fermentation clades shown in Figure 6 were correlated with the slow and fast evolution lineages defined by these authors. Branches were in agreement with phylogenetic classifications that were based on ribosomal genes [19]. It might be interesting to use the group as an evolution model to determine the mechanism by which the fermentation group diverged separately from the fruit group [36], giving less species diversity probably due to slow evolution mechanisms. Further work will be needed to understand whether the process might be an example of domestication, as has been proposed for *S. cerevisiae* wine and beer strains [64].

Previous studies have compared the fermentation capacity of two species belonging to the fruit group: *H. guillermondi* and *H.uvarum* [27,29], and the work presented here is the first assessment of a member of the fermentation group of *Hanseniaspora*.

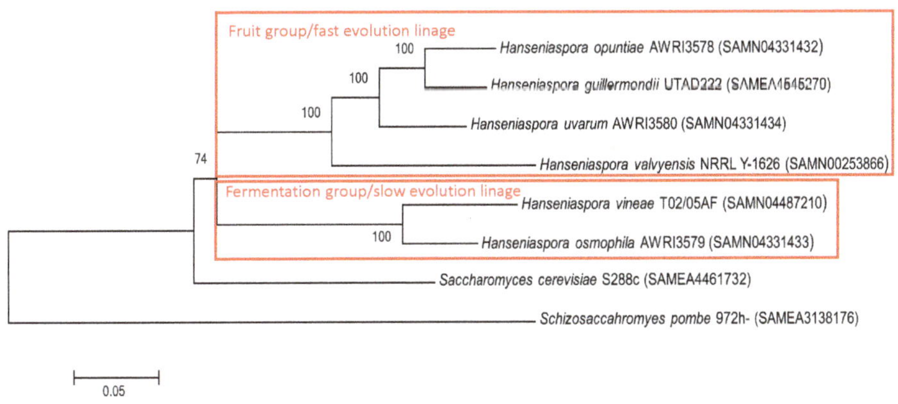

Figure 6. Dendrogram of seven concatenated DNA sequences from *Hanseniaspora* species constructed using the neighbour-joining method. The robustness of branching is indicated by bootstrap values (%) calculated for 1000 subsets. The entries in brackets correspond to NCBI BioSample identifiers.

4. Conclusions

The results suggest that *H. vineae* is clearly better adapted to the fermentation niche compared to what we named as the *Hanseniaspora* fruit clade. These results are in agreement with a separately evolution divergence between the two clades of the genus *Hanseniaspora* as was proposed previously. Phenotypic behavior of *H. vineae* growth, ethanol tolerance, and fermentation kinetics are in agreement with the genetic and transcriptomic data provided. The results obtained demonstrate that *H. vineae* and a genetically closely related species, *H. osmophila*, behave similarly. Homologies of glycolytic and alcoholic fermentation enzyme sequences of both species were compared to *S. cerevisiae*, and the similarities observed allowed the differentiation of *H. uvarum* from *H. osmophila* and *H. vineae*. High sequence homology in these latter two species was observed for key genes involved in glycolysis such as *HXK2*, which encodes hexokinase, *PFK1/PFK2* subunits of phosphofructokinase, and *CDC19* that encodes pyruvate kinase. This homology could explain the improved fermentative performance observed for *H. vineae* compared with other *Hanseniaspora* species. The elevated number of copies of *ADH* genes in *H. vineae* might be associated with increased ethanol tolerance in the species. The presence of active genes typically related to wine fermentation capacities in *H. vineae* and *H. osmophila* such as sulfite tolerance (*SSU1*) and sucrose hydrolyzing invertase (*SUC2*) differentiate both species from the other sequenced species of the genus. Taken together, findings reported here support the characterization of the *Hanseniaspora* genus into two different groups that are adapted to two different niches, fruit and juice fermentation. These results have contributed to the improved characterization of the genus and furthermore might support the importance of it as a model for further studies related to the genetic and evolutionary phenomena of yeast domestication processes.

Author Contributions: M.J.V., E.B., E.D., and F.C. conceived the study and its design; M.J.V. and F.C. wrote the manuscript; M.J.V. performed laboratory experiments and data analysis; E.B. carried out statistical analysis. All authors read and approved the manuscript.

Funding: This research was funded by Agencia Nacional de Investigación e Innovación (ANII), Application of *Hanseniaspora vineae* Project ALI_2_2019_1_155314 with Lage y Cia-Lallemand, Uruguay.

Acknowledgments: We wish to thank our Universidad de la Republica for basic support of this work: CSIC Group Project 802 and Facultad de Quimica, Uruguay.

Conflicts of Interest: The authors declare no conflict of interest.

References

1. Rainieri, S.; Pretorius, I.S. Selection and improvement of wine yeasts. *Ann. Microbiol.* **2000**, *50*, 15–31.
2. Fleet, G.H. Wine yeasts for the future. *FEMS Yeast Res.* **2008**, *8*, 979–995. [CrossRef] [PubMed]
3. Jolly, N.P.; Varela, C.; Pretorius, I.S. Not your ordinary yeast: Non-*Saccharomyces* yeasts in wine production uncovered. *FEMS Yeast Res.* **2014**, *14*, 215–237. [CrossRef] [PubMed]
4. Mas, A.; Guillamón, J.M.; Beltran, G. Non-conventional Yeast in the Wine Industry. *FEMS Yeast Res.* **2016**, *7*, 1–2. [CrossRef]
5. Comitini, F.; Capece, A.; Ciani, M.; Romano, P. New insights on the use of wine yeasts. *Curr. Opin. Food Sci.* **2017**, *13*, 44–49. [CrossRef]
6. Carrau, F.; Gaggero, C.; Aguilar, P.S. Yeast diversity and native vigor for flavor phenotypes. *Trends Biotechnol.* **2015**, *33*, 148–154. [CrossRef] [PubMed]
7. Lleixà, J.; Manzano, M.; Mas, A.; Portillo, M.d.C. *Saccharomyces* and non-*Saccharomyces* competition during microvinification under different sugar and nitrogen conditions. *Front. Microbiol.* **2016**, *7*, 1959. [CrossRef]
8. Masneuf-Pomarede, I.; Bely, M.; Marullo, P.; Albertin, W. The genetics of non-conventional wine yeasts: Current knowledge and future challenges. *Front. Microbiol.* **2016**, *6*, 1563. [CrossRef]
9. Varela, C. The impact of non-*Saccharomyces* yeasts in the production of alcoholic beverages. *Appl. Microbiol. Biotechnol.* **2016**, *100*, 9861–9874. [CrossRef]
10. Esteve-Zarzoso, B.; Gostíncar, A.; Bobet, R.; Uruburu, F.; Querol, A. Selection and molecular characterization of wine yeasts isolated from the "El Penedes" area (Spain). *Food Microbiol.* **2000**, *17*, 553–562. [CrossRef]
11. Degre, R. Selection and Commercial Cultivation of Wine Yeast and Bacteria. In *Wine Microbiology and Biotechnology*; Fleet, G.H., Ed.; Taylor and Francis: New York, NY, USA, 1993.
12. Romano, P.; Capece, A.; Jespersen, L. Taxonomic and ecological diversity of food and beverage yeasts. In *Yeasts in Foods and Beverages*; Springer: Berlin/Heidelberg, Germany, 2006; pp. 13–53.
13. Jolly, N.P.; Augustyn, O.P.H.; Pretorius, I.S. The Effect of Non-*Saccharomyces* Yeasts on Fermentation and Wine Quality. *S. Afr. J. Enol. Vitic.* **2003**, *24*, 55–62. [CrossRef]
14. Zott, K.; Miot-Sertier, C.; Claisse, O.; Lonvaud-Funel, A.; Masneuf-Pomarede, I. Dynamics and diversity of non-*Saccharomyces* yeasts during the early stages in winemaking. *Int. J. Food Microbiol.* **2008**, *125*, 197–203. [CrossRef] [PubMed]
15. Pina, C.; Santos, C.; Couto, J.A.; Hogg, T. Ethanol tolerance of five non-*Saccharomyces* wine yeasts in comparison with a strain of *Saccharomyces cerevisiae*—Influence of different culture conditions. *Food Microbiol.* **2004**, *21*, 439–447. [CrossRef]
16. Albergaria, H.; Francisco, D.; Gori, K.; Arneborg, N.; Gírio, F. *Saccharomyces cerevisiae* CCMI 885 secretes peptides that inhibit the growth of some non-*Saccharomyces* wine-related strains. *Appl. Microbiol. Biotechnol.* **2010**, *86*, 965–972. [CrossRef] [PubMed]

17. Wang, K.; Dang, W.; Xie, J.; Zhu, R.; Sun, M.; Jia, F.; Zhao, Y.; An, X.; Qiu, S.; Li, X.; et al. Antimicrobial peptide protonectin disturbs the membrane integrity and induces ROS production in yeast cells. *Biochim. Biophys. Acta Biomembr.* **2015**, *1848*, 2365–2373. [CrossRef] [PubMed]

18. Moreira, N.; Pina, C.; Mendes, F.; Couto, J.A.; Hogg, T.; Vasconcelos, I. Volatile compounds contribution of *Hanseniaspora guilliermondii* and *Hanseniaspora uvarum* during red wine vinifications. *Food Control* **2011**, *22*, 662–667. [CrossRef]

19. Martin, V.; Valera, M.J.; Medina, K.; Boido, E.; Carrau, F. Oenological impact of the *Hanseniaspora/Kloeckera* yeast genus on wines—A review. *Fermentation* **2018**, *4*, 76. [CrossRef]

20. Fleet, G.H. Yeast interactions and wine flavour. *Int. J. Food Microbiol.* **2003**, *86*, 11–22. [CrossRef]

21. Medina, K.; Boido, E.; Fariña, L.; Gioia, O.; Gomez, M.E.; Barquet, M.; Gaggero, C.; Dellacassa, E.; Carrau, F. Increased flavour diversity of Chardonnay wines by spontaneous fermentation and co-fermentation with *Hanseniaspora vineae*. *Food Chem.* **2013**, *141*, 2513–2521. [CrossRef]

22. Giorello, F.; Valera, M.J.; Martin, V.; Parada, A.; Salzman, V.; Camesasca, L.; Fariña, L.; Boido, E.; Medina, K.; Dellacassa, E.; et al. Genomic and transcriptomic basis of *Hanseniaspora vineae*'s impact on flavor diversity and wine quality. *Appl. Environ. Microbiol.* **2019**, *75*, 1–20. [CrossRef]

23. Valera, M.J.; Boido, E.; Ramos, J.C.; Manta, E.; Radi, R.; Dellacassa, E.; Carrau, F. The mandelate pathway, an alternative to the PAL pathway for the synthesis of benzenoids in yeast. *Appl. Environ. Microbiol.* **2020**, *86*, 701–720. [CrossRef] [PubMed]

24. Giorello, F.M.; Berna, L.; Greif, G.; Camesasca, L.; Salzman, V.; Medina, K.; Robello, C.; Gaggero, C.; Aguilar, P.S.; Carrau, F. Genome Sequence of the Native Apiculate Wine Yeast *Hanseniaspora vineae* T02/19AF. *Genome Announc.* **2014**, *2*, e00530-14. [CrossRef] [PubMed]

25. Sternes, P.R.; Lee, D.; Kutyna, D.R.; Borneman, A.R. Genome Sequences of Three Species of *Hanseniaspora* Isolated from Spontaneous Wine Fermentations. *Genome Announc.* **2016**, *4*, e01287-16. [CrossRef] [PubMed]

26. Riley, R.; Haridas, S.; Wolfe, K.H.; Lopes, M.R.; Hittinger, C.T.; Göker, M.; Salamov, A.A.; Wisecaver, J.H.; Long, T.M.; Calvey, C.H.; et al. Comparative genomics of biotechnologically important yeasts. *Proc. Natl. Acad. Sci. USA* **2016**, *113*, 9882–9887. [CrossRef]

27. Seixas, I.; Barbosa, C.; Mendes-Faia, A.; Güldener, U.; Tenreiro, R.; Mendes-Ferreira, A.; Mira, N.P. Genome sequence of the non-conventional wine yeast *Hanseniaspora guilliermondii* UTAD222 unveils relevant traits of this species and of the *Hanseniaspora* genus in the context of wine fermentation. *DNA Res.* **2019**, *26*, 67–83. [CrossRef]

28. Walker, M.E.; Nguyen, T.D.; Liccioli, T.; Schmid, F.; Kalatzis, N.; Sundstrom, J.F.; Gardner, J.M.; Jiranek, V. Genome-wide identification of the Fermentome; genes required for successful and timely completion of wine-like fermentation by *Saccharomyces cerevisiae*. *BMC Genom.* **2014**, *15*, 552. [CrossRef]

29. Langenberg, A.; Bink, F.J.; Wolff, L.; Walter, S.; Grossmann, M.; Schmitz, H. Glycolytic Functions Are Conserved in the Genome of the Wine Yeast. *Appl. Environ. Microbiol.* **2017**, *83*, 1–20. [CrossRef]

30. Stanke, M.; Keller, O.; Gunduz, I.; Hayes, A.; Waack, S.; Morgenstern, B. AUGUSTUS: Ab initio prediction of alternative transcripts. *Nucleic Acids Res.* **2006**, *34*, W435–W439. [CrossRef]

31. Cherry, J.M.; Hong, E.L.; Amundsen, C.; Balakrishnan, R.; Binkley, G.; Chan, E.T.; Christie, K.R.; Costanzo, M.C.; Dwight, S.S.; Engel, S.R.; et al. *Saccharomyces* Genome Database: The genomics resource of budding yeast. *Nucleic Acids Res.* **2012**, *40*, D700–D706. [CrossRef]

32. Kimura, M. A simple method for estimating evolutionary rates of base substitutions through comparative studies of nucleotide sequences. *J. Mol. Evol.* **1980**, *16*, 111–120. [CrossRef]

33. Tamura, K.; Dudley, J.; Nei, M.; Kumar, S. MEGA4: Molecular Evolutionary Genetics Analysis (MEGA) software version 4.0. *Mol. Biol. Evol.* **2007**, *24*, 1596–1599. [CrossRef] [PubMed]

34. Carrau, F.M.; Medina, K.; Boido, E.; Farina, L.; Gaggero, C.; Dellacassa, E.; Versini, G.; Henschke, P.A. De novo synthesis of monoterpenes by *Saccharomyces cerevisiae* wine yeasts. *FEMS Microbiol. Lett.* **2005**, *243*, 107–115. [CrossRef] [PubMed]

35. Henschke, P.A.; Jiranek, V. Yeast: Metabolism of nitrogen compounds. In *Wine Microbiology and Biotechnology*; Taylor and Francis: New York, NY, USA, 1993; ISBN 0-415-27850-3.

36. Steenwyk, J.L.; Opulente, D.A.; Kominek, J.; Shen, X.X.; Zhou, X.; Labella, A.L.; Bradley, N.P.; Eichman, B.F.; Čadež, N.; Libkind, D.; et al. Extensive loss of cell-cycle and DNA repair genes in an ancient lineage of bipolar budding yeasts. *PLoS Biol.* **2019**, *17*, e3000255. [CrossRef] [PubMed]

37. Gallone, B.; Steensels, J.; Prahl, T.; Soriaga, L.; Saels, V.; Herrera-Malaver, B.; Merlevede, A.; Roncoroni, M.; Voordeckers, K.; Miraglia, L.; et al. Domestication and Divergence of *Saccharomyces cerevisiae* Beer Yeasts. *Cell* **2016**, *166*, 1397–1410.e16. [CrossRef]

38. Albertin, W.; Setati, M.E.; Miot-Sertier, C.; Mostert, T.T.; Colonna-Ceccaldi, B.; Coulon, J.; Girard, P.; Moine, V.; Pillet, M.; Salin, F.; et al. *Hanseniaspora uvarum* from winemaking environments show spatial and temporal genetic clustering. *Front. Microbiol.* **2016**, *6*, 1569. [CrossRef]

39. Forsberg, H.; Ljungdahl, P.O. Sensors of extracellular nutrients in *Saccharomyces cerevisiae*. *Curr. Genet.* **2001**, *40*, 91–109. [CrossRef]

40. Jordan, P.; Choe, J.Y.; Boles, E.; Oreb, M. Hxt13, Hxt15, Hxt16 and Hxt17 from *Saccharomyces cerevisiae* represent a novel type of polyol transporters. *Sci. Rep.* **2016**, *6*, 23502. [CrossRef]

41. Rodríguez, A.; De La Cera, T.; Herrero, P.; Moreno, F. The hexokinase 2 protein regulates the expression of the GLK1, HXK1 and HXK2 genes of *Saccharomyces cerevisiae*. *Biochem. J.* **2001**, *355*, 625–631. [CrossRef]

42. Reifenberger, E.; Freidel, K.; Ciriacy, M. Identification of novel *HXT* genes in *Saccharomyces cerevisiae* reveals the impact of individual hexose transporters on qlycolytic flux. *Mol. Microbiol.* **1995**, *16*, 157–167. [CrossRef]

43. Tondini, F.; Lang, T.; Chen, L.; Herderich, M.; Jiranek, V. Linking gene expression and oenological traits: Comparison between *Torulaspora delbrueckii* and *Saccharomyces cerevisiae* strains. *Int. J. Food Microbiol.* **2019**, *294*, 42–49. [CrossRef]

44. Ozcan, S.; Johnston, M. Function and regulation of yeast hexose transporters. *Microbiol. Mol. Biol. Rev.* **1999**, *63*, 554–569. [CrossRef] [PubMed]

45. Ferreira, C.; Van Voorst, F.; Martins, A.; Neves, L.; Oliveira, R.; Kielland-Brandt, M.C.; Lucas, C.; Brandt, A. A member of the sugar transporter family, Stl1p is the glycerol/H + symporter in *Saccharomyces cerevisiae*. *Mol. Biol. Cell* **2005**, *16*, 2068–2076. [CrossRef] [PubMed]

46. Casal, M.; Paiva, S.; Andrade, R.P.; Gancedo, C.; Leão, C. The lactate-proton symport of *Saccharomyces cerevisiae* is encoded by *JEN1*. *J. Bacteriol.* **1999**, *181*, 2620–2623. [CrossRef] [PubMed]

47. Akita, O.; Nishimori, C.; Shimamoto, T.; Fujii, T.; Iefuji, H. Transport of pyruvate in *Saccharomyces cerevisiae* and cloning of the gene encoded pyruvate permease. *Biosci. Biotechnol. Biochem.* **2000**, *64*, 980–984. [CrossRef] [PubMed]

48. Cortassa, S.; Aon, J.C.; Aon, M.A. Fluxes of carbon, phosphorylation, and redox intermediates during growth of *Saccharomyces cerevisiae* on different carbon sources. *Biotechnol. Bioeng.* **1995**, *47*, 193–208. [CrossRef] [PubMed]

49. Aguilera, A.; Zimmermann, F.K. Isolation and molecular analysis of the phosphoglucose isomerase structural gene of *Saccharomyces cerevisiae*. *MGG Mol. Gen. Genet.* **1986**, *202*, 83–89. [CrossRef]

50. Aguilera, A. Deletion of the phosphoglucose isomerase structural gene makes growth and sporulation glucose dependent in *Saccharomyces cerevisiae*. *MGG Mol. Gen. Genet.* **1986**, *204*, 310–316. [CrossRef]

51. McAlister, L.; Holland, M.J. Differential expression of the three yeast glyceraldehyde-3-phosphate dehydrogenase genes. *J. Biol. Chem.* **1985**, *260*, 15019–15027.

52. McAlister, L.; Holland, M.J. Isolation and characterization of yeast strains carrying mutations in the glyceraldehyde-3-phosphate dehydrogenase genes. *J. Biol. Chem.* **1985**, *260*, 15013–15018.

53. Grauslund, M.; Rønnow, B. Carbon source-dependent transcriptional regulation of the mitochondrial glycerol-3-phosphate dehydrogenase gene, *GUT2*, from *Saccharomyces cerevisiae*. *Can. J. Microbiol.* **2000**, *46*, 1096–1100. [CrossRef]

54. Boles, E.; Hollenberg, C.P. The molecular genetics of hexose transport in yeasts. *FEMS Microbiol. Rev.* **1997**, *21*, 85–111. [CrossRef] [PubMed]

55. Jurica, M.S.; Mesecar, A.; Heath, P.J.; Shi, W.; Nowak, T.; Stoddard, B.L. The allosteric regulation of pyruvate kinase by fructose-1,6-bisphosphate. *Structure* **1998**, *6*, 195–210. [CrossRef]

56. Kellermann, E.; Seeboth, P.G.; Hollenberg, C.P. Analysis of the primary structure and promoter function of a pyruvate decarboxylase gene (PDCI) from *Saccharomyces cerevisiae*. *Nucleic Acids Res.* **1986**, *14*, 8963–8977. [CrossRef] [PubMed]

57. Hohmann, S.; Cederberg, H. Autoregulation may control the expression of yeast pyruvate decarboxylase structural genes *PDC1* and *PDC5*. *Eur. J. Biochem.* **1990**, *188*, 615–621. [CrossRef]

58. König, S. Subunit structure, function and organisation of pyruvate decarboxylases from various organisms. *Biochim. Biophys. Acta Protein Struct. Mol. Enzymol.* **1998**, *1385*, 271–286. [CrossRef]

59. Muller, E.H.; Richards, E.J.; Norbeck, J.; Byrne, K.L.; Karlsson, K.A.; Pretorius, G.H.J.; Meacock, P.A.; Blomberg, A.; Hohmann, S. Thiamine repression and pyruvate decarboxylase autoregulation independently control the expression of the *Saccharomyces cerevisiae PDC5* gene. *FEBS Lett.* **1999**, *449*, 248–250. [CrossRef]

60. Eberhardt, I.; Cederberg, H.; Li, H.; König, S.; Jordan, F.; Hohmann, S. Autoregulation of yeast pyruvate decarboxylase gene expression requires the enzyme but not its catalytic activity. *Eur. J. Biochem.* **1999**, *262*, 196–201. [CrossRef]

61. Bataillon, M.; Rico, A.; Sablayrolles, J.M.; Salmon, J.M.; Barre, P. Early thiamin assimilation by yeasts under enological conditions: Impact on alcoholic fermentation kinetics. *J. Ferment. Bioeng.* **1996**, *82*, 145–150. [CrossRef]

62. Molina, A.M.; Swiegers, J.H.; Varela, C.; Pretorius, I.S.; Agosin, E. Influence of wine fermentation temperature on the synthesis of yeast-derived volatile aroma compounds. *Appl. Microbiol. Biotechnol.* **2007**, *77*, 675–687. [CrossRef]

63. Rossouw, D.; Bauer, F.F. Comparing the transcriptomes of wine yeast strains: Toward understanding the interaction between environment and transcriptome during fermentation. *Appl. Microbiol. Biotechnol.* **2009**, *84*, 937. [CrossRef]

64. Gonçalves, M.; Pontes, A.; Almeida, P.; Barbosa, R.; Serra, M.; Libkind, D.; Hutzler, M.; Gonçalves, P.; Sampaio, J.P. Distinct Domestication Trajectories in Top-Fermenting Beer Yeasts and Wine Yeasts. *Curr. Biol.* **2016**, *26*, 2750–2761. [CrossRef] [PubMed]

 *fermentation*

Article

Alterations in Yeast Species Composition of Uninoculated Wine Ferments by the Addition of Sulphur Dioxide

Kathleen Cuijvers [1,†], **Steven Van Den Heuvel** [1,†], **Cristian Varela** [1,2], **Mark Rullo** [1], **Mark Solomon** [1], **Simon Schmidt** [1] and **Anthony Borneman** [1,3,*]

[1] The Australian Wine Research Institute, PO Box 197, Glen Osmond, SA 5064, Australia; kate.cuijvers@awri.com.au (K.C.); steven.vandenheuvel@awri.com.au (S.V.D.H.); cristian.varela@awri.com.au (C.V.); mark.rullo@adelaide.edu.au (M.R.); mark.solomon@awri.com.au (M.S.); simon.schmidt@awri.com.au (S.S.)

[2] School of Agriculture, Food & Wine, Faculty of Sciences, University of Adelaide, Adelaide, SA 5005, Australia

[3] School of Biological Sciences, University of Adelaide, Adelaide, SA 5005, Australia

* Correspondence: anthony.borneman@awri.com.au; Tel.: +618-8313-6600; Fax: +618-8313-6601

† Contributed equally.

Received: 3 June 2020; Accepted: 19 June 2020; Published: 23 June 2020

Abstract: Uninoculated wine fermentations are conducted by a consortium of wine yeast and bacteria that establish themselves either from the grape surface or from the winery environment. Of the additives that are commonly used by winemakers, sulphur dioxide (SO_2) represents the main antimicrobial preservative and its use can have drastic effects on the microbial composition of the fermentation. To investigate the effect of SO_2 on the resident yeast community of uninoculated ferments, Chardonnay grape juice from 2018 and 2019 was treated with a variety of SO_2 concentrations ranging up to 100 mg/L and was then allowed to undergo fermentation, with the yeast community structure being assessed via high-throughput meta-barcoding (phylotyping). While the addition of SO_2 was shown to select against the presence of many species of non-*Saccharomyces* yeasts, there was a clear and increasing selection for the species *Hanseniaspora osmophila* as concentrations of SO_2 rose above 40 mg/L in fermentations from both vintages. Chemical analysis of the wines resulting from these treatments showed significant increases in acetate esters, and specifically the desirable aroma compound 2-phenylethyl acetate, that accompanied the increase in abundance of *H. osmophila*. The ability to modulate the yeast community structure of an uninoculated ferment and the resulting chemical composition of the final wine, as demonstrated in this study, represents an important tool for winemakers to begin to be able to influence the organoleptic profile of uninoculated wines.

Keywords: wine; uninoculated fermentation; yeast; sulphur dioxide

1. Introduction

Wine is a complex beverage, produced through the interplay of grape and microbial metabolomes during the process of fermentation. While the majority of modern commercial wine fermentation is performed using inoculated commercial strains of the major wine yeast *Saccharomyces cerevisiae*, a significant proportion of commercial wine fermentations are now being performed using uninoculated grape must. In these situations, the fermentation is conducted by a consortium of wine yeast and bacteria that establish themselves either from the grape surface or from the winery via shared equipment or other vectors such as insects [1].

In the very early stages of fermentation, apiculate yeasts, and yeast-like fungi which reside on the surface of intact grape berries or winery equipment and include the genera *Aureobasidium*, *Rhodotorula*,

Pichia, Candida, Hanseniaspora and *Metschnikowia*, represent the majority of the microbiota [2]. However, the majority of these yeasts succumb very early after fermentation commences. Mildly fermentative yeasts, such as *Hanseniaspora uvarum, Candida stellata, Metschnikowia pulcherrima, Torulaspora delbrueckii* and *Lachancea thermotolerans* have been shown to proliferate and survive well into the fermentation, but reduce in numbers as ethanol levels increase above 6% [3–6].

Vineyard geography, environment and management practices, and harvest, juice/must processing and fermentation conditions can all affect yeast population dynamics during wine fermentation [7–13]. Of those fermentation conditions that are readily modulated by winemakers, the addition of the antimicrobial sulphur dioxide (SO_2) represents the most broadly available intervention practice. Previous microbiological studies have shown that species (and strains) of the major wine yeasts can respond differently to the application of SO_2. Typically, commercial strains of *S. cerevisiae* display high tolerance to SO_2, while "wild" yeasts display lower tolerances and are therefore thought to be broadly selected against through the application of moderate amounts of SO_2 prior to the start of fermentation [14–16].

In order to explore the effect of SO_2 addition on the yeast microbiota during uninoculated Chardonnay wine, meta-barcoding (phylotyping) analysis was used to assess the population dynamics of wine produced across two successive vintages using a range of pre-ferment SO_2 levels.

2. Results

In order to evaluate the effect of SO_2 addition on the yeast population structure, triplicate uninoculated fermentations were established in Chardonnay grape juice across two consecutive vintages. The effect of these different SO_2 concentrations on wine volatile composition was also evaluated.

2.1. SO_2 Addition Affects Yeast Population Structure

In vintage 2018, the grape juice was treated with one of five different concentrations of total SO_2 (0, 40, 60, 80 and 100 mg/L). In addition to its antimicrobial effect, SO_2 is also a powerful antioxidant [17]. To differentiate between the antimicrobial and antioxidant effects of the SO_2 addition, an alternate antioxidant, glutathione (GSH, 250 mg/L), was also assessed for its effects on the yeast community structure.

The progress of each ferment was tracked via sugar consumption (Figure S1), with samples taken immediately after SO_2 or GSH addition (T1), at 90% of sugar remaining (T2), 50% sugar remaining (T3) and 10% sugar remaining (T4), for meta-barcoding analysis using the fungal Internal Transcribed Spacer (ITS) region [18,19]. The addition of GSH did not affect the duration of fermentation, however, SO_2 had a significant impact on the length of time required for the fermentation to reach completion, with two of the 100 mg/mL treatments requiring five to seven days longer than the control ferments (26 day fermentation) and one of the 100 mg/mL treatments becoming stuck with 13 g/L of residual sugar.

Across the 18 samples from 2018 (6 treatments in triplicate), Operational Taxonomic Units (OTUs) that could be assigned to a total of 26 fungal genera were detected that exceeded 0.01% of the total abundance in at least one sample (Figure 1; full results in Table S1). Triplicate samples were shown to be highly concordant for each combination of SO_2 concentration and timepoint (Figure S2). The highest level of fungal diversity was observed at the T1 timepoint, while *Hanseniaspora, Metschnikowia, Saccharomyces* and *Torulaspora* dominated the fermentations from T2 through T4, accounting over 95% of the total ITS reads (Figure 1).

As seen for the fermentation kinetics, the GSH addition did not affect the overall population structure relative to the control samples, however, the addition of SO_2 had a significant, but differential effect on the four main genera observed across the samples (Figure 1). *Metschnikowia* displayed the highest sensitivity to SO_2, with 40 mg/L completely inhibiting the detection of this genus by the T2 timepoint. *Torulaspora* was shown to have a higher abundance at 40 mg/L relative to 0 mg/L, however, this genus was progressively inhibited by higher concentrations of SO_2 in a gradient from 60 through to 100 mg/L, at which point it was completely inhibited at timepoint T2. *Hanseniaspora* and *Saccharomyces* were both shown to be tolerant across all the tested SO_2 concentrations, with *Hanseniaspora* increasing in its total proportion relative to the other genera as the concentration of SO_2 was increased.

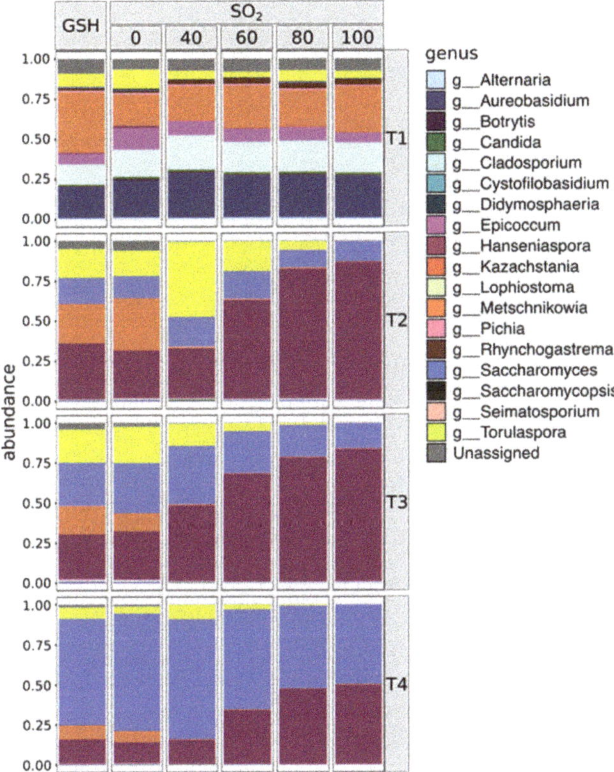

Figure 1. Genus-level metabarcoding analysis of community response to SO_2 addition. Vintage 2018 Chardonnay juice was treated with increasing concentrations of total SO_2 (mg/L) or glutathione (GSH, 250 mg/L) as an alternate antioxidant. Ferment samples were taken at four timepoints (T1, at crush; T2, 10% sugar utilization; T3, 50% sugar utilization; T4, 90% sugar utilization) and subjected to ITS metabarcoding. Only genera that exceeded 0.1% abundance in at least one sample are shown.

As fungal ITS sequencing generally affords the ability to define OTUs to the species level, the genus level counts were partitioned into species-level units to determine the effect of SO_2 concentration on the abundance of individual species. There were 29 species that exceeded 0.1% of the total abundance in any sample, with the genus *Hanseniaspora* displaying the highest number of individual species ($n = 4$). While the addition of SO_2 was shown to increase the overall abundance of *Hanseniaspora* at the genus level, there was a far more complex response profile when species designations were taken into account (Figure 2). Rather than a general increase in all species of *Hanseniaspora*, two species, *H. uvarum* and *H. opuntiae*, were the dominant species when SO_2 was absent (GSH and SO_2 0 mg/L treatments). However, the addition of 40 mg/L of SO_2 resulted in a drastic shift in the species composition such that *H. osmophila* was the sole representative of this genus at 40 mg/L of SO_2. The relative abundance of this species increased substantially as SO_2 levels were raised, producing the overall increase in *Hanseniaspora* that was observed at the genus level.

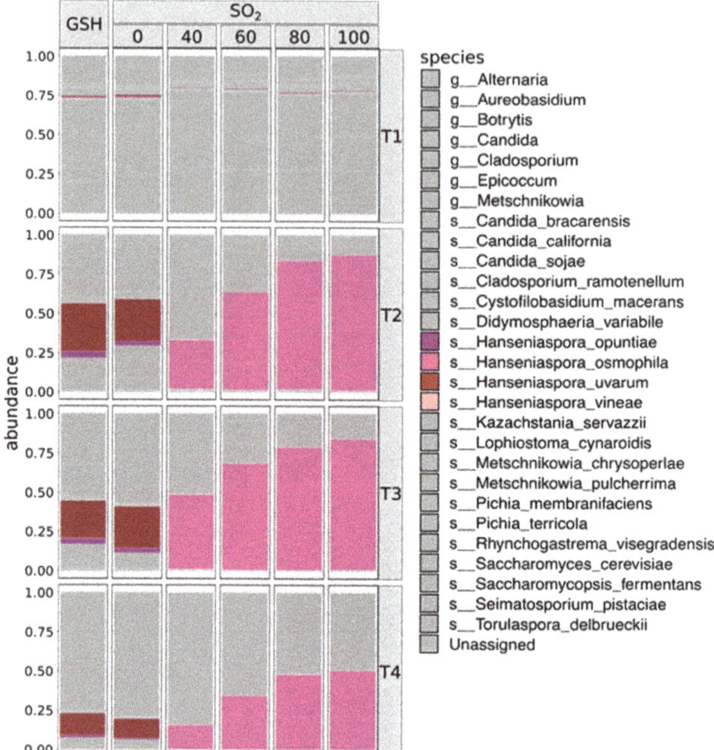

Figure 2. Species-level meta-barcoding analysis of community response to the SO$_2$ addition. Vintage 2018 Chardonnay juice was treated with increasing concentrations of total SO$_2$ (mg/L) or glutathione (GSH, 250 mg/L) as an alternate antioxidant. Ferment samples were taken at four timepoints (T1, at crush; T2, 10% sugar utilization; T3, 50% sugar utilization; T4, 90% sugar utilization) and subjected to ITS metabarcoding. Only species that exceeded 0.1% abundance in at least one sample are shown. For those OTU where a species-level designation was not possible, the genus-level taxonomic classification of the OTU was used. Only members of the genus *Hanseniaspora* are colored.

A second set of fermentations were established in the subsequent year (2019) using a finer set of SO$_2$ treatment intervals (0, 10, 20 and 40 mg/L). Consistent with the observations from the 2018 vintage, the SO$_2$ addition affected fermentation kinetics, particularly for the 40 mg/L treatments (Figure S1). The 2019 ferments displayed a different overall yeast diversity compared with the 2018 samples, with a lack of OTUs that could be assigned to *Metschnikowia* and prominent contributions from OTUs assigned to *Candida* spp., which increased over the range of SO$_2$ concentration used, and *Kazachstania* spp. that were present at up to 20 mg/L of SO$_2$ (Figure 3). However, when the species level contributions were investigated, there were clear similarities between the two vintages in the dynamics of the OTUs assigned to the genus *Hanseniaspora* (Figure 4). As also seen in 2018, *Hanseniaspora* was represented by the greatest number of species designations in 2019. The 2019 ferments also displayed a clear species shift that was associated with the use of SO$_2$, with the proliferation of *H. uvarum*, *H. opuntiae* and *H. vineae* all being inhibited by SO$_2$ in a concentration-dependent manner, while the proportion of *H. osmophila* was shown to be enhanced by the addition of 40 mg/L of SO$_2$.

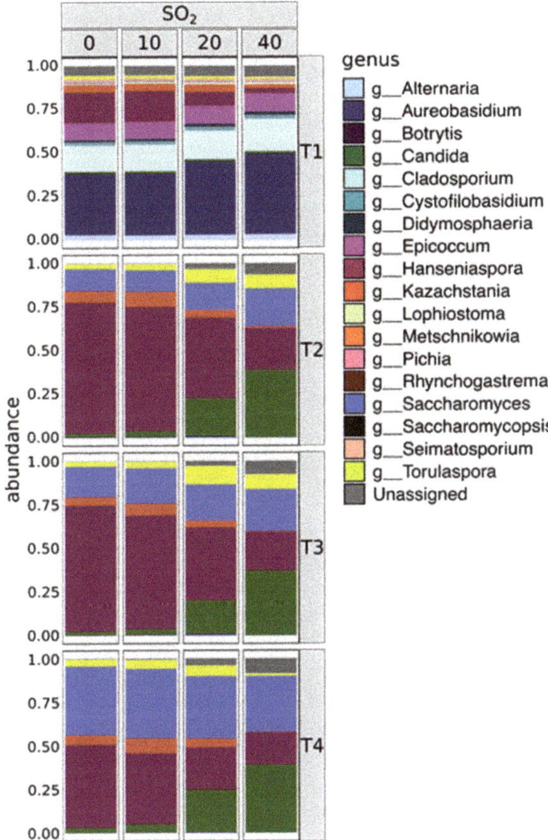

Figure 3. Genus-level meta-barcoding analysis of community response to the SO$_2$ addition. Vintage 2019 Chardonnay juice was treated with increasing concentrations of total SO$_2$ (mg/L). Ferment samples were taken at four timepoints (T1, at crush; T2, 10% sugar utilization; T3, 50% sugar utilization; T4, 90% sugar utilization) and subjected to ITS metabarcoding. Only genera that exceeded 0.1% abundance in at least one sample are shown.

In order to compare the 2018 and 2019 data, the metabarcoding time course results were analyzed by Bray–Curtis dissimilarity analysis (Figure 5). The T1 samples from across both vintages were broadly similar, with the ferments characterized by the presence of non-fermentative genera such as *Aureobasidium*, *Cladosporium* and *Epicoccum*. As observed in the abundance plots, ferments progressed towards being dominated by *S. cerevisiae* (Axis 1), however, there was a clear division between samples with 40 mg/L or more of added SO$_2$, which deviated along Axis 2 towards *H. osmophila*, while the samples with less than 40 mg/L were dominated by the signal from *H. uvarum*. Thus, despite differences in the overall microbial populations of the wild fermentations performed across the two vintages, both displayed consistent alterations in the microbial community due to the amount of SO$_2$ addition.

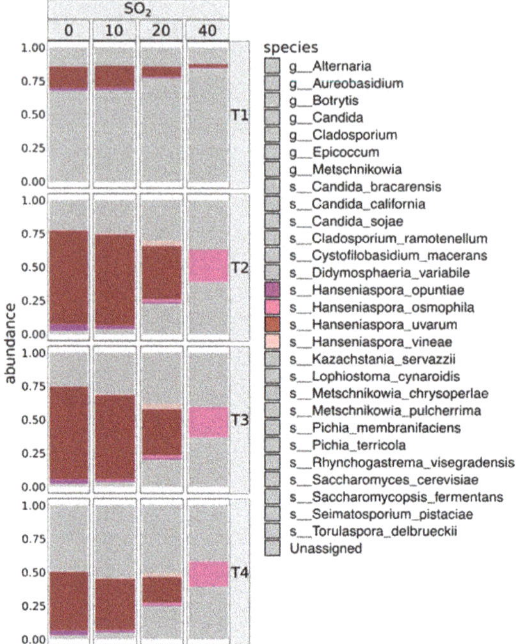

Figure 4. Species-level metabarcoding analysis of community response to the SO$_2$ addition. Vintage 2019 Chardonnay juice was treated with increasing concentrations of total SO$_2$ (mg/L). Ferment samples were taken at four timepoints (T1, at crush; T2, 10% sugar utilization; T3, 50% sugar utilization; T4, 90% sugar utilization) and subjected to ITS metabarcoding. Only species that exceeded 0.1% abundance in at least one sample are shown. For those OTU where a species-level designation was not possible, the genus-level taxonomic classification of the OTU was used. Only members of the genus *Hanseniaspora* are colored.

2.2. SO$_2$ Addition Influences Wine Volatile Composition

Given the significant effect of SO$_2$ addition on the microbial community structure, it was of interest to understand whether these different SO$_2$ treatments were also associated with changes to the chemical composition of the wine. This was assessed through an analysis of the volatile yeast metabolites known to contribute to the aromatic profile of wine. Of the 39 aroma compounds analyzed, 18 displayed a significant difference (ANOVA, $p \leq 0.001$) in concentration across one of the SO$_2$ regimes in either 2018 or 2019 (Table S2). Of these, ten analytes displayed more than a 1.5-fold decrease in at least one of the SO$_2$ treatments, while seven displayed an increase of the same magnitude (Figure 6). Three analytes, 2-methylpropanol (decreasing in response to SO$_2$) and 2-phenylethyl acetate and hexyl acetate (increasing), displayed the same effect across the 2018 and 2019 vintages. In all cases, there was a significant effect at 40 mg/L SO$_2$. Furthermore, in most situations in which a significant difference in analyte concentration was observed across multiple SO$_2$ regimes, there was a correlation between SO$_2$ concentration and the magnitude of change. The largest change in analyte concentration was observed for the desirable aroma compound 2-phenylethyl-acetate, with the 100 mg/L SO$_2$ treatment in 2018 displaying over two orders of magnitude more of this metabolite than the control, and the 2019 40 mg/L SO$_2$ treatment having over nine times as much 2-phenylethyl-acetate as the control (Table S2). More generally, higher SO$_2$ concentrations resulted in decreases in short chain acetates and higher alcohols and increases in 6-carbon and 8-carbon esters and acids. There was no effect of SO$_2$ concentration on low molecular weight volatile sulphur compound production.

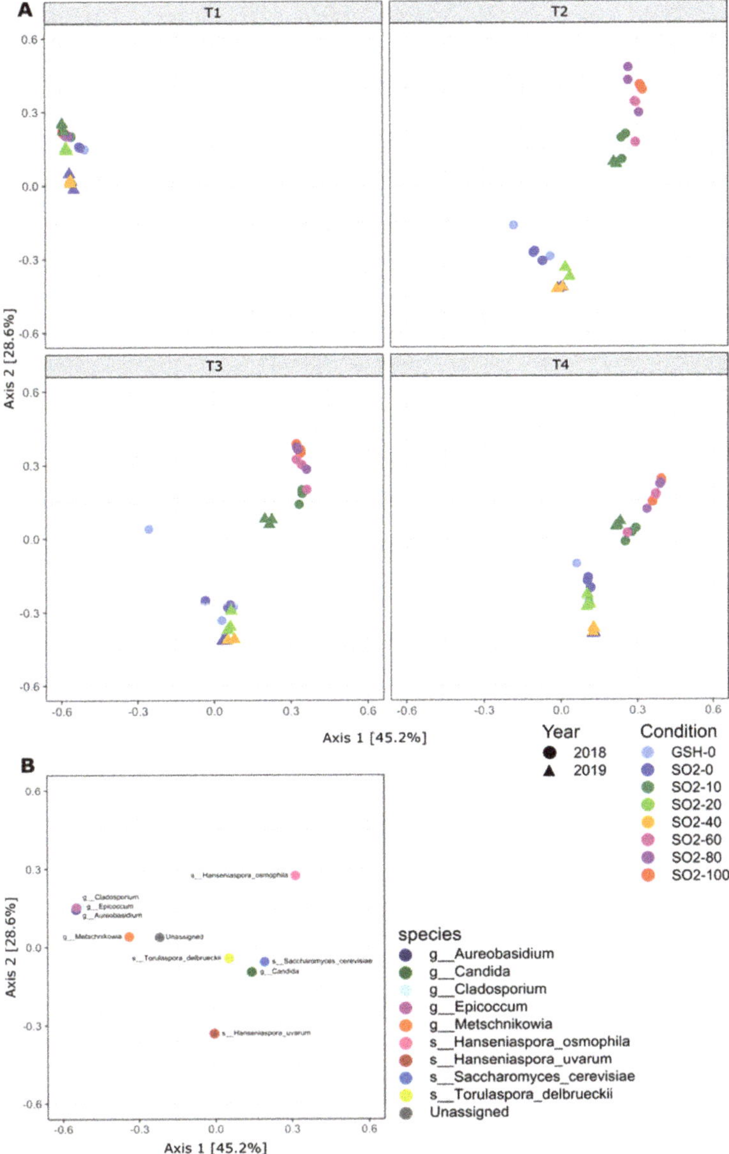

Figure 5. Dissimilarity analysis of ITS-amplicon abundance from vintage 2018 and 2019 fermentations. (**A**) Triplicate samples from each time point were subjected to Bray–Curtis dissimilarity analysis (clustered by PCoA) based upon the top 10 most abundant species and are shaded by treatment condition. (**B**) The weightings of the top 10 most abundant species relative to the plots in part (**A**). Points are shaded by species. For those OTU where a species-level designation was not possible, the genus-level taxonomic classification of the OTU was used.

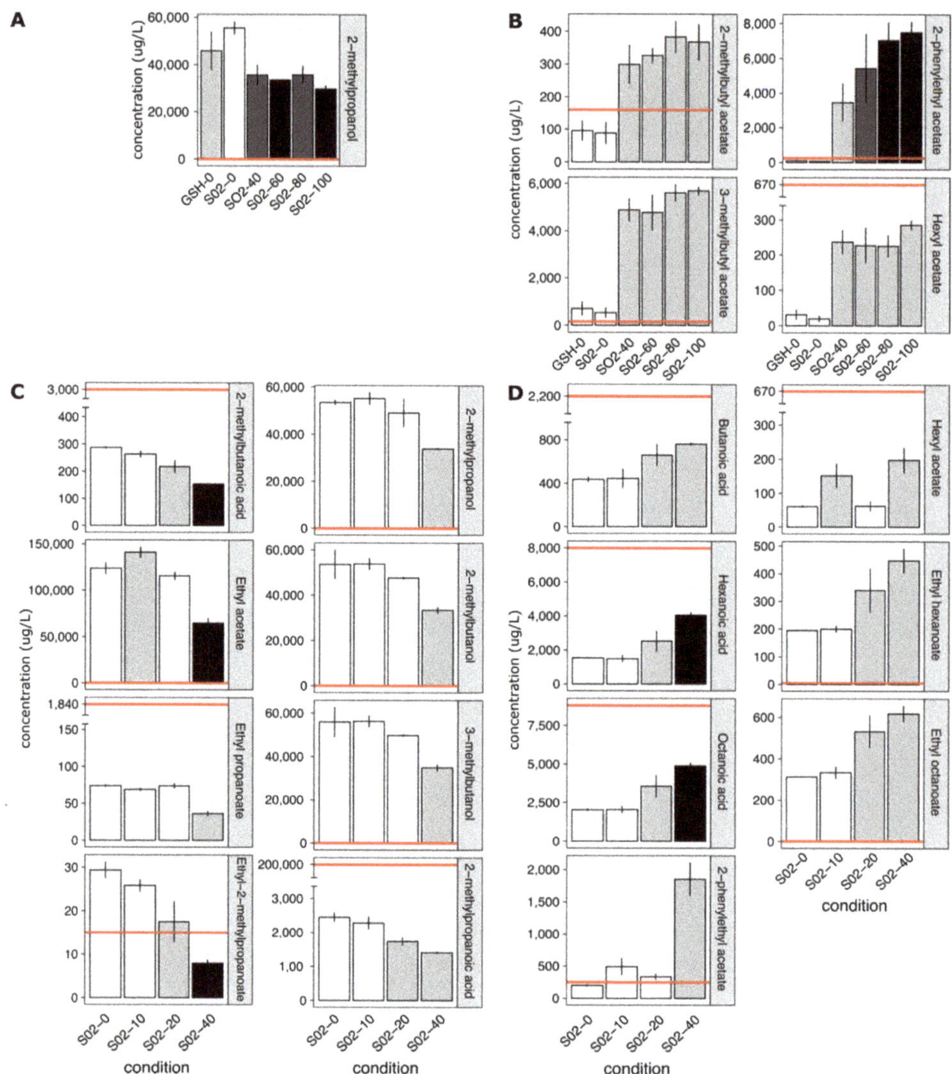

Figure 6. Concentration differences in aroma compounds due to the addition of SO$_2$. (**A**) Analytes with significantly reduced (ANOVA, $p \leq 0.001$) concentrations in 2018. (**B**) Analytes with significantly increased (ANOVA, $p \leq 0.001$) concentrations in 2018. (**C**) Analytes with significantly decreased (ANOVA, $p \leq 0.001$) concentrations in 2019. (**D**) Analytes with significantly increased (ANOVA, $p \leq 0.001$) concentrations in 2019. Individual bars are shaded according to their significance group and the estimated aroma thresholds (see Materials and Methods) are indicated in red.

3. Discussion

Winemakers are limited in their ability to influence the native microbial population of grape juice, with SO$_2$ addition representing the main available intervention. Previous microbiological studies have shown that species (and strains) of the major wine yeasts can respond differently to the application of SO$_2$, with commercial strains of *S. cerevisiae* displaying diverse but higher tolerance to SO$_2$ [20], while "wild" yeasts display lower tolerances and are therefore thought to be broadly selected against through the application of moderate amounts of SO_2 prior to the start of fermentation [14–16,21]. Through the application of ITS meta-barcoding, this study demonstrated that the addition of over 40 mg/L of total SO_2 favored the presence of the non-*Saccharomyces* species *H. osmophila* at the expense of other genera such as *Metschnikowia*, *Torulaspora* and *Kazachstania*. *H. osmophila* has previously been shown to be resistant to SO_2 concentrations of over 40 mg/L [22,23]. While previous research into the effects of SO_2 on grape juice consortia did observe the antagonistic effect of SO_2 against non-*Saccharomyces* yeasts, the presence of *H. osmophila* was not specifically reported, although *Hanseniaspora* yeasts were observed at levels of SO_2 above 40 mg/L [16,24,25].

Much is known regarding the molecular basis of SO_2 tolerance in *S. cerevisiae*, where the sulfite efflux pump *SSU1* provides the main mode of resistance [26,27], but very little information is available on the main genetic determinants of SO_2 resistance in non-*Saccharomyces* species, although *SSU1* has been suggested to have a major role in *Brettanomyces bruxellensis* [28]. However, recent comparative genomic studies of *Hanseniaspora* spp. have shown that there is a clear differentiation of this genus into two well-defined phylogenetic clades, in which one of the differentiating factors is a homolog of *SSU1*, which is absent in the large clade containing *H. uvarum*, but present in the clade containing *H. osmophila* and *H. vineae* [29–31]. While this may explain the different response to SO_2 of *H. osmophila* versus *H. uvarum*, there are likely many other factors that impact the response of a specific species, as *T. delbrueckii*, which also possesses an *SSU1* homolog and displayed increased abundance at up to 40 mg/L SO_2, and is less tolerant at higher SO_2 concentrations than *H. osmophila*. Likewise, *H. vineae*, which was present at levels similar to *H. osmophila* in the control ferments and contains an *SSU1* homolog, did not increase in abundance in response to increases in the concentrations of SO_2.

Detailed chemical analysis showed that the addition of SO_2 resulted in a significant increase in the concentration of key esters and particularly the aroma compound 2-phenylethyl-acetate, which increased over 9-fold, to levels well above the sensory threshold for this compound, even under modest SO_2 additions (40 mg/L). Given the microbiological shift that was observed, it is likely that this change in ester production is due to the increasing prevalence of *H. osmophila* in these ferments with higher SO_2 levels. This is supported by published data from fermentations established with purified non-*Saccharomyces* strains, in which mixtures of *H. osmophila* and *S. cerevisiae* (90:10 ratio) were shown to produce higher concentrations of acetate esters (with the exception of isoamyl acetate) and concentrations of 2-phenylethyl-acetate almost 10-fold greater than those observed using *S. cerevisiae* alone [32,33].

In summary, uninoculated fermentations can provide desirable complexity, however, the process lacks the ability to introduce specific fermentation characteristics through the use of commercial starter strains with distinct fermentation aroma and/or flavor profiles. The ability to modulate the yeast community structure of an uninoculated ferment, and the resulting chemical composition of the final wine, demonstrated in this study represents an important tool for winemakers to begin to be able to influence the organoleptic profile of uninoculated wines.

4. Materials and Methods

4.1. Fermentation

Commercial, high-solids Chardonnay juice (not pre-treated with SO_2 during harvesting or processing) was obtained directly after destemming and crushing from Yalumba wineries during the 2018 and 2019 vintages (Table 1). The juice was transferred to 2 L Schott bottles and then treated with either potassium metabisulfite (ACE Chemical Company, Camden Park, Australia) to the appropriate final total SO_2 concentration or 250 mg of reduced glutathione (GSH, Sigma-Aldrich Sydney, Australia) as indicated. Each experiment was performed in triplicate. Bottles were sealed with airlocks and incubated at 18 °C. Ferments were assessed at least every 24 h by refractometry and sugar analysis (see below), with samples taken for meta-barcoding at four approximate sugar levels (T1, directly after treatment; T2, 90% sugar remaining; T3, 50% sugar remaining; T4, 10% sugar remaining) from an in-built sampling port.

Table 1. Juice composition.

2018	
pH	3.32
Total soluble solids	22.3°Brix
Yeast assimilable nitrogen	249 mg/L
Ammonia	87 mg/L
Alpha amino nitrogen	177 mg/L
Titratable acidity pH 7.0	5.7 g/L
Titratable acidity pH 8.2	6.0 g/L
2019	
pH	3.41
[Glucose + Fructose]	229.6 g/L
Yeast assimilable nitrogen	412 mg/L
Ammonia	147 mg/L
Alpha amino nitrogen	291 mg/L
Titratable acidity pH 7.0	5.7 g/L
Titratable acidity pH 8.2	6.0 g/L

4.2. Meta-Barcoding

DNA was isolated using the DNeasy PowerFood Microbial DNA Isolation Kit (Qiagen Hilden, Germany) following the manufacturer's instructions. Bead-beating was carried out using a combination of 0.1 mm and 0.5 mm zirconia/silica beads (BioSpec Products, Butlersville, Oklahoma) in a Precellys Evolution homogenizer (Bertin Instruments, Montigny-le-Bretonneux, France) at 8000 RPM for 4×60 s. In order to prepare amplicons for sequencing, a two-step PCR was performed using sequences designed to amplify the fungal ITS region, while adding experiment-specific inline barcodes and appropriate adaptors for the Illumina sequencing platform. Briefly, first-round amplification of the ITS region was performed using the fungal-specific primers BITS (ACCTGCGGARGGATCA) and B58S3 (GAGATCCRTTGYTRAAAGTT) [18] which were modified to include both an inline barcode and Illumina adaptor sequences [19]. Second-round amplification added sequences required for Illumina dual-indexed sequencing via overhang PCR. Sequencing was performed using 2×300 bp chemistry (Ramaciotti Centre for Functional Genomics, Sydney, Australia). Paired-end reads were quality trimmed (Trimmomatic v0.38 [34]), adaptor trimmed (cutadapt v1.16 [35]) and merged into single synthetic reads (FLASH2 v2.2.00 [36]). Merged reads were de-replicated (USEARCH v10.0.240 [37]) and clustered (Swarm v2.2.2 [38]) into operational taxonomic units (OTUs) as presented previously [19]. Taxonomic annotation was performed against the UNITE database (qiime_ver8_dynamic_02.02.2019) using a 98% similarity cut off (assign_taxonomy.py module of QIIME v1.9.1 [39]). All sequence reads have been lodged in the NCBI database under the Bioproject accession PRJNA634973.

4.3. Chemical Analysis

Titratable acidity and pH were determined using a TitraLab 840 (Radiometer) and the yeast assimilable nitrogen concentration was estimated by the NOPA + enzymatic ammonia method [40] on a Gallery Discrete Analyser (ThermoFischer, Waltham, USA) by AWRI Commercial Services (Australia).

During fermentation, [Glucose + Fructose] concentrations were determined spectrophotometrically using a Randox kit (Randox Laboratories Ltd., Crumlin, Antrim, UK) with adaptations for use in a 96-well microplate format [41]. Volatile acetates, esters and higher alcohol concentrations in the finished wines were determined using large-volume, stable-isotope dilution headspace–GC/MS analysis (Metabolomics Australia, Adelaide Australia) as adapted from [42] and as described by [43]. Volatile sulphur compounds contributing sulfidic off-aromas, were determined by gas chromatography with sulphur chemiluminescence detection (GC/SCD) [42,44]. Aroma thresholds were in wine estimated using data from Siebert et al., [42]

Free SO_2 was measured in grape juice supplemented with freshly prepared PMS using the aspiration/titration method [45].

4.4. Statistical Analysis

An analysis of variance (ANOVA) was conducted using the formula aov (analyte ~ treatment) in R (version 3.6.3) to determine whether mean aroma-active compound concentrations ($n = 3$ for all treatments) differed with regard to SO_2 treatment. If ANOVA p values were less than 0.05, a multiple comparison of the analyte concentration with respect to treatment was undertaken using the function HSD.test (agricolae) to determine the grouping of the treatments at alpha = 0.05. ANOVA $F_{3,8}$ values, p values, treatment means, standard deviations and treatment group are reported in Table S2.

Supplementary Materials: The following are available online at http://www.mdpi.com/2311-5637/6/2/62/s1, Figure S1: Sugar-consumption kinetics, Figure S2: Replicate fermentation comparisons, Table S1: OTU abundance measurements, Table S2: Chemical analysis results.

Author Contributions: Conceptualization, A.B.; methodology, K.C., S.V.D.H., M.S., C.V.; investigation, K.C.; S.V.D.H.; M.R., M.S., C.V.; formal analysis, S.S., A.B.; writing—original draft preparation, A.B., S.S.; writing—review and editing, all authors. All authors have read and agreed to the published version of the manuscript.

Funding: The AWRI is supported by Australia's grape growers and winemakers through their investment body Wine Australia with matching funds from the Australian Government.

Acknowledgments: Thank you to Caroline Bartel for performing SO_2 analysis and Ross Sanders for performing the volatile sulphur compound analysis and Markus Herderich for critical review of the manuscript. The AWRI is a member of the Wine Innovation Cluster in Adelaide.

Conflicts of Interest: The authors declare that they have no conflict of interest.

References

1. Varela, C.; Borneman, A.R. Yeasts found in vineyards and wineries. *Yeast Chichester Engl.* **2017**, *34*, 111–128. [CrossRef]
2. Fleet, G.H. Wine yeasts for the future. *FEMS Yeast Res.* **2008**, *8*, 979–995. [CrossRef] [PubMed]
3. Fleet, G.H.; Lafon-Lafourcade, S.; Ribéreau-Gayon, P. Evolution of yeasts and Lactic acid bacteria during fermentation and storage of Bordeaux wines. *Appl. Environ. Microbiol.* **1984**, *48*, 1034–1038. [CrossRef]
4. Fleet, G.H. Growth of yeasts during wine fermentations. *J. Wine Res.* **1990**, *1*, 211–223. [CrossRef]
5. Beltran, G.; Torija, M.J.; Novo, M.; Ferrer, N.; Poblet, M.; Guillamón, J.M.; Rozès, N.; Mas, A. Analysis of yeast populations during alcoholic fermentation: A six year follow-up study. *Syst. Appl. Microbiol.* **2002**, *25*, 287–293. [CrossRef] [PubMed]
6. Combina, M.; Elía, A.; Mercado, L.; Catania, C.; Ganga, A.; Martinez, C. Dynamics of indigenous yeast populations during spontaneous fermentation of wines from Mendoza, Argentina. *Int. J. Food Microbiol.* **2005**, *99*, 237–243. [CrossRef] [PubMed]
7. Epifanio, S.I.; Gutierrez, A.R.; Santamaría, M.P.; López, R. The influence of enological practices on the selection of wild Yeast strains in spontaneous fermentation. *Am. J. Enol. Vitic.* **1999**, *50*, 219–224.
8. Torija, M.J.; Rozès, N.; Poblet, M.; Guillamón, J.M.; Mas, A. Yeast population dynamics in spontaneous fermentations: Comparison between two different wine-producing areas over a period of three years. *Antonie Van Leeuwenhoek* **2001**, *79*, 345–352. [CrossRef]
9. Jemec, K.P.; Raspor, P. Initial *Saccharomyces cerevisiae* concentration in single or composite cultures dictates bioprocess kinetics. *Food Microbiol.* **2005**, *22*, 293–300. [CrossRef]
10. Sturm, J.; Grossmann, M.; Schnell, S. Influence of grape treatment on the wine yeast populations isolated from spontaneous fermentations. *J. Appl. Microbiol.* **2006**, *101*, 1241–1248. [CrossRef]
11. Albertin, W.; Miot-Sertier, C.; Bely, M.; Marullo, P.; Coulon, J.; Moine, V.; Colonna-Ceccaldi, B.; Masneuf-Pomarede, I. Oenological prefermentation practices strongly impact yeast population dynamics and alcoholic fermentation kinetics in Chardonnay grape must. *Int. J. Food Microbiol.* **2014**, *178*, 87–97. [CrossRef] [PubMed]

12. Bokulich, N.A.; Thorngate, J.H.; Richardson, P.M.; Mills, D.A. Microbial biogeography of wine grapes is conditioned by cultivar, vintage, and climate. *Proc. Natl. Acad. Sci. USA* **2014**, *111*, E139–E148. [CrossRef] [PubMed]

13. Maturano, Y.P.; Mestre, M.V.; Combina, M.; Toro, M.E.; Vazquez, F.; Esteve-Zarzoso, B. Culture-dependent and independent techniques to monitor yeast species during cold soak carried out at different temperatures in winemaking. *Int. J. Food Microbiol.* **2016**, *237*, 142–149. [CrossRef] [PubMed]

14. Andorrà, I.; Landi, S.; Mas, A.; Guillamón, J.M.; Esteve-Zarzoso, B. Effect of oenological practices on microbial populations using culture-independent techniques. *Food Microbiol.* **2008**, *25*, 849–856. [CrossRef]

15. Morgan, S.C.; Scholl, C.M.; Benson, N.L.; Stone, M.L.; Durall, D.M. Sulfur dioxide addition at crush alters *Saccharomyces cerevisiae* strain composition in spontaneous fermentations at two Canadian wineries. *Int. J. Food Microbiol.* **2017**, *244*, 96–102. [CrossRef]

16. Morgan, S.C.; Tantikachornkiat, M.; Scholl, C.M.; Benson, N.L.; Cliff, M.A.; Durall, D.M. The effect of sulfur dioxide addition at crush on the fungal and bacterial communities and the sensory attributes of Pinot gris wines. *Int. J. Food Microbiol.* **2019**, *290*, 1–14. [CrossRef]

17. Danilewicz, J.C.; Seccombe, J.T.; Whelan, J. Mechanism of interaction of polyphenols, oxygen, and sulfur dioxide in model wine and wine. *Am. J. Enol. Vitic.* **2008**, *59*, 128–136.

18. Bokulich, N.A.; Mills, D.A. Improved selection of internal transcribed spacer-specific primers enables quantitative, ultra-high-throughput profiling of fungal communities. *Appl. Environ. Microbiol.* **2013**, *79*, 2519–2526. [CrossRef]

19. Sternes, P.R.; Lee, D.; Kutyna, D.R.; Borneman, A.R. A combined meta-barcoding and shotgun metagenomic analysis of spontaneous wine fermentation. *GigaScience* **2017**, *6*, gix040. [CrossRef]

20. Nardi, T.; Corich, V.; Giacomini, A.; Blondin, B. A sulphite-inducible form of the sulphite efflux gene *SSU1* in a *Saccharomyces cerevisiae* wine yeast. *Microbiol. Read. Engl.* **2010**, *156*, 1686–1696. [CrossRef]

21. Henick-Kling, T.; Edinger, W.; Daniel, P.; Monk, P. Selective effects of sulfur dioxide and yeast starter culture addition on indigenous yeast populations and sensory characteristics of wine. *J. Appl. Microbiol.* **1998**, *84*, 865–876. [CrossRef]

22. Granchi, L.; Ganucci, D.; Messini, A.; Vincenzini, M. Oenological properties of and from wines produced by spontaneous fermentations of normal and dried grapes. *FEMS Yeast Res.* **2002**, *2*, 403–407. [CrossRef]

23. Hierro, N.; Gonzalez, A.; Mas, A.; Guillamon, J.M. Diversity and evolution of non-*Saccharomyces* yeast populations during wine fermentation: Effect of grape ripeness and cold maceration. *FEMS Yeast Res.* **2006**, *6*, 102–111. [CrossRef] [PubMed]

24. Bagheri, B.; Bauer, F.F.; Cardinali, G.; Setati, M.E. Ecological interactions are a primary driver of population dynamics in wine yeast microbiota during fermentation. *Sci. Rep.* **2020**, *10*, 1–12. [CrossRef] [PubMed]

25. Bokulich, N.A.; Swadener, M.; Sakamoto, K.; Mills, D.A.; Bisson, L.F. Sulfur dioxide treatment alters wine microbial diversity and fermentation progression in a dose-dependent fashion. *Am. J. Enol. Vitic.* **2015**, *66*, 73–79. [CrossRef]

26. Divol, B.; du Toit, M.; Duckitt, E. Surviving in the presence of sulphur dioxide: Strategies developed by wine yeasts. *Appl. Microbiol. Biotechnol.* **2012**, *95*, 601–613. [CrossRef]

27. Park, H.; Bakalinsky, A.T. SSU1 mediates sulphite efflux in *Saccharomyces cerevisiae*. *Yeast* **2000**, *16*, 881–888. [CrossRef]

28. Varela, C.; Bartel, C.; Roach, M.; Borneman, A.; Curtin, C. *Brettanomyces bruxellensis SSU1* haplotypes confer different levels of sulfite tolerance when expressed in a *Saccharomyces cerevisiae SSU1* null mutant. *Appl. Environ. Microbiol.* **2019**, *85*, e02429-18. [CrossRef]

29. Sternes, P.R.; Lee, D.; Kutyna, D.R.; Borneman, A.R. Genome sequences of three species of *Hanseniaspora* isolated from spontaneous wine fermentations. *Genome Announc.* **2016**, *4*, e01287-16. [CrossRef]

30. Seixas, I.; Barbosa, C.; Mendes-Faia, A.; Güldener, U.; Tenreiro, R.; Mendes-Ferreira, A.; Mira, N.P. Genome sequence of the non-conventional wine yeast *Hanseniaspora guilliermondii* UTAD222 unveils relevant traits of this species and of the *Hanseniaspora* genus in the context of wine fermentation. *DNA Res.* **2019**, *26*, 67–83. [CrossRef]

31. Čadež, N.; Bellora, N.; Ulloa, R.; Hittinger, C.T.; Libkind, D. Genomic content of a novel yeast species *Hanseniaspora gamundiae* sp. nov. from fungal stromata (*Cyttaria*) associated with a unique fermented beverage in Andean Patagonia, Argentina. *PLoS ONE* **2019**, *14*, e0210792. [CrossRef]

32. Viana, F.; Gil, J.V.; Genovés, S.; Vallés, S.; Manzanares, P. Rational selection of non-*Saccharomyces* wine yeasts for mixed starters based on ester formation and enological traits. *Food Microbiol.* **2008**, *25*, 778–785. [CrossRef] [PubMed]

33. Viana, F.; Gil, J.V.; Vallés, S.; Manzanares, P. Increasing the levels of 2-phenylethyl acetate in wine through the use of a mixed culture of *Hanseniaspora osmophila* and *Saccharomyces cerevisiae*. *Int. J. Food Microbiol.* **2009**, *135*, 68–74. [CrossRef] [PubMed]

34. Bolger, A.M.; Lohse, M.; Usadel, B. Trimmomatic: A flexible trimmer for Illumina sequence data. *Bioinform. Oxf. Engl.* **2014**, *30*, 2114–2120. [CrossRef] [PubMed]

35. Martin, M. Cutadapt removes adapter sequences from high-throughput sequencing reads. *EMB-Net. J.* **2011**, *17*, 10–12. [CrossRef]

36. Magoč, T.; Salzberg, S.L. FLASH: Fast length adjustment of short reads to improve genome assemblies. *Bioinform. Oxf. Engl.* **2011**, *27*, 2957–2963. [CrossRef] [PubMed]

37. Edgar, R.C. Search and clustering orders of magnitude faster than BLAST. *Bioinform. Oxf. Engl.* **2010**, *26*, 2460–2461. [CrossRef] [PubMed]

38. Mahé, F.; Rognes, T.; Quince, C.; de Vargas, C.; Dunthorn, M. Swarm: Robust and fast clustering method for amplicon-based studies. *PeerJ* **2014**, *2*, e593. [CrossRef]

39. Caporaso, J.G.; Kuczynski, J.; Stombaugh, J.; Bittinger, K.; Bushman, F.D.; Costello, E.K.; Fierer, N.; Peña, A.G.; Goodrich, J.K.; Gordon, J.I.; et al. QIIME allows analysis of high-throughput community sequencing data. *Nat. Methods* **2010**, *7*, 335–336. [CrossRef]

40. Gump, B.H.; Zoecklein, B.W.; Fugelsang, K.C.; Whiton, R.S. Comparison of analytical methods for prediction of prefermentation nutritional status of grape juice. *Am. J. Enol. Vitic.* **2002**, *53*, 325–329.

41. Vermeir, S.; Nicolaï, B.M.; Jans, K.; Maes, G.; Lammertyn, J. High-throughput microplate enzymatic assays for fast sugar and acid quantification in Apple and Tomato. *J. Agric. Food Chem.* **2007**, *55*, 3240–3248. [CrossRef] [PubMed]

42. Siebert, T.E.; Smyth, H.E.; Capone, D.L.; Neuwöhner, C.; Pardon, K.H.; Skouroumounis, G.K.; Herderich, M.J.; Sefton, M.A.; Pollnitz, A.P. Stable isotope dilution analysis of wine fermentation products by HS-SPME-GC-MS. *Anal. Bioanal. Chem.* **2005**, *381*, 937–947. [CrossRef] [PubMed]

43. Holt, H.; Cozzolino, D.; McCarthy, J.; Abrahamse, C.; Holt, S.; Solomon, M.; Smith, P.; Chambers, P.J.; Curtin, C. Influence of yeast strain on Shiraz wine quality indicators. *Int. J. Food Microbiol.* **2013**, *165*, 302–311. [CrossRef] [PubMed]

44. Bekker, M.Z.; Day, M.P.; Holt, H.; Wilkes, E.; Smith, P.A. Effect of oxygen exposure during fermentation on volatile sulfur compounds in Shiraz wine and a comparison of strategies for remediation of reductive character: Macro-oxygenation: Effect on sulfurous off-odours. *Aust. J. Grape Wine Res.* **2016**, *22*, 24–35. [CrossRef]

45. Rankine, B.; Pocock, K. Alkalimetric determination of sulphur dioxide in wine. *Aust. Wine Brew. Spirit Rev.* **1970**, *88*, 40–44.

 *fermentation*

Article

Do Non-*Saccharomyces* Yeasts Work Equally with Three Different Red Grape Varieties?

Rocío Escribano-Viana, Patrocinio Garijo, Isabel López-Alfaro, Rosa López, Pilar Santamaría, Ana Rosa Gutiérrez and Lucía González-Arenzana *

ICVV, Instituto de Ciencias de la Vid y el Vino, Universidad de La Rioja, Gobierno de La Rioja, CSIC. Finca La Grajera, Ctra LO-20- salida, 13, 26071 Logroño, Spain; rocio.escribano@icvv.es (R.E.-V.); pgarijo@larioja.org (P.G.); isabel.lopez@unirioja.es (I.L.-A.); rosa.lopez@icvv.es (R.L.); psantamaria@larioja.org (P.S.); ana-rosa.gutierrez@unirioja.es (A.R.G.)
* Correspondence: lucia.gonzalez@icvv.es; Tel.: +34-941894980

Received: 4 November 2019; Accepted: 27 December 2019; Published: 31 December 2019

Abstract: The present study aimed to investigate the oenological changes induced by non-*Saccharomyces* yeasts in three red grape varieties from the Rioja Qualified Designation of Origin. Pilot plants fermentation of three different varieties, were conducted following early inoculations with *Metschnikowia pulcherrima* and with mixed inoculum of *Lachancea thermotolerans-Torulaspora delbrueckii* from La Rioja and compared to a wine inoculated with *Saccharomyces cerevisiae*. The microbiological and physicochemical characteristics of vinifications were analysed. Results showed that most of the variations due to inoculation strategies were observed in Tempranillo just after the alcoholic fermentation, probably because of the better adaptation of the inocula to the must's oenological properties. Finally, after the malolactic fermentation the inoculation with the mix of *Lachancea thermotolerans* and *Torulaspora delbrueckii* caused more changes in Tempranillo and Grenache wines while the early inoculation with *Metschnikowia pulcherrima* had more effects on Grenache wines. Therefore, the study was aimed to identify the fermentation effects of each inoculation strategy by using different non-*Saccharomyces* yeasts and different grape varieties.

Keywords: *Metschnikowia pulcherrima*; *Lachancea thermotolerans*; *Torulaspora delbrueckii*; Grenache; Graciano

---

1. Introduction

Grapes hold a diverse microbial population consisting of bacteria and yeasts that meet the microorganisms located in the winery facilities after the harvest. During the initial stages of the spontaneous alcoholic fermentation (AF), this pool of microbes achieves a balance until *Saccharomyces* (*S.*) *cerevisiae* becomes the main yeast in the fermentative process.

Early AF is characterized by a diverse yeast population, with low frequency of detection of *S. cerevisiae*, but with a high presence of non-*Saccharomyces* yeasts. The presence of unknown microbiota makes it a risky and unpredictable practice. Therefore, the inoculation of commercial *S. cerevisiae* strains has been widespread in the modern wine industry all over the World. Indeed, the non-*Saccharomyces* yeasts have not been well-regarded by oenologists and these have tended to make efforts to avoid their involvement in AF [1]. These traditional and conservative oenological practices have led to a homogenization and globalization of winemaking, a sameness in the taste and flavours of finished wines [2].

A general strategy to increase the diversification of wines has made oenology return to its origins of natural and diverse microbial populations. For this purpose, the employment of non-*Saccharomyces* yeast species has shown promising results. This new trend has triggered the studies and published results of non-*Saccharomyces* yeasts which has led to some of them being used as commercial culture starters [1].

The use of mixed starter cultures of selected non-*Saccharomyces* combined with *S. cerevisiae* to avoid any stuck fermentations is thought to be a solution for ensuring AF completion, while various organoleptic characteristics involved in the quality of the final products are improved [2,3]. Furthermore, mixed cultures composed of more than one non-*Saccharomyces* species in combination with *S. cerevisiae* have been employed with the aim of simulating this complex yeast community present in spontaneous AF [4,5]. In general terms, the early inoculation of *Metschnikowia* (*M.*) *pulcherrima* has been aimed to improve flavour of wines [6]. In the case of *Lachancea* (*L.*) *thermotolerans*, the objective is the increase of lactic acid that would have also an impact in the aromatic profile of wines [7]. Moreover, *Torulaspora* (*T.*) *delbrueckii* has been initially employed for reducing the alcohol after the AF and for improving the aroma profile of wines [8].

The current study aims to describe the oenological effects of the sequential early inoculation of a pure culture of *M. pulcherrima* and a mixed culture of *L. thermotolerans* and *T. delbrueckii* in the vinification of Tempranillo, Grenache and Graciano grape varieties. With this purpose, the impact dependent on the specific grape variety in semi-industrial conditions was analysed. To this end, the kinetics of AF, implantation rate, variation of the oenological, colour and aromatic parameters after AF and clustering after malolactic fermentation (FML) were individually performed for each grape variety.

2. Material and Methods

2.1. Grapes and Initial Must Samples of the Three Varieties

Grapes of the three red grape varieties from the D.O.Ca. Rioja, Tempranillo, Grenache and Graciano were employed to perform this study. These grape varieties were chosen for being important in the region where this study was developed but also, they are very present in international winemaking areas. When the grapes had reached an average probable alcohol by volume (APBV) of approximately 13%, around 225 k of each one were individually harvested, crushed and destemmed (Figure S1).

Samples of the three must were physicochemical characterized. APBV, pH and total acidity were analysed according to official ECC methods [9]. Malic acid was determined also by the official method [9]. by an enzymatic method carried out with an automated clinical chemistry analyser (Miura One, TDI, Madrid, Spain). The yeast assailable nitrogen (YAN) was measured following the protocol described by Aerny [10].

The three musts were also microbiologically characterised by plating the appropriate dilution on Chloramphenicol Glucose Agar (CGA 05% yeast extract, 20% glucose, 005% chloramphenicol, 17% agar,) plates, incubated at 28 °C for 48 h. Ten yeast colonies were isolated from each plate containing between 30 and 300 colony forming units per millilitre (CFU/mL). DNA was then extracted from fresh culture following the protocol determined by López et al. [11]. Then, a partial region of the 26S rDNA gene was amplified with PCR using the primers and conditions established by Cocolin et al. [12]. PCR amplicons were purified and sequenced by Macrogen Inc. (Seoul, South Korea). The sequences were compared to the GenBank nucleotide database using the Basic Local Alignment Search Tool (BLAST) [13]. The identification was considered appropriate if gene sequences showed identities of at least 98%.

2.2. Yeast Species

This study was performed with four oenological yeast species, *M. pulcherrima*, *L. thermotolerans*, *T. delbrueckii* and *S. cerevisiae* (VRB commercial yeast from Lallemand Bio S.L., Toronto, Canada). *M. pulcherrima* and *S. cerevisiae* were pure cultures while *L. thermotolerans* and *T. delbrueckii* (*L&T*) were combined in percentages of 30% and 70%, respectively, following the natural combination of the two species observed in other studies of non-*Saccharomyces* population in Rioja red wines [14,15]. All these yeasts were selected in the Rioja Qualified Designation of Origin (D.O. Ca. Rioja) from Spain, and they are in the last stage of the selection program. Furthermore, they are stored in the Instituto de Ciencias de la Vid y del Vino (ICVV) collection. These yeast were identified by Macrogen Inc. with the amplified region D1 of the 26S rRNA gene using the primers NL1GC and LS2 [16].

2.3. Inoculation Procedure and Alcoholic Fermentation

The must of each variety were put into nine 30 L tanks that were kept at 25 °C to carry out the AF (Figure S1). When the tanks were filled, potassium metabisulphite was added to the samples to achieve a total SO_2 concentration of 50 mg/L.

After this, the 27 tanks were inoculated with the different yeasts following three different inoculation strategies. For each variety, three out of the nine tanks (n = 3) made up the control sample (C) and were inoculated with the commercial *S. cerevisiae* starter culture VRB^{TM} following the producer's instructions, another three made up the sample early inoculated with *M. pulcherrima* (n = 3) (M) and the last three (n = 3) the sample early inoculated with a 30/70 mixture of *L. thermotolerans* and *T. delbrueckii* (L&T). The non-*Saccharomyces* yeasts had been pre-cultured in YPD liquid medium at 25 °C for 48 h with orbital shaking until the stationary phase. The concentration of cells/mL was counted with the Neubauer chamber. *M. pulcherrima* pure culture was inoculated in a concentration of 10^6 cells/mL counted while the mixed culture contained 3×10^5 cells/mL of *L. thermotolerans* and 7×10^5 cells/mL of *T. delbrueckii*. Three days later, all the 27 tanks were inoculated with the *S. cerevisiae* starter culture VRB^{TM} at a concentration of 1×10^6 cells/mL.

The kinetics of AF was monitored by daily determination of the Brix degree and density decrease. Samples for implantation control were taken under aseptic conditions at three different moments. The first one was three days after harvest and initial inoculation with *Saccharomyces* and non-*Saccharomyces* yeasts (day 3). The second one was at the fourth day (day 4) when the 27 tanks had been inoculated with *S. cerevisiae* VRB^{TM}. Eventually, the third control of implantation was performed one week after the first inoculation (day 7) (Figure S1). At these three moments, serial dilutions were carried out and the samples were microbiologically characterized as described above (Section 2.1). With the sequencing results, the percentage of each species composing each replicate was determined.

When the 27 wines had reached about 990 g/L density, they were pressed and fermented to dryness. The AF was complete when reducing sugars were lower than 2 g/L. Then, the wines were characterized by measuring the alcohol by volume (ABV), pH, total acidity, volatile acidity, colour intensity and hue according to official ECC methods [9]. Moreover, the malic and lactic acids, glycerol and acetaldehyde contents were determined by an enzymatic method carried out by an automated clinical chemistry analyser (Miura One) and tartaric acid by the Rebelein method [17]. Furthermore, total anthocyanins were measured by decolouring using SO_2 [18] and total phenolics were determined as the total polyphenol index by spectrophotometric absorbance at 280 nm after dilution of samples. Ionized anthocyanins were determined according to Glories [19] and the polymerization index was calculated according to Ruiz [20].

2.4. Analytical Techniques

The analysis of fermentative volatile or aromatic compounds after the AF was performed using the method described by Ortega et al. [21] with some modifications. The extraction was carried out by 4 mL of sample, 9 mL of $(NH_4)_2SO_4$ saturated solution, 40 µL of internal standard solution (2-butanol, 4-methyl-2-pentanol, 4-hydroxy-4-methyl-2-pentanone, 2-octanol, and heptanoic acid, 40 mg of each of them in 100 mL of ethanol) and 300 µL of dichloromethane in tubes. The tubes were shaken for 1 h at 32 × g and then centrifuged at 3220× *g* for 10 min. Once the phases were separated, the dichloromethane phase was recovered. Two µL was injected onto a Hewlett-Packard (Palo Alto, California, CA, USA) 6890 series II gas chromatograph. Separation was carried out with a DB-Wax capillary column (60 m × 0.32 mm I.D. × 0.5 µm film thickness; J&W Scientific, Folsom, CA, USA).

2.5. Malolactic Fermentation

After AF, the wines were drawn off the lees and transferred to 15 L containers that were inoculated with the commercial LAB *Uvaferm alpha*® (Lallemand Bio S.L., Toronto, Canada) to carry out the MLF, at a temperature of 20 °C. The evolution of the fermentation was controlled by periodic determination of the malic acid content (g/L). After this, the wines were sulphited again and bottled. One month after MLF had ended, the wines were again analysed in terms of oenological and colour parameters, including the parameters described above for the AF (Section 2.2).

2.6. Statistical Treatment

The statistical analysis of physicochemical data consisted of two multivariate analysis performed with discriminant analysis and classification by a hierarchical cluster. The analysis of the discriminate capacity of the oenological parameters was assessed for each replicate (n = 3) of must and wines after AF. The hierarchical cluster was built with the averages of every oenological and colour parameter assessed by triplicates (n = 3) for the oenological parameters of samples after FML. Both analyses were carried out by using the statistical package IBM SPSS Statistic 20.0 (Chicago, IL, USA). Raw data of replicates employed for statistical analysis could be consulted in the Spreadsheet S1.

3. Results

3.1. Musts Physicochemical Characterization

Results of the statistical canonical discriminant analysis (CDA) of oenological parameters of must samples of Tempranillo, Grenache and Graciano are shown in Figure 1. The 100% of the variability between the three musts was explained by two possible canonical functions (F). F1 explained over 96.1% of variability and F2 3.9%, with both being significant.

A)

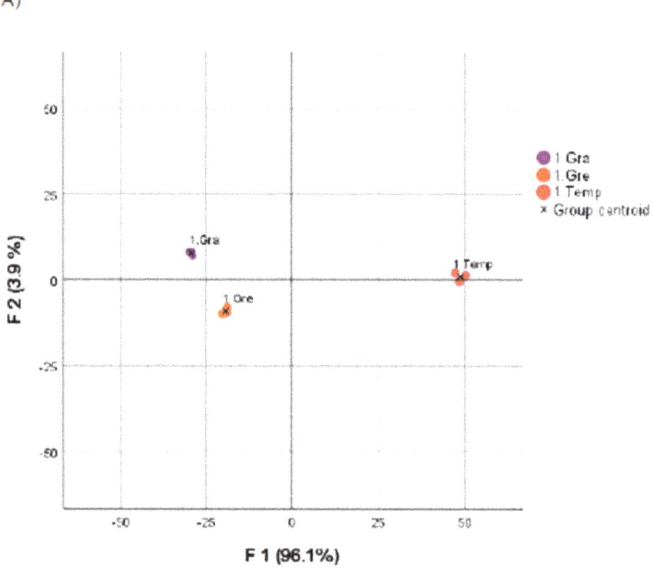

Figure 1. *Cont.*

B)

Standardized canonical coefficients

Oenological parameters	F 1	F 2
NFA	2.792	0.784
APBV (%)	- 4.247	-0.804
pH	2.056	-0.130
Total acidity (acetic acid g/l)	- 1.045	0.802
Malic acid (g/l)	0.444	1.198

Figure 1. (**A**) Canonical discriminant analysis of control initial must (1.) of Tempranillo (Temp), Grenache (Gre) and Graciano (Gra). (**B**) Standardized canonical coefficients of the two main discriminant functions (F1 and F2) obtained for oenological parameters.

All five analysed parameters contributed to the separation along F1, but APBV loading was the most dominant. F2 was also employed by the statistical software to construct the graph being mainly loaded by the malic acid content. The Tempranillo must sample was separated along F1 from the other two varieties. Grenache must was placed in the negative part of the F2 axis, and Graciano must on the positive F2 axis. Tempranillo had low APBV and the high pH and malic acid content while the Graciano must also was characterised also by low ABPV and high total acidity (data shown in Spreadsheet S1). The Grenache must had high APBV and low malic acid content.

3.2. Control of Yeast Populations and AF Kinetics in Each Grape Variety

3.2.1. Tempranillo

Results of the yeast population found in Tempranillo are shown in Figure 2A. The initial indigenous yeast population of Tempranillo must (day 0) was composed of 50% *Hanseniaspora* (*H.*) *uvarum*, 31% *S. cerevisiae* and the 19% remaining *T. delbrueckii* and *Cryptococcus* (*Cr.*) *laurenti*. The control sample of Tempranillo (C) before the second *S. cerevisiae* inoculation (day 3), was 30% *S. cerevisiae* and 70% *M. pulcherrima*. One day later (day 4), it was 77% *S. cerevisiae* and 23% *M. pulcherrima* and after a week (day 7) it was 100% *S. cerevisiae*. For samples early inoculated with *M. pulcherrima* (M), at day 3, the yeast community was 60% *M. pulcherrima* and 30% *S. cerevisiae*, 7% *H. uvarum* and 3% *L. thermotolerans*. One day later (day 4), the yeast community was composed of 70% *S. cerevisiae* and 30% *M. pulcherrima*. Eventually, after a week (day 7), all the identified yeasts were *S. cerevisiae*. In the case of Tempranillo grapes initially inoculated with *L. thermotolerans* and *T. delbrueckii* (L&T), three days after their inoculation (day 3, the yeast community was 87% *T. delbrueckii* and 13% *L. thermotolerans*. One day after *S. cerevisiae* was inoculated (day 4), it reached 40% of the yeast community and 47% *T. delbrueckii* and 13% *L. thermotolerans* was found. Finally, a week after the first inoculation (day 7), all identified yeasts were *S. cerevisiae*.

Figure 2. Percentage of yeast species in (**A**)Tempranillo, (**B**) Grenache and (**C**) Graciano, initial must (day 0), and control (C) samples and samples early inoculated with *M. pulcherrima* (M) and with a mix of *L. thermotolerans* and *T. delbrueckii* (*L&T*) during days 3, 4 and 7.

Considering the AF completed when the Brix degree had values between five and seven, the control AF of Tempranillo was completed in six days and the other two (M and *L&T*) lasted a day longer (Figure 3A).

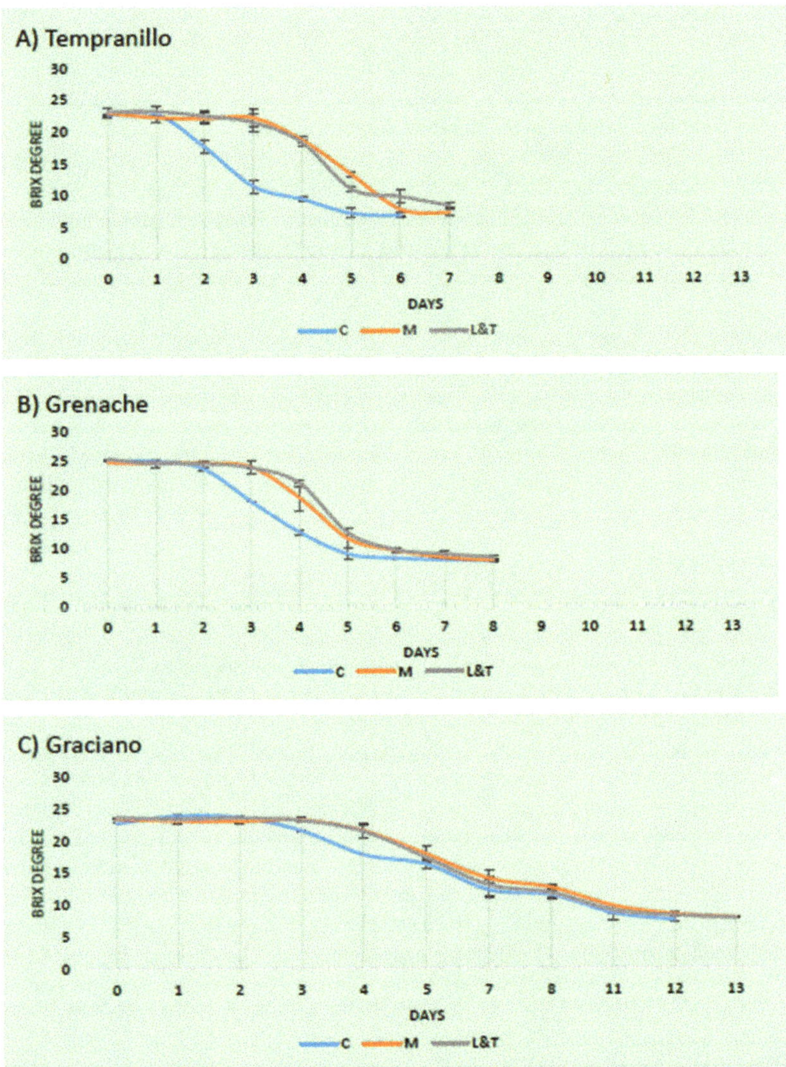

Figure 3. Brix degree measurement during alcoholic fermentation of control (C) samples and samples early inoculated with *M. pulcherrima* (M) and with a mix of *L. thermotolerans* and *T. delbrueckii* (L&T); (**A**)Tempranillo, (**B**) Grenache and (**C**) Graciano.

3.2.2. Grenache

The yeast population identified in Grenache are shown in Figure 2B. The initial Grenache must was composed of 70% *H. uvarum*, 20% *S. cerevisiae*, and 10% of *M. pulcherrima*, *Pichia (P.) membranaefaciens*, *Aureobasidium (A.) pullulans* and *Williopsis (W.) anomalous (day 0)*. In control (C) samples analysed three, four and seven days after the first inoculation of *S. cerevisiae*, all yeast isolates belonged to this species. Samples of Grenache inoculated with *M. pulcherrima* (M) were composed of 53% *M. pulcherrima*, 30% *S. cerevisiae* and 21% of *H. uvarum* and *Starmerella (St.) bacillaris* at day 3. One day later (day 4), 87% was *S. cerevisiae* and 13% *M. pulcherrima* and a week after the first inoculation (day 7), the entire yeast community was identified as *S. cerevisiae*. Grenache inoculated with *L. thermotolerans* and *T. delbrueckii*

(*L&T*) were composed of 27% *H. uvarum*, 33% *S. cerevisiae*, 23% *T. delbrueckii* and 17% *L. thermotolerans* at day 3. The other two checks of implantation (days 4 and 7) showed that 100% of the yeast community was *S. cerevisiae*. In the case of the Grenache variety (Figure 3B), the control sample either ended AF in six days and the others in seven days. The kinetics of the AF control sample were quicker than the AF of the samples early inoculated with the non-*Saccharomyces* yeasts.

3.2.3. Graciano

The species identified in Graciano samples are shown in Figure 2B. The initial Graciano must (day 0) had 75% *H. uvarum*, 12% *S. cerevisiae*, and 13% of *T. delbrueckii* and *H. osmophila*. Control samples were composed of 100% *S. cerevisiae* at each sampling checked. The samples inoculated with *M. pulcherrima* (M) at the third day had 17% *M. pulcherrima* and 83% *S. cerevisiae*. One day later (day 4), *S. cerevisiae* was 93% and one week later (day 7) all the samples were composed of *S. cerevisiae*. The samples inoculated with *L. thermotolerans* and *T. delbrueckii* (L&T) three days later (day 3) were made up of 7% *H. uvarum*, 43% *S. cerevisiae*, 43% *T. delbrueckii* and 6% *L. thermotolerans*. One day later (day 4), samples had 87% *S. cerevisiae* and one week later (day 7), they were composed solely of *S. cerevisiae*.

The Graciano AF kinetics (Figure 3C) of control samples and samples early inoculated with non-*Saccharomyces* yeasts were similar regardless of the inoculated yeasts used and took thirteen days to complete.

3.3. Characterisation of Wines

The statistical CDA of the oenological parameters of the samples of Tempranillo, Grenache and Graciano wines after AF, are shown in Figure 4. The variability between the samples (n = 3) was explained by four possible canonical functions (F) with statistical significance. F1 explained over 78.9% of variability and F2 15.2%, both explaining the 94.1% of the variance. Four out of the five assessed parameters contributed to the separation along F1, but the pH was the most influencer. F2 was mainly loaded by the total acidity, F3 and F4 (not included in the graph) were loaded by the volatile acidity. The three samples of Graciano wines stayed close in the negative part of axis F1 and Grenache and Tempranillo wines were separated by the axis 2. The three Grenache wines were clustered together. The sample of Tempranillo early inoculated with *L. thermotolerans* and *T. delbrueckii* (LT) was separated from the other two types of Tempranillo wines (control –C- and inoculated with *M. pulcherrima* –M-).

Results of statistical CDA of the colour parameters of the samples of Tempranillo, Grenache and Graciano after AF, are shown in Figure 5. The variability between the samples (n = 3) was explained by four possible canonical functions (F) with statistical significance. F1 explained over 89.2% of variability and F2 8.3%, both explaining the 98.5% of the variance. The six colour parameters analysed contributed to the separation along F1 and F2, but the most important one was the total polyphenol index. F3 and F4 (not included in the graph) were loaded by the hue and the colour intensity, respectively. The three samples of Graciano wines stayed together in the positive part of axis F1 and Grenache and Tempranillo wines were in negative part of axis F1 and separated by the axis 2. The three Grenache wines were clustered together in the positive part of axis F2. The samples of Tempranillo were placed in the negative part of both axis and the samples early inoculated with *L. thermotolerans* and *T. delbrueckii* (LT)were separated from the other two types of Tempranillo wines (control –C- and inoculated with *M. pulcherrima* –M-).

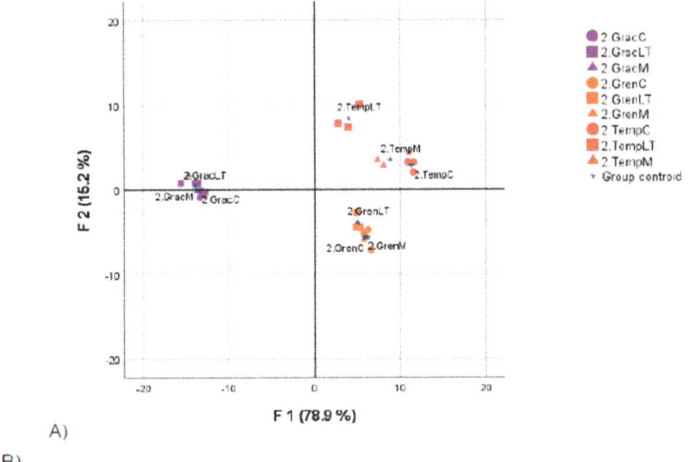

A)

B)

Standardized canonical coefficients

Oenological parameters	F 1	F 2	F3	F4
pH	0.690	0.755	-0.224	0.343
Total acidity (acetic acid g/l)	-0.481	0.978	-0.402	0.115
Volatile acidity (acetic acid g/l)	-0.116	-0.217	0.636	0.835
Lactic acid (g/l)	0.167	0.141	1.044	-0.254

Figure 4. (**A**) Canonical discriminant graph of oenological parameters in control samples (C), samples early inoculated with *M. pulcherrima* (M) and samples early inoculated with a mix of *L. thermotolerans* & *T. delbrueckii* (LT), after alcoholic fermentation (2.) of Tempranillo (Temp), Grenache (Gre) and Graciano (Gra) (**B**) Standardized canonical coefficients of the four main discriminant variables in functions (F1 and F2) for oenological parameters.

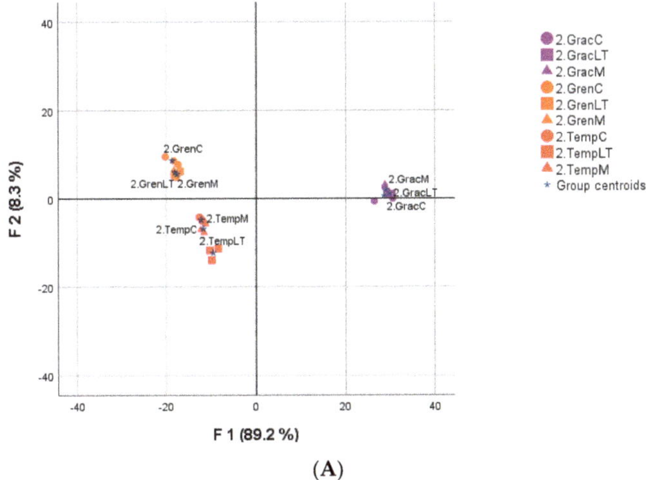

(**A**)

Figure 5. *Cont.*

Standardized Canonical Coefficients.				
Colour Parameters	F 1	F 2	F3	F4
Colour intensity	1.380	−0.203	−0.164	−1.787
Hue	−0.136	0.303	1.255	0.080
Anthocyanins (mg/l)	0.658	−1.461	0.716	1.017
Total poliphenol index	−1.616	1.773	0.002	0.267
Ionization index	0.509	0.668	0.551	0.722
Polymerization index	−0.052	−0.246	−0.230	0.499

(B)

Figure 5. (**A**) Canonical discriminant graph of colour parameters in control samples (C), samples early inoculated with *M. pulcherrima* (M) and samples early inoculated with a mix of *L. thermotolerans* & *T. delbrueckii* (LT), after alcoholic fermentation (2.) of Tempranillo (Temp), Grenache (Gre) and Graciano (Gra). (**B**) Standardized canonical coefficients of the four main discriminant variables in functions (F1 and F2) for colour parameters.

The statistical CDA of the aromatic compounds of samples of Tempranillo, Grenache and Graciano after AF, are shown in Figure 6. The variability between the samples (n = 3) was explained by four possible canonical functions (F) with statistical significance. F1 explained over 99.1% of variability and F2 0.4%, explaining the 99.5% of the variance. 12 alcohols and six esters out of the 34 aromatic compounds measured, contributed to the separation along F1 that was mainly loaded by propanol-1 compound and F2 by the hexyl acetate contents. The F3 by 2-phenylacetate and F4 by ethyl-3-hidroxibutyrate although not included in the graph. The three samples of Graciano wines stayed together in the negative part of axis F1, being separated the control sample (C) of the other two samples. Grenache and Tempranillo wines were separated by the axis 1 but in the positive part. In this case, wines were separated, being the control samples of Tempranillo and Grenache very close while the samples of both varieties but early inoculated with *L. thermotolerans* and *T. delbrueckii* (LT) were quite distant.

The MLF of each wine was completed without problems (data no shown). Six months after completion of MLF and bottling, the wines were analytically analysed in the colour and oenological parameters described for samples after AF and the hierarchical cluster built with the average data is shown in Figure 7.

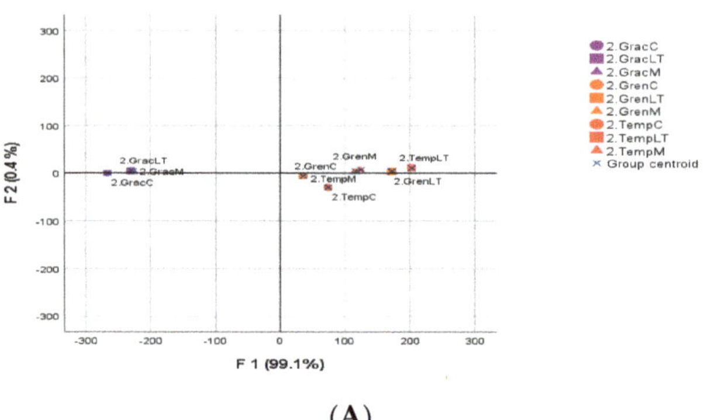

(A)

Figure 6. *Cont.*

Standardized Canonical Coefficients				
Aromatic Compounds	**F1**	**F2**	**F3**	**F4**
Propanol-1	22.50	1.130	1.052	0.426
1-Butanol	5.032	0.673	0.317	−0.075
Isobutanol	5.265	3.127	−1.571	−0.358
Amyl alcohols	9.093	−0.583	1.999	−2.660
2-Phenylethanol	−19.06	1.024	−1.460	3.235
1-Hexanol	−18.62	−3.436	0.393	−0.780
Benzyl alcohol	4.95	−1.644	0.809	−0.204
Methionol	20.36	0.404	0.566	−1.000
Cis-3-hexanol	16.92	−1.313	−0.198	−0.318
Isoamyl acetate	−19.78	−7.869	−2.783	−0.868
Hexyl acetate	0.879	9.344	−0.553	1.507
2-phenylacetate	21.38	−0.245	2.894	−0.175
Ethyl propionate	6.045	2.237	−0.493	−0.493
Ethyl-3-hidroxibutyrate	19.44	3.535	−0.014	3.737
Ethyl isobutyrate	−13.88	−1.270	0.264	0.838
Ethyl butyrate	−7.602	−2.158	1.274	−0.366
Ethyl hexanoate	5.535	2.814	0.109	1.004
Ethyl octanoate	−14.04	−1.859	0.277	−2.057

(B)

Figure 6. (**A**) Canonical discriminant graph of aromatic compounds in control samples (C), samples early inoculated with *M. pulcherrima* (M) and samples early inoculated with a mix of *L. thermotolerans* & *T. delbrueckii* (LT), after alcoholic fermentation (2.) of Tempranillo (Temp), Grenache (Gre) and Graciano (Gra) (**B**) Standardized canonical coefficients of the four main discriminant variables in functions (F1 and F2) for aromatic compounds.

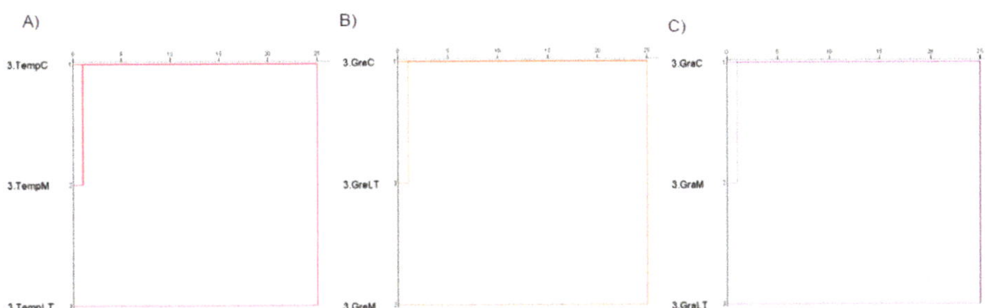

Figure 7. Hierarchical clusters assessed with average oenological and colour parameters of control samples (C), samples early inoculated with *M. pulcherrima* (M) and samples early inoculated with a mix of *L. thermotolerans* & *T. delbrueckii* (LT), after the malolactic fermentation (3.) of (**A**) Tempranillo (Temp); (**B**) Grenache (Gre); and (**C**) Graciano (Gra).

Tempranillo and Graciano grapes inoculated early with *L. thermotolerans* and *T. delbrueckii* (LT) were separated from the control samples and from the samples inoculated early with *M. pulcherrima* (M) that were clustered together. In contrast, in Grenache wines after MLF the samples inoculated early with *M. pulcherrima* (M) were separated from the other two samples that stayed clustered together.

4. Discussion

This study was focused on individual pilot plant vinifications of Tempranillo, Grenache and Graciano inoculated with non-*Saccharomyces* yeast inocula for responding if every non-*Saccharomyces* yeast would cause similar physicochemical and aromatic profiles in different grape varieties. The initial must of three grape varieties musts were separated according to parameters of APBV and acidity. Moreover, their indigenous yeast communities were also different. These initial differences fitted with a standard winemaking of non-sterile grapes [22].

4.1. Yeasts Establishment and Fermentation Kinetics

Tempranillo grapes had low APBV and high malic acid content, which was initially positive for the establishment of yeasts and bacteria populations and consequently for the evolution of AF and MLF. The presence of *S. cerevisiae* in the grape surface and must is usually low and this was corroborated in this study [23]. The initial must had *S. cerevisiae* as residual yeast and a high population of *H. uvarum*, *H. osmophila* and *T. delbrueckii* that were naturally present. The control sample inoculated only with *S. cerevisiae* had a large population of *M. pulcherrima* after three days, which might be due to an external contamination of the tanks with the *M. pulcherrima* inoculated vinification that coexisted in the experimental winery. Nonetheless, indigenous *T. delbrueckii* and *H. osmophila* were not detected and AF proceeded without problems; it was rapid and lasted only six days. In Tempranillo samples, early inoculated with non-*Saccharomyces* and then with *S. cerevisiae*, the establishment of the different yeast species happened as it was expected, probably due to the preadaptation of the strains to the grape variety because they had been isolated from this same variety.

The microbial composition of the Grenache must was characterised by a large population of *H. uvarum*, with *S. cerevisiae* as a minority strain. A diverse indigenous population characterized the initial must. Furthermore, indigenous *M. pulcherrima* was found in Grenache grapes, although with low percentage. *S. cerevisiae* inoculated in the control sample was able to achieve total implantation in spite of the high APBV of the must, and of the ecological pressure that other initial yeast species could have exerted. Indeed, the AF was not as rapid as it was in Tempranillo. The establishment of inoculated yeast species in Grenache must sample were not so successful that the observed in Tempranillo samples, in effect, the diversity of indigenous and inoculated non-*Saccharomyces* stayed until the day 4 and after this, *S. cerevisiae* became the majority.

The Graciano must had a similar microbial composition to that observed for Tempranillo. *H. uvarum* was the most frequently detected species and *T. delbrueckii* was initially present in the must sample. Similarly, to what was observed in Grenache, the implantation of *S. cerevisiae* in the control sample was total in spite of the high acidity and low pH of the must although the AF kinetics was very slow and lasted thirteen days. Similar to the described in Grenache, the establishment of inoculated non-*Saccharomyces* species was even less successful in percentages of identification.

4.2. Discriminant Analysis of Wines after AF

4.2.1. Statistical Analysis of Oenological Parameters

In order to know how the wine samples were separated depending only on the must inoculation strategy, the statistical analysis was performed without the ABV and the malic acid content that separated the must samples in the discriminant analysis.

The early inoculation of *S. cerevisiae*, *M. pulcherrima* and the mix of *L. thermotolerans* and *T. delbrueckii* did not provide enough changes in the oenological parameters of Graciano and Grenache wine samples, so that they appeared together regardless the inoculation strategy in the representation of the two main canonical discriminate functions. Only Tempranillo samples early inoculated with *L. thermotolerans* and *T. delbrueckii* was separated in the graph, from control wine samples and from wine samples early inoculated with *M. pulcherrima*. These Tempranillo wine samples early inoculated with the mix of *L. thermotolerans* and *T. delbrueckii* were characterized by a low pH and a high total acidity. As far as we know, this is the first time that the mixed inocula of *L&T* (30/70) has been tested in a pilot plant in three different grape varieties. Results showed that in the Tempranillo must, both yeasts achieved a total implantation maintaining a ratio of 13/87. Post AF, the inoculated wine had interesting increased acidity parameters due to the capacity of *L. thermotolerans* to produce lactic acid [5], which could achieve balance in a grape variety generally characterised by high pH and low acidity.

4.2.2. Statistical Analysis of Colour Parameters

Analysing statistically the colour parameters of the wine samples of the three grape varieties early inoculated with *S. cerevisiae*, *M. pulcherrima* and a mix of *L. thermotolerans* and *T. delbrueckii* provided similar results to the described for oenological parameters in the later Section 4.2.1. Again, Graciano and Grenache wine samples were separated only for being different grape varieties, but not because of the three different yeast inoculation strategies. Moreover, the Tempranillo control wine samples and the samples early inoculated with *M. pulcherrima* reached high values of total polyphenol index, while samples early inoculated with a mix of *L. thermotolerans* and *T. delbrueckii* yeasts reached lower values what make them stay separated in the graph of the two main canonical functions extracted from the discriminant analysis. In one previous study, of this same mix of *L. thermotolerans* and *T. delbrueckii* was tested for oenological parameters and anthocyanins and stilbenes and similar results were described [14]. In general terms, the reduction of the total polyphenol index is not a good result for wine quality, but observing this effect only on Tempranillo that is a grape variety characterized for normal anthocyanins content, might not be so negative than if it happened in Grenache that has a low anthocyanins content [24].

4.2.3. Statistical Analysis of Aromatic Profile

Results of the aromatic profile of the three varieties inoculated with different strategies showed interesting results. For instance, wine samples were mainly separated in the graph of the two main discriminant functions by the content of propanol-1 compound that provide alcoholic and mature fruit notes. Any other aromatic compound was able to discriminate samples. Graciano wine control samples, with lower propanol-1 concentrations, were separated from samples that had been early inoculated with *M. pulcherrima* and with the mix of *L. thermotolerans* and *T. delbrueckii*, this would mean that early inoculation of non-*Saccharomyces* yeast in Graciano must samples led to a more alcoholic profile than the samples inoculated only with *Saccharomyces*. Grenache wine samples were also separated by propanol content but in this case, the samples inoculated with *S. cerevisiae* had lower concentrations than the early inoculate with *M. pulcherrima* and these ones than the early inoculated with a mix of *L. thermotolerans* and *T. delbrueckii*. Just this same result was observed for Tempranillo wine samples. Giudici et al. [25] published that the higher alcohol n-propanol was directly related with the ability of some yeast strains to metabolise methionine and threonine aminoacids and depended on their initial content in wine, what could explain why the same inoculation strategy led to different concentration of propanol in wines depending on the variety. In any case, odour threshold for propanol was established by Peinado et al. [26] in 306 mg/L that was very high comparing concentrations obtained in the current study. This means that the different concentration between wine samples observed in the current research would not probably led to a differentiation in sensory terms.

Furthermore, the three Tempranillo wine samples were slightly differenced by the hexyl acetate content. In this way, the control wine sample was the one with the highest content of hexyl acetate compared to the early inoculated with *M. pulcherrima* and with a mix of *L. thermotolerans* and *T. delbrueckii*. The hexyl acetate aromatic compound is related to apple, cherry, pear and floral aromas and the odour threshold is 1.5 mg/L [27]. Only Tempranillo samples inoculated with *S. cerevisiae* overcame this threshold so that it would be fruitier than the Tempranillo samples early inoculated non-*Saccharomyces*.

4.3. Discriminant Analysis of Wines after MLF

Aromatic composition of wines after MLF was not considered because this fermentation was seeded with one commercial strain of *O. oeni*, so that differences in aroma could probably be due to the effect of this strain but not to the different inoculation strategies. Multivariate statistical analysis of oenological and colour parameters of samples of the three varieties showed clearly that early inoculation of Tempranillo and Grenache varieties with a mix of *L. thermotolerans* and *T. delbrueckii* caused separation of control wine samples while the early inoculation of Graciano with *M. pulcherrima* was the wine that was differenced of control Grenache wine samples.

5. Conclusions

To sum up, most of the oenological differentiations due to inoculation strategies were observed in Tempranillo wines while in Graciano and Grenache changes due to different inoculation strategies were scarce in many cases. Nevertheless, wine elaborated with different wine varieties were perfectly identified considering the grape variety. This would be indicating that one of the most important previous step in obtaining not homogenous wines is the winemaking of different grape varieties. Non-*Saccharomyces* early inoculated had been isolated from Tempranillo musts, so that a preadaptation to these grape variety properties might be expected. Therefore, changes in Tempranillo wines might be linked to the implantation or establishment rates of the inoculated yeasts. This research indicated, for the first time, that early inoculation with non-*Saccharomyces* should be carefully adjusted to the properties and features of a specific grape variety in order to increase the heterogeneity of the final product.

Supplementary Materials: The following are available online at http://www.mdpi.com/2311-5637/6/1/3/s1, One Figure S1 and one spreadsheet S1 has been included in this submission. The Figure S1 aims to clarify the sampling with a schematic representation and the spreadsheet S1 contains every data employed in statistical analysis.

Author Contributions: R.E.-V. was in charge of the methodology and the original draft preparation. P.G. was also responsible of the methodology. I.L.-A. was part of the research equipment. R.L. was a researcher of the project and also collaborated in the review and editing of the draft. P.S. researched and was in charge of resources. A.R.G. was responsible for the finding acquisition and for the project administration, and developed the formal analysis, resources and reviewed and edited the draft and eventually L.G.-A. submitted the manuscripts after reviewing and editing the draft. All authors have read and agreed to the published version of the manuscript.

Funding: This research received no external funding.

Acknowledgments: This work was supported by the Instituto Nacional de Investigaciones Agrarias.

Conflicts of Interest: The authors declare not conflict of interest.

References

1. Masneuf-Pomarede, I.; Bely, M.; Marullo, P.; Albertin, W. The Genetics of Non-Conventional Wine Yeasts: Current Knowledge and Future Challenges. *Front. Microbiol.* **2015**, *6*, 1563. [CrossRef] [PubMed]
2. Medina-Trujillo, L.; González-Royo, E.; Sieczkowski, N.; Heras, J.; Canals, J.M.; Zamora, F. Effect of Sequential Inoculation (Torulaspora delbrueckii /Saccharomyces cerevisiae) in the First Fermentation on the Foaming Properties of Sparkling Wine. *Eur. Food Res. Technol.* **2017**, *243*, 681–688. [CrossRef]
3. Whitener, M.E.B.; Stanstrup, J.; Carlin, S.; Divol, B.; Toit, M.D.; Vrhovsek, U. Effect of Non- Saccharomyces Yeasts on the Volatile Chemical Profile of Shiraz Wine. *Aust. J. Grape Wine Res.* **2017**, *23*, 179–192. [CrossRef]
4. Sadineni, V.; Kondapalli, N.; Obulam, V.S.R. Effect of Co-Fermentation with Saccharomyces cerevisiae and Torulaspora delbrueckii or Metschnikowia pulcherrima on the Aroma and Sensory Properties of Mango Wine. *Ann. Microbiol.* **2012**, *62*. [CrossRef]
5. Gobbi, M.; Comitini, F.; Domizio, P.; Romani, C.; Lencioni, L.; Mannazzu, I.; Ciani, M. Lachancea thermotolerans and Saccharomyces cerevisiae in Simultaneous and Sequential Co-Fermentation: A Strategy to Enhance Acidity and Improve the Overall Quality of Wine. *Food Microbiol.* **2013**, *33*. [CrossRef]
6. Jolly, N.P.; Varela, C.; Pretorius, I.S. Not Your Ordinary Yeast: Non-Saccharomyces Yeasts in Wine Production Uncovered. *FEMS Yeast Res.* **2014**, *14*, 215–237. [CrossRef]
7. Benito, Á.; Calderón, F.; Palomero, F.; Benito, S. Combine Use of Selected Schizosaccharomyces pombe and Lachancea thermotolerans Yeast Strains as an Alternative to the Traditional Malolactic Fermentation in Red Wine Production. *Molecules* **2015**, *20*, 9510–9523. [CrossRef]
8. Comitini, F.; Gobbi, M.; Domizio, P.; Romani, C.; Lencioni, L.; Mannazzu, I.; Ciani, M. Selected Non-Saccharomyces Wine Yeasts in Controlled Multistarter Fermentations with Saccharomyces Cerevisiae. *Food Microbiol.* **2011**, *28*, 873–882. [CrossRef]

9. Commission Regulation. Commission Regulation (EC) No 606/2009 of 10 July 2009 Laying down Certain Detailed Rules for Implementing Council Regulation (EC) No 479/2008 as Regards the Categories of Grapevine Products, Oenological Practices and the Applicable Restrictions. *Off. J. Eur. Union* **2009**, *193*, 1–59.

10. Aerny, J. Composés Azotés Des Moûts et Des Vins. *Rev. Suisse Vitic. Arboric. Hortic.* **1996**, *28*, 161–165.

11. López, I.; Torres, C.; Ruiz-Larrea, F. Genetic Typification by Pulsed-Field Gel Electrophoresis (PFGE) and Randomly Amplified Polymorphic DNA (RAPD) of Wild Lactobacillus plantarum and Oenococcus oeni Wine Strains. *Eur. Food Res. Technol.* **2008**, *227*, 547–555. [CrossRef]

12. Cocolin, L.; Dolci, P.; Rantsiou, K.; Urso, R.; Cantoni, C.; Comi, G. Lactic acid bacteria ecology of three traditional fermented sausages produced in the North of Italy as determined by molecular method. *Meat Sci.* **2009**, *82*, 125–132. [CrossRef] [PubMed]

13. Altschul, S.F.; Gish, W.; Miller, W.; Myers, E.W.; Lipman, D.J. Basic Local Alignment Search Tool. *J. Mol. Biol.* **1990**, *215*, 403–410. [CrossRef]

14. Escribano-Viana, R.; Portu, J.; Garijo, P.; López, R.; Santamaría, P.; López-Alfaro, I.; Gutiérrez, A.R.; González-Arenzana, L. Effect of the Sequential Inoculation of Non-Saccharomyces/Saccharomyces on the Anthocyans and Stilbenes Composition of Tempranillo Wines. *Front. Microbiol. 10 APR* **2019**. [CrossRef] [PubMed]

15. Escribano-Viana, R.; González-Arenzana, L.; Portu, J.; Garijo, P.; López-Alfaro, I.; López, R.; Santamaría, P.; Gutiérrez, A.R. Wine Aroma Evolution throughout Alcoholic Fermentation Sequentially Inoculated with Non- Saccharomyces/Saccharomyces Yeasts. *Food Res. Int.* **2018**, *112*, 17–24. [CrossRef] [PubMed]

16. Cocolin, L. Direct Profiling of the Yeast Dynamics in Wine Fermentations. *FEMS Microbiol. Lett.* **2000**, *189*, 81–87. [CrossRef] [PubMed]

17. Lipka, Z.; Tanner, H. Une Nouvelle Méthode de Dosage Rapide de l'acide Tartrique Dans Les Moûts, Les Vins et Autres Boissons (Selon Rebelein). *Rev. Suisse de Vitic. Arboric. Hortic.* **1974**, *6*, 5–10.

18. Somers, T.C.; Evans, E. Wine Quality: Correlations with Colour Density and Anthocyanin Equilibria in a Group of Young Red Wines. *J. Sci. Food Agric.* **1974**, *25*, 1369–1379. [CrossRef]

19. Glories, Y. Recherches Sur La Matière Colorantes Des Vins Rouges. Thesis, Université de Bourdeaux II, Bordeaux, French, 1978.

20. Ruiz, M. *La Cata y El Conocimento de Los Vinos*; Madrid Mundi Prensa: Madrid, Spain, 1999.

21. Ortega, C.; López, R.; Cacho, J.; Ferreira, V. Fast Analysis of Important Wine Volatile Compounds. Development and Validation of a New Method Based on Gas Chromatographic-Flame Ionisation Detection Analysis of Dichloromethane Microextracts. *J. Chromatogr. A* **2001**, *923*, 205–214. [CrossRef]

22. Portillo, M.C.; Franquès, J.; Araque, I.; Reguant, C.; Bordons, A. Bacterial Diversity of Grenache and Carignan Grape Surface from Different Vineyards at Priorat Wine Region (Catalonia, Spain). *Int. J. Food Microbiol.* **2016**, *219*, 56–63. [CrossRef]

23. Barata, A.; Malfeito-Ferreira, M.; Loureiro, V. The Microbial Ecology of Wine Grape Berries. *Int. J. Food Microbiol.* **2012**, *153*, 243–259. [CrossRef] [PubMed]

24. García-Beneytez, E.; Revilla, E.; Cabello, F. Anthocyanin Pattern of Several Red Grape Cultivars and Wines Made from Them. *Eur. Food Res. Technol.* **2002**, *215*, 32–37. [CrossRef]

25. Giudici, P.; Zambonelli, C.; Kunkee, R.E. Increased Production of N-Propanol in Wine by Yeast Strains Having an Impaired Ability to Form Hydrogen Sulfide. *Am. J. Enol. Vitic.* **1993**, *44*, 17–21.

26. Peinado, R.A.; Moreno, J.; Bueno, J.E.; Moreno, J.A.; Mauricio, J.C. Comparative Study of Aromatic Compounds in Two Young White Wines Subjected to Pre-Fermentative Cryomaceration. *Food Chem.* **2004**, *84*, 585–590. [CrossRef]

27. Etievant, P.X. Wine. In *Volatile Compounds in Food*; Maarse, Ed.; Food Science and Technology: New York, NY, USA, 1991.

Article

Pilot Scale Fermentations of Sangiovese: An Overview on the Impact of *Saccharomyces* and Non-*Saccharomyces* Wine Yeasts

Cristina Romani [1], Livio Lencioni [1], Alessandra Biondi Bartolini [2], Maurizio Ciani [3], Ilaria Mannazzu [4,*] and Paola Domizio [1,*]

[1] Department of Agriculture, Food, Environment and Forestry (DAGRI), University of Florence, 50144 Firenze, Italy; criromani@supereva.it (C.R.); livio.lencioni@unifi.it (L.L.)

[2] R&D Wine and Sensory Consultant, 51017 Pescia, Italy; abiondibartolini@tiscali.it

[3] Department of Life and Environmental Sciences, Polytechnic University of Marche, 60121 Ancona, Italy; m.ciani@staff.univpm.it

[4] Department of Agriculture, University of Sassari, 07100 Sassari, Italy

* Correspondence: imannazzu@uniss.it (I.M.); paola.domizio@unifi.it (P.D.)

Received: 30 May 2020; Accepted: 22 June 2020; Published: 30 June 2020

Abstract: The production of wines with peculiar analytical and sensorial profiles, together with the microbiological control of the winemaking process, has always been one of the main objectives of the wine industry. In this perspective, the use of oenological starters containing non-*Saccharomyces* yeasts can represent a valid tool for achieving these objectives. Here we present the results of seven pilot scale fermentations, each of which was inoculated with a different non-*Saccharomyces* yeast strain and after three days with a commercial *Saccharomyces cerevisiae* starter. The fermentations were carried out in double on 70 L of Sangiovese grape must, the most widely planted red grape variety in Italy and particularly in Tuscany, where it is utilized for the production of more than 80% of red wines. Fermentations were monitored by assessing both the development of the microbial population and the consumption of sugars at the different sampling times. The impact of the different starters was assessed after stabilization through the evaluation of the standard analytical composition of the resulting wines, also taking into account polysaccharides and volatile compounds. Moreover, quantitative descriptive sensory analyses were carried out. Compared to the control wines obtained by inoculating the *S. cerevisiae* starter strain, those inoculated with non-*Saccharomyces*/*Saccharomyces* mixed starters presented a significant differentiation in the chemical-analytical composition. Moreover, sensory analysis revealed differences among wines mainly for intensity of color, astringency, and dryness mouthfeel perception.

Keywords: non-*Saccharomyces* yeasts; wine; mixed starter cultures; fermentation; Sangiovese; sensory analysis

1. Introduction

Non-*Saccharomyces* yeasts have attracted increasing attention in recent years, with several studies providing evidence of their impact on the organoleptic characteristics and chemical-physical stability of wines. Despite their large intraspecific biodiversity, non-*Saccharomyces* yeasts often show species-specific metabolic features that contribute to the specific imprint of the resulting wines, when inoculated in mixed fermentation with *Saccharomyces cerevisiae* [1]. For instance, among the non-*Saccharomyces* yeasts, those belonging to the species *Torulaspora delbrueckii* result in the production of low volatile acidity, high terpenols, and 2-phenylethanol when utilized in mixed culture with *S. cerevisiae* [2–7]. In addition, the release of higher concentrations of thiols, with consequent increase of varietal characters, was reported for *T. delbrueckii/S. cerevisiae* mixed starters [8,9]. *Starmerella bacillaris* (synonym *Candida zemplinina*) contributes to reduce the amount of acetic acid in mixed fermentation with *S. cerevisiae* [5,10–12]. Moreover, this yeast is usually characterized by high glycerol production [11,13–17] and low ethanol yield [16,18–20] making it an interesting tool to increase the wine sweetness and modulate the ethanol content. *Lachancea thermotolerans* strains produce lactic acid during the alcoholic fermentation causing a decrease of wine pH while reducing its volatile acidity [4,21–24]. Moreover, an increase of 2-phenylethanol, glycerol, and polysaccharides in mixed fermentation *L. thermotolerans/S. cerevisiae* was reported [4,21,22,25]. Regarding *Metschnikowia pulcherrima*, some studies showed its high β-glucosidase activity [26–28] with consequent increase of volatile terpene content from glycosylated flavorless precursors present in grapes. Moreover, because of a high β-lyase activity, yeasts belonging to the species *M. pulcherrima* release high quantity of varietal thiols from grape precursors conjugated to cysteine or glutathione [29,30]. In the last few years there has been also a renewed interest in yeasts belonging to the genus *Schizosaccharomyces*. Indeed, besides reducing malic acid in grape juice and/or wine, these yeasts produce high quantities of pyruvic acid [31,32] and polysaccharides during the course of alcoholic fermentation [33–36], positively contributing to the chemical-physical stability of wines. Finally, among yeasts typically considered as potential spoilage, those belonging to the genus *Zygosaccharomyces* have also started to attract attention [37,38]. In particular, the species *Z. florentina* contributes to increase esters and glycerol concentration when used in co-culture with *S. cerevisiae*, thus producing wines with higher floral notes and lower perception of astringency [39].

A Scopus database search with the combination of terms "wine and non-*Saccharomyces*" as query statement to highlight the relevant literature in the last decade, indicates that an increasing number of peer-reviewed publications have considered the use of non-*Saccharomyces* yeasts as starters together with *S. cerevisiae*. In particular, of a total of 458 peer reviewed scientific articles published from 2010 to 2019, the average number of publications/year on this topic was 26 during the period in between 2010 and 2014, and reached 65 in the following years (2015–2019). It is worth pointing out that most of these publications refer to laboratory scale fermentations. In particular, considering publications starting from 2015, 76% of the fermentations were carried out at the laboratory scale (50% in up to 1 L, 15% in 1.2–5 L and 11% in 10–20 L). Instead, pilot plant and industrial scale fermentations, that regarded 24% of the trials, are still quite limited. Of these, 18% were carried out in grape must volumes ranging from 30 to 200 L (pilot plant fermentations) and 5% in 700 to 1000 L (industrial scale fermentations). One fermentation was performed in a 100,000 L vessel.

Moreover, while 12% of the publications starting from 2015 describe fermentations carried out in synthetic media, the majority of works report on the utilization of different grape varieties to evaluate the impact of the non-*Saccharomyces* yeasts on the chemical and physical characteristics of the relevant wine. Among these, Shiraz, Sauvignon Blanc, Barbera, Cabernet Sauvignon, Chardonnay and Merlot were the most frequently utilized. Few articles (4%) describe mixed fermentations in Sangiovese grape must despite the importance of this grape variety that in Italy represents 90% of total world Sangiovese vineyard area (http://www.oiv.int/public/medias/5888/en-distribution-of-the-worlds-grapevine-varieties.pdf). In order to avoid any metabolic interference by other microorganisms, half of the studies evaluated the impact of pure and mixed starters on the final wine by using sterile grape juice.

Indeed, laboratory scale fermentation and synthetic media or sterile grape juice are important conditions to evaluate the specific metabolic traits of the non-*Saccharomyces* yeasts included in the mixed starter and also to establish their possible interactions with *S. cerevisiae*. However, the results obtained under these conditions are likely far away from those obtainable under technological conditions, also due to the unpredictability of the interactions that the inoculated starters may establish with wild grape must microflora.

Based on these observations, in the present work seven different non-*Saccharomyces/S. cerevisiae* mixed starters were inoculated in Sangiovese grape must at the pilot plant scale and their impact on the final product was evaluated through chemical and sensory analyses of the resulting wines after stabilization. Sangiovese is the most widely planted red grape variety in Italy, particularly in Tuscany where it represents the obligatory variety in the production of wines with a protected and guaranteed designation of origin (DOCG) such as Chianti Classico and Brunello di Montalcino.

2. Material and Methods

2.1. Yeast Strains

Seven non-*Saccharomyces* strains from the yeast culture collection of the Department of Agriculture, Food, Environment and Forestry (DAGRI, University of Florence, Italy) and of the Department of Life and Environmental Sciences (DiSVA, Polytechnic University of Marche, Ancona, Italy) were used (Table 1). The yeast strains were sub-cultured on YPD (1% yeast extract, 2% peptone, 2% glucose, 2% agar) at six months intervals, and maintained at 4 °C.

Table 1. Origin of the seven non-*Saccharomyces* strains and the commercial strain of *S. cerevisiae* used in this study.

Strain	Species	Origin
# 4	*Pichia fermentans*	DiSVA [a]
# 22	*Starmerella bacillaris*	DAGRI [b]
# 32	*Hanseniaspora osmophila*	DAGRI [b]
# 42	*Zygotorulaspora florentina*	DAGRI [b]
# 46	*Metschnikowia pulcherrima*	DiSVA [a]
# 92	*Torulaspora delbrueckii*	DAGRI [b]
# 103	*Lachancea thermotolerans*	DiSVA [a]
# EC1118	*Saccharomyces cerevisiae*	Lallemand [c]

[a] Department of Life and Environmental Sciences of the Polytechnic University of Marche (Italy), [b] Department of Agriculture, Food, Environment and Forestry, University of Florence (Italy). [c] Lallemand Inc. (Montreal, QC, Canada).

The strains reported in Table 1 were isolated from grapes and musts of different origins and characterized for their enological performances in mixed fermentations carried out in grape juice at laboratory scale [4,37,38]. A commercial *S. cerevisiae* starter, Lalvin EC1118 (Lallemand Inc., Montreal, QC, Canada), was used as reference strain and for comparison determination.

2.2. Pilot Scale Fermentation

The fermentation trials were carried out in 100 L steel tanks containing 70 L of Sangiovese grape must with the following characteristics: pH 3.66, 234 ±7 g/L sugars, 4.0 g/L total acidity (as tartaric acid), 1.2 g/L malic acid. Non-*Saccharomyces* yeasts were inoculated in 12 L filtered sterilized commercial red grape must, aliquoted within 2 L flasks, each containing 1.5 L and grown for 48-h at 25 °C under shaking conditions (150 rpm). Cell concentration was determined by microscope counting. Each tank was inoculated with 10^7 cell/mL of the non-*Saccharomyces* yeast strain. After 3 days of fermentation, *S. cerevisiae* EC1118 was inoculated as active dry yeast (ADY) at the final concentration of 10^7 cell/mL. Control trials were inoculated with 10^7 cell/mL of *S. cerevisiae* EC1118. Skin cap was punched down twice a day and fermenting must was sampled during the fermentation process immediately after inoculation (T0) and 3, 5, and 10 days after inoculation (T3, T5, T10, respectively) to evaluate the evolution of the yeast populations as viable cell counts and to determine the residual sugars. Alcoholic fermentation was monitored by periodically measuring the density by a double scale hydrometer (density and Baumé). All trials were fermented at 25 °C, in duplicate. After completion of fermentation, the wines were naturally fined by three successive rackings over a month at 16–18 °C and added with SO_2 up to 100 mg/L before bottling (0.75 L cork-capped glass bottles).

2.3. Analyses

2.3.1. Evaluation of Cell Growth

During the fermentation, cell concentrations were evaluated as viable cell counts and expressed as number of colony forming units (cfu)/mL at the different sampling times. Viable plate count was carried out both on lysine agar (LA medium; Oxoid Unipath, Hampshire, UK), and Wallerstein Laboratory nutrient agar medium (WL medium; Oxoid Unipath, Hampshire, UK) [40] to estimate the non-*Saccharomyces* yeast and the total yeast population, respectively.

2.3.2. Analytical Determinations

Residual sugar, organic acid, total and volatile acidity and pH were determined according to Official EU Methods (EC 2000). Total polysaccharides concentration was evaluated by HPLC [35], using a Varian instrument equipped with auto-sample injector (loop 20 μL) and coupled with refractive index detector. For the separation of total polysaccharides, a column Progel-TSK G-OLIGO PW (Supelco 808031) and a TSK-gel PW (Supelco 808034) precolumn were used with isocratic elution (NaCl 0.2 M; 0.8 mL/min; 40 °C). Before injection, the samples were filtered (1.2 μm) and purified on polyamide SC6 (Macherey-Nagel, Dylan, Germany). Polysaccharides quantification was performed by comparison with an external calibration curve of mannans from *S. cerevisiae* (M7504, Sigma-Aldrich, St. Louis, MO, USA), at concentrations ranging from 50 mg/L to 500 mg/L.

Total polyphenols in wines were determined by the Folin–Ciocalteu method according to Di Stefano [41] and expressed as catechin equivalents in mg of catechin/L, whereas total anthocyanins were determined by direct reading of the absorbance at 540 nm of wine in hydrochloric ethanol solution [42] and expressed in g malvidin/L [43].

Higher alcohols and acetaldehyde were determined by a GC method with flame ionization detector (FID) detection at 250 °C, on a Carlo Erba HRGC 5300 instruments equipped with a glass column (2 m; 2 mm ID) packaged with Carbopack C + 0.2% Carbowax 1500, 80–100 meshes (Supelco). The other chromatographic conditions were as follows: temperature gradient from 45 °C to 160 °C (3 °C/min), held to 160 °C for 20 min; inj. temperature 220 °C; carrier: Helium 2 mL/min; injection volume: 1 μL of distilled sample spiked with 3-methyl-2-butanol as internal standard. The acquisition and elaboration of the FID signal was carried out by means of Galaxy software (Varian Inc., Walnut Creek, CA, USA).

Minor volatile compounds were evaluated by capillary gas-liquid chromatography as previously reported [36]. In particular, the analyses were carried out on a Carlo Erba HRGC 5300 instrument, injecting an ether/hexane extracts (1/1, *v/v*) of the wine samples previously spiked with 3-octanol as internal standard. The chromatographic conditions were as follow: glass capillary column 0.25 μm Supelcowax 10 (60 m length, 0.32 mm internal diameter, 0.25 μm film thickness). One μL was injected in split-splitless mode (60 s splitless); carrier gas: helium at 2.5 mL/min flow rate; injection temperature: 220 °C; elution temperature gradient: from 50 °C (held 5 min) to 220 °C (3 °C/min); detection by flame ionization detector (FID) at 250 °C. The acquisition and integration of the FID signals were carried out using the Galaxy software (Varian Inc., Walnut Creek, CA, USA) and the content of each compound was evaluated in respect to an external standard curve. All analyses were carried out in double from each fermentation tank.

2.3.3. Sensory Evaluation of Wines

The wines were left to mellow for about four months after bottling, before sensory evaluation. Wine tasting was performed by an 11 member trained and formed panel, in two sessions. Organoleptic evaluations were conducted by quantitative descriptive analysis, using a pre-defined protocol and descriptive terminology, previously developed by the tasting group. In particular, every sample was tasted twice by each taster, within completely randomized blocks, and the panelists were asked to express their judgment by quantification of each sensory descriptor (color intensity, floral, fruity, preserved fruit, spicy, candy, sulfur, chemical, earthy, mouthfeel volume, acidity, tannic intensity, astringency, dryness, bitterness) on the basis of a four point structured scale, from 0 to 3 for the olfactory descriptors (no presence of the perception to high intensity), or from 1 to 4 (low to high intensity) for color intensity and mouthfeel descriptors. The olfactive descriptors were chosen previously in a round table session of the panel basing on a free profile description of the same wines and the judges were subsequently trained on them.

For evaluation of gustative descriptors, the method used was that developed by ICV (Institute Cooperative du Vin, Lattes, France) for the Quantitative Descriptive Analysis of red wine [44,45].

2.4. Statistical Analysis

Data from chemical analysis of the wines were subjected to one-way ANOVA using STATISTICA 7 (Statsoft, Tulsa, OK, USA) software. Duncan test was carried out to compare mean values and evaluate significant differences. Mean values of volatile compounds were analyzed by principal component analysis (PCA), using JMP® 11 statistical software.

The sensory scores were statistically analyzed and compared according to analysis of variance (ANOVA) using a mixed effect model considering as fixed factors those related to the experimental thesis and as random factors the deviations due to the effect of the judge from the general average of each parameter.

3. Results and Discussion

Growth kinetics of *S. cerevisiae* EC1118 pure culture (control fermentation) and of non-*Saccharomyces*/*S. cerevisiae* mixed cultures are reported in Figure 1. In all mixed cultures the initial concentration of the non-*Saccharomyces* yeasts ranged from 10^6 to 10^7 cell/mL. *S. cerevisiae* was able to dominate in most of the mixed fermentations and showed, in mixed culture, a growth kinetics that was similar to that of control. In particular, in spite of the presence of the non-*Saccharomyces* yeasts, *S. cerevisiae* reached a cell concentration of about 10^8 CFU/mL (T5) and maintained it until the end of the fermentation (T10). When in mixed culture with *H. osmophila*, *S. cerevisiae* reached a lower cell concentration (2.5×10^7 cell/mL at T5) which remained unvaried for 5 further days of fermentation (T10). Similarly, *L. thermotolerans*, even if to a lower extent, affected the growth of *S. cerevisiae*, which reached a cell concentration of 4.8×10^7 cell/mL and 6.7×10^7 cell/mL at T5 and T10, respectively. Conversely, and contrary to that found by Englezos et al. [20], *S. bacillaris* did not affect *S. cerevisiae* growth. Similar to *H. osmophila* and *L. thermotolerans*, *Z. florentina* and *T. delbrueckii* showed a higher level of competitiveness being still present at concentrations ranging from 1×10^4 to 5×10^5 cells/mL at T10. On the contrary, *P. fermentans*, *S. bacillaris*, and *M. pulcherrima* persisted at high concentration up to T5 and they almost disappeared at T10.

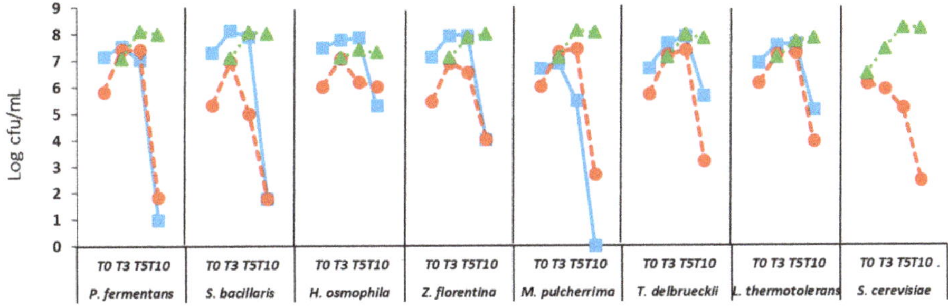

Figure 1. Growth of inoculated non-*Saccharomyces* (■), other non-*Saccharomyces* yeasts (•) and *S. cerevisiae* (▲). Viable plate counts were done immediately after (T0), and 3, 5, and 10 days after inoculation (T3, T5, and T10, respectively). Data are means ±SD (*n* = 2).

Sugar consumption was consistent with growth kinetics (Figure 2). As expected, in control fermentations (*S. cerevisiae* pure culture), it started soon after grape crushing and proceeded faster compared to that observed in all mixed fermentation trials. Among these, the combination *H. osmophila/S. cerevisiae* showed an impairment of sugar consumption with 7.50 g/L residual sugar at T10, while the other mixed cultures left from 1 to 1.4 g/L residual sugar.

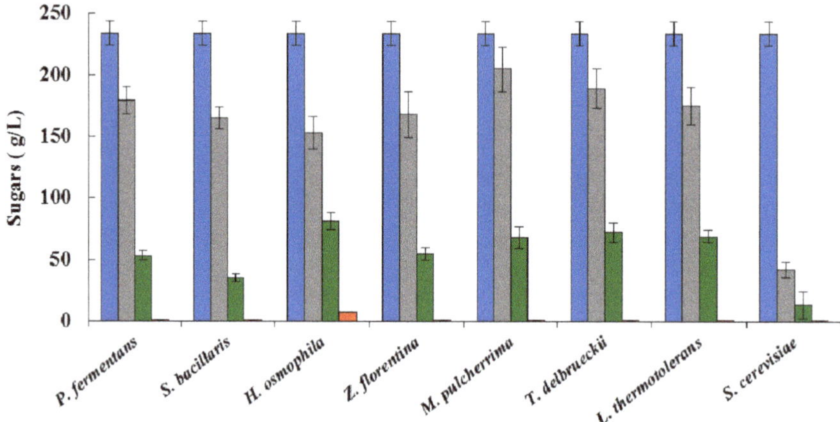

Figure 2. Non-*Saccharomyces* and *S. cerevisiae* sugar consumption in each fermentation trial. Different colors indicate different sampling times; T0 (■), T3 (■), T5 (■), T10 (■). Data are means ±SD (*n* = 2).

In most of the mixed cultures ethanol concentrations were comparable to that of the control (Table 2). Mixed starters involving *Z. florentina* and *T. delbrueckii* were exceptions and produced less ethanol than the control. Lencioni et al. [39], also found slightly lower ethanol concentrations in *Z. florentina/S. cerevisiae* fermentations performed at laboratory scale in white grape must, with respect to the pure *S. cerevisiae* culture. Regarding the mixed starter *T. delbrueckii/S. cerevisiae*, decreases in ethanol concentrations, ranging from 0.3% to 0.5%, were also reported by other authors at the end of pilot-scale fermentations [46,47]. Non-*Saccharomyces/S. cerevisiae* mixed starters have already been proposed as a tool for the possible reduction of ethanol content in wine [19,48–54]. Indeed, the lower ethanol concentration is a consequence of some features of non-*Saccharomyces* yeasts, such as reduced ethanol yield, low fermentation efficiency, and respiro-fermentative metabolism.

Table 2. Main analytical parameters of the wines evaluated before bottling.

	pH	Ethanol% (v/v)	Volatile Acidity (g/L)	Total Acidity (g/L)	Malic Acid (g/L)	Lactic Acid (g/L)	Free SO_2 (mg/L)	Total Polyphenols (mg catechin/L)	Total Anthocyanins (mg malvidin/L)
P. fermentans	3.37 ± 0.03 [ab]	12.80 ± 0.01 [abc]	0.34 ± 0.02 [bc]	5.50 ± 0.05 [bc]	1.08 ± 0.02 [abc]	0.11 ± 0.03 [cd]	18 ± 0.8 [bc]	1536 ± 27.32 [a]	194 ± 3.55 [ab]
S. bacillaris	3.34 ± 0.02 [ab]	12.85 ± 0.00 [bc]	0.37 ± 0.02 [b]	5.80 ± 0.06 [ab]	1.06 ± 0.02 [bc]	0.11 ± 0.00 [d]	13 ± 2.50 [c]	1556 ± 30.50 [a]	204 ± 0.50 [a]
H. osmophila	3.41 ± 0.05 [a]	12.88 ± 0.13 [ab]	0.53 ± 0.02 [a]	5.50 ± 0.33 [bc]	0.96 ± 0.23 [cd]	0.24 ± 0.09 [b]	16 ± 1.00 [bc]	1381 ± 66.00 [ab]	173 ± 12.50 [bc]
Z. florentina	3.34 ± 0.01 [ab]	12.63 ± 0.03 [bc]	0.26 ± 0.01 [c]	6.15 ± 0.00 [a]	1.34 ± 0.01 [a]	0.22 ± 0.01 [bc]	14 ± 0.00 [bc]	1394 ± 44.50 [ab]	172 ± 2.00 [bc]
M. pulcherrima	3.34 ± 0.02 [ab]	12.78 ± 0.03 [abc]	0.25 ± 0.04 [c]	5.87 ± 0.06 [ab]	1.24 ± 0.01 [ab]	0.15 ± 0.00 [bcd]	12 ± 3.00 [c]	1518 ± 53.50 [ab]	187 ± 0.50 [abc]
T. delbrueckii	3.32 ± 0.00 [b]	12.70 ± 0.15 [c]	0.30 ± 0.03 [bc]	6.07 ± 0.02 [a]	1.13 ± 0.02 [bc]	0.17 ± 0.02 [bcd]	14 ± 0.50 [bc]	1350 ± 42.50 [ab]	181 ± 11.50 [abc]
L. thermotolerans	3.40 ± 0.02 [a]	12.98 ± 0.03 [a]	0.37 ± 0.02 [b]	5.58 ± 0.07 [bc]	1.04 ± 0.02 [d]	0.78 ± 0.01 [a]	19 ± 1.00 [ab]	1421 ± 7.50 [ab]	188 ± 2.00 [ab]
S. cerevisiae	3.40 ± 0.02 [ab]	13.00 ± 0.00 [a]	0.28 ± 0.06 [bc]	5.36 ± 0.10 [c]	1.15 ± 0.09 [abc]	0.16 ± 0.01 [bcd]	24 ± 3.00 [a]	1300 ± 92.00 [b]	163 ± 12.50 [c]

Data are media ± semi-difference of two independent experiments. Data with different superscript letters within each column are significantly different (Duncan test; $p \leq 0.05$).

Mixed starters, including *P. fermentans*, *S. bacillaris*, and *L. thermotolerans*, determined an increase of volatile acidity of about 0.1 g/L, as compared to the control (Table 2). Although non-significant, *H. osmophila/S. cerevisiae* resulted in a more marked increase in volatile acidity which reached values of 0.5 g/L. This is in contrast with previous results obtained with the same yeast strain in co-fermentation with *S. cerevisiae*, but at laboratory scale and with a commercial white grape must [38]. Instead, the increase in volatile acidity observed in the fermentation inoculated with *S. bacillaris/S. cerevisiae* is in agreement with the results obtained by Whitener et al. [55] but in contrast with previously published works showing *C. zemplinina* (synonym *S. bacillaris*) able to reduce the amount of acetic acid when in mixed culture with *S. cerevisiae* [5,10–12]. Nisiotou et al. [56] found lower acetic acid concentration in sequential fermentation *S. bacillaris/S. cerevisiae* carried out as a pilot plant as compared to those performed at laboratory scale fermentation. These discrepancies might be due to the significant strain diversity within this species, as already observed by Englezos et al. [11], but also to a strain specific response to the different fermentation conditions, including the grape variety utilized. In our experimental trials, the presence of other microorganisms, starting from the beginning of the alcoholic fermentations performed at pilot scale, might have interfered with the metabolic activity of the *S. bacillaris* yeast strain. Similar observations can be extended to the mixed fermentation conducted with *L. thermotolerans* that is usually recognized for low volatile acidity production in wine [57].

The utilization of *P. fermentans* and *S. bacillaris* resulted also in a significantly higher amount of both total polyphenols and anthocyanins, with respect to the control (Table 2). Recent works indicate that wine color and anthocyanin composition may benefit from the fermentative activity of non-*Saccharomyces* yeasts [58,59]. In particular, it was shown that the inoculation of non-*Saccharomyces/Saccharomyces* mixed starters results in higher acetaldehyde production, with effects on anthocyanin-derived pigments [60]. Here, no differences in acetaldehyde content were observed at T10 although its increase at T3 and T5 cannot be excluded in fermentations carried out by mixed starters including *P. fermentans* and *S. bacillaris*.

Analyses of total polysaccharides, glycerol and volatile compounds, together with the sensory analyses were performed four months after bottling.

Polysaccharides, in particular mannoproteins, impact wine sensorial features by decreasing astringency, improving the mouthfeel and fullness, adding complexity and aromatic persistence, and increasing roundness and sweetness [61–64]. With the exception of those including *H. osmophila* and *Z. florentina*, all mixed starters produced significantly higher polysaccharides concentrations in respect to the control (Figure 3). In particular, the increase ranged from 2.5% to 33%. In this respect, the most interesting association was *L. thermotolerans/S. cerevisiae* with a final content of total polysaccharides of 732 mg/L versus 550 mg/L of the control, in agreement with the result obtained by Gobbi et al. [22] with a different *L. thermotolerans* strain. The release of polysaccharides by non-*Saccharomyces* yeasts is not new and a wide intraspecific biodiversity for this characteristic was observed in *Hanseniaspora*, *Zygosaccharomyces* [4,35,37,38] and *Schizosaccharomyces* yeasts [34].

The concentration of glycerol, responsible for the sweetness of red and white wines [65,66], was significantly higher in most of the wines produced by mixed starters, apart from those including *H. osmophila*, *P. fermentans* and *Z. florentina* (Figure 3). As expected, the association *S. bacillaris/S. cerevisiae* resulted in the highest glycerol concentration (11.4 g/L), in accordance with that already observed for the species *S. bacillaris* [25,67,68].

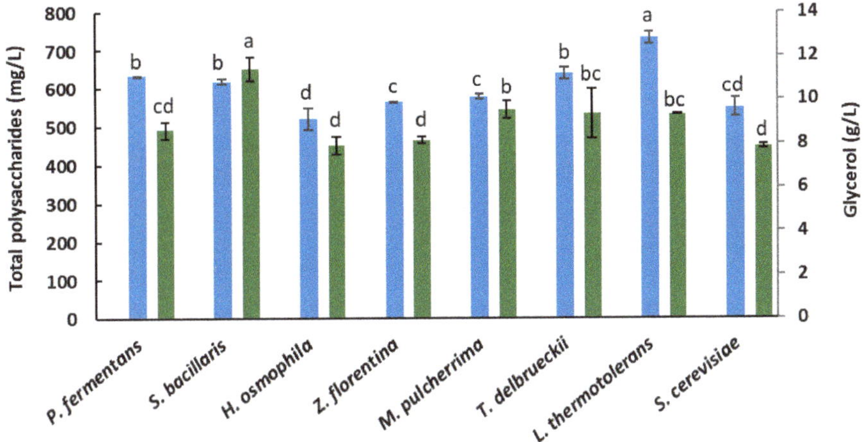

Figure 3. Total polysaccharides (■) and glycerol (■) in wines obtained four months after bottling. Data are means ±SD (*n* = 2). Values displaying different letters (a, b, c, d) are significantly different according to the Duncan test (*p* ≤ 0.05).

The concentrations of the main volatile compounds are reported in Table 3. The concentrations of acetaldehyde, propanol, and hexanol produced by mixed starter cultures were comparable to that of the control. In contrast, mixed starters resulted in higher production of some of the higher alcohols. In particular, significant increases of 2-methyl-1-propanol (isobutanol) were observed in respect to control fermentation (47 ± 2 mg/L). This compound ranged from a minimum of 66 mg/L (for the associations including *H. osmophila* and *Z. florentina*) to a maximum of 123 ± 8 mg/L in the wine produced by *S. bacillaris/S. cerevisiae*. However, the sum of amylic alcohols (i.e., 2-methyl-1-butanol, 3-methyl-1-butanol) was significantly higher in wines produced with the mixed starters including *M. pulcherrima* (328 mg/L), *T. delbrueckii* (299 mg/L) and *L. thermotolerans* (294 mg/L) than in the control wine (266 mg/L). In agreement with that reported by other authors [69–72], mixed starters including *Hanseniaspora*, *Pichia* and *Zygosaccharomyces* showed lower production of higher alcohols. Interestingly, all the wines obtained with mixed starters presented significantly higher concentrations of 2-phenylethanol (8–9.5 mg/L) (which provides a rose-like flavor) compared to the control (6.1 mg/L). In particular, the highest concentrations of 2-phenylethanol were reached in mixed fermentations including *M. pulcherrima* (9.2 mg/L) and *T. delbrueckii* (9.4 mg/L). These results agree with those found by other authors showing that *M. pulcherrima* and *T. delbrueckii* produce high level of 2-phenylethanol [3,73,74]. Ethyl acetate was the main ester produced. At high concentrations (>100–150 mg/L) ethyl acetate determines a solvent-like aroma. Interestingly, with the exception of the associations including *M. pulcherrima* and *H. osmophila* that nearly doubled the amount produced by *S. cerevisiae* starter culture (54 mg/L and 51 mg/L, respectively), the other mixed starters determined slight increases in ethyl acetate in respect to the control. In any case, ethyl acetate concentration was always below the perception threshold (Table 3). These findings confirm those already observed in other studies [37,38], where some non-*Saccharomyces* yeasts, generally considered spoilage yeasts, produced in mixed culture ethyl acetate concentrations below those normally produced by the relevant pure culture. On the other hand, many studies report that most non-*Saccharomyces* yeasts can produce high amounts of ethyl acetate [71,75]. However, this discrepancy may be due to the wide inter-generic and intra-generic variability observed for the production of this ester compound. Accordingly, Domizio et al. [38] found, by analyzing eleven yeast strains of *Hanseniaspora* (belonging to four different species), that ethyl-acetate production ranged from 27 to 333 mg/L. It is also worth underlining that this compound, at low concentration, might contribute to wine fruity aroma.

Table 3. Volatile compounds (mg/L) of wines four months after bottling.

	P. fermentans	*S. bacillaris*	*H. osmophila*	*Z. florentina*	*M. pulcherrima*	*T. delbrueckii*	*L. thermotolerans*	*S. cerevisiae*
Acetaldehyde	71 ± 2 [a]	66 ± 0 [a]	61 ± 3 [a]	62 ± 1 [a]	72 ± 8 [a]	61 ± 3 [a]	62 ± 5 [a]	64 ± 7 [a]
1-propanol	34 ± 2 [b]	38 ± 1 [ab]	33 ± 3 [b]	41 ± 2 [a]	35 ± 2 [b]	33 ± 1 [b]	41 ± 2 [a]	36 ± 2 [ab]
2-Methyl-1-propanol	72 ± 2 [d]	123 ± 8 [a]	66 ± 3 [d]	66 ± 1 [d]	92 ± 1 [b]	74 ± 3 [cd]	84 ± 1 [bc]	47 ± 2 [e]
2-Methyl-1-butanol	54 ± 0 [bc]	42 ± 2 [d]	45 ± 5 [cd]	53 ± 1 [bcd]	68 ± 7 [a]	58 ± 1 [ab]	63 ± 2 [ab]	61 ± 3 [ab]
3-Methyl-1-butanol	212 ± 9 [bcd]	184 ± 5 [d]	213 ± 6 [bcd]	187 ± 1 [d]	260 ± 14 [a]	241 ± 7 [ab]	231 ± 5 [abc]	205 ± 3 [cd]
Hexanol	0.108 ± 0.012 [bc]	0.096 ± 0.001 [c]	0.109 ± 0.001 [bc]	0.113 ± 0.003 [b]	0.111 ± 0.007 [bc]	0.115 ± 0.002 [ab]	0.129 ± 0.002 [a]	0.117 ± 0.001 [ab]
2-Phenylethanol	7.355 ± 0.140 [ab]	7.995 ± 0.015 [ab]	8.920 ± 0.5 [a]	6.995 ± 0.045 [ab]	9.240 ± 0.250 [a]	9.455 ± 0.355 [a]	8.440 ± 0.430	6.125 ± 0.095 [b]
Ethyl acetate	35 ± 0 [bc]	39 ± 2 [b]	54 ± 1 [a]	31 ± 0 [cd]	51 ± 3 [a]	38 ± 2 [b]	39 ± 1 [b]	28 ± 2 [d]
Isoamyl acetate	0.027 ± 0.005 [b]	0.020 ± 0.001 [b]	0.042 ± 0.003 [a]	0.028 ± 0.001 [ab]	0.034 ± 0.002 [ab]	0.034 ± 0.003 [ab]	0.030 ± 0.001 [ab]	0.040 ± 0.001 [a]
Phenylethyl acetate	0.003 ± 0.001 [b]	0.003 ± 0.001 [b]	0.016 ± 0.005 [a]	0.004 ± 0 [b]	0.005 ± 0.001 [b]	0.006 ± 0.001 [b]	0.004 ± 0 [b]	0.003 ± 0 [b]
Ethyl lactate	3.4 ± 0.6 [b]	3.7 ± 0 [b]	5.5 ± 2.1 [b]	5.6 ± 0.4 [b]	5.5 ± 0.2 [b]	4.4 ± 0.2 [b]	16.5 ± 1 [a]	4.7 ± 0.6 [b]
Ethyl butyrate	0.181 ± 0.01 [abc]	0.192 ± 0.006 [ab]	0.149 ± 0.009 [c]	0.205 ± 0.009 [a]	0.166 ± 0.01 [bc]	0.171 ± 0.021 [bc]	0.194 ± 0.001 [ab]	0.172 ± 0.001 [abc]
Ethyl hexanoate	0.013 ± 0.002 [bc]	0.008 ± 0.001 [d]	0.007 ± 0.001 [d]	0.010 ± 0.001 [cd]	0.014 ± 0.001 [b]	0.014 ± 0.001 [b]	0.014 ± 0.001 [b]	0.023 ± 0.002 [a]
Ethyl octanoate	0.013 ± 0.003 [b]	0.008 ± 0.001 [cd]	0.003 ± 0 [e]	0.011 ± 0.001 [bc]	0.013 ± 0 [b]	0.010 ± 0.001 [bcd]	0.007 ± 0.001 [de]	0.024 ± 0.001 [a]

Data with different superscript letters (a, b, c, d) within each line are significantly different (Duncan test; $p \leq 0.05$).

Among other acetates analyzed, 2-phenylethyl acetate (with a fruity and flowery flavor) was significantly higher only in wines fermented by *H. osmophila/S. cerevisiae* (0.016 mg/L) in comparison with the control wine (0.003 mg/L). This result is in agreement with the known capacity of this yeast species to release high levels of 2-phenylethyl acetate [70–72,76].

Other ethyl esters compounds, such as ethyl lactate, ethyl butyrate, ethyl hexanoate and ethyl octanoate, were present in all the wines with similar or slightly lower concentrations in comparison to those present in the control wine. An exception, regarding ethyl lactate, was made for wines produced by the association of *L. thermotolerans/S. cerevisiae* that reached a concentration about 3-fold higher than that measured in the control wine (4.7 mg/L). This result is in accordance with that observed in previous studies [22,58,77], and is compatible with lactic acid production by *L. thermotholerans* [78].

PCA analysis showed evident differences among the strains tested as a function of volatile compounds production and this reflects the ability of each strain to give a specific aromatic imprint to the final wines (Figure 4). Based on volatile compounds content in the resulting wines, *H. osmophila* and *M. pulcherrima* were positioned in the upper left quadrant and characterized by acetate esters and 2-phenyl ethanol. *S. bacillaris* and *P. fermentans* were placed in bottom left quadrant characterized by the production of isobutanol. *T. delbrueckii* and *L. thermotolerans* were placed in the upper right quadrant due to the production of isoamyl alcohol and isoamyl acetate, while *Z. florentina* and *S. cerevisiae* control strain were positioned in the right bottom quadrant and were characterized by the production of ethyl esters.

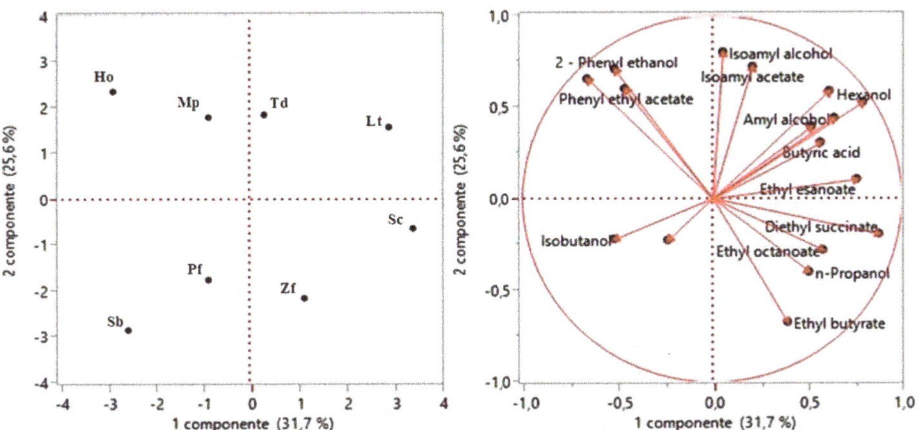

Figure 4. Principal component analysis (PCA) based on the production of volatile compounds.

According to the results of sensory analysis carried out four months after bottling, all the wines obtained with mixed fermentation starters were perceived as significantly more provided in color intensity, in respect to the control wine. This was particularly true for wines obtained with associations including *S. bacillaris* and *M. pulcherrima* (Figure 5). This result is in accordance with the higher amounts of total polyphenols and anthocyanins found in the relevant wines, and in respect to the control. Moreover, it agrees with the findings of other authors pointing out that many non-*Saccharomyces* yeasts in sequential fermentation with *S. cerevisiae* may enhance color intensity of wines, promoting the formation of derivatives with more stable color than anthocyanins [79–82]. This is particularly important for Sangiovese wine that, being rich in unstable and oxidizable phenols, is characterized by limited color stability [83] and suggests that the utilization of mixed starters, including non-*Saccharomyces* yeasts, might represent an option for the management of Sangiovese color stability.

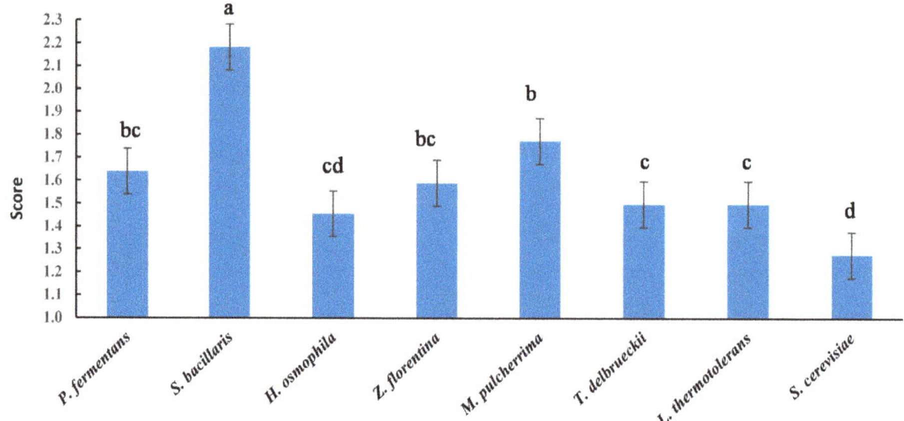

Figure 5. Sensory perception of color in wines 4 months after bottling (QDA score: scale 1–4). Values displaying different letters (a, b, c, d) are significantly different according to the Duncan test ($p \leq 0.001$).

Despite the differences found in the relevant volatile compounds profile, no significant differences among the wines were found in the aromatic profile (descriptors: floral, fruity, canned fruits, spicy, candy, chemical, earthy).

Concerning the taste descriptors used in the organoleptic assessment of wines, significant differences resulted only regarding astringency ($p \leq 0.01$) and mouth dryness ($p \leq 0.001$) perception (Figure 6). In particular, while the association including *P. fermentans* resulted in a more astringent wine, in respect to the control, that including *T. delbrueckii* emerged as less astringent with respect to the control wine and all the other wines.

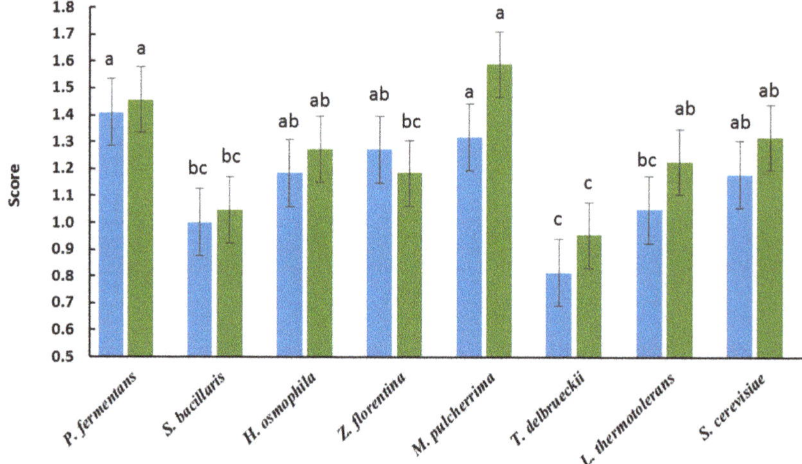

Figure 6. Sensory perception of astringency (■) and dryness (■) in wines 4 months after bottling (QDA: scale 1–4) Values displaying different letters (a, b, c, d) are significantly different according to the Duncan test (for astringency at $p \leq 0.01$ and for dryness at $p \leq 0.001$).

The perception of mouth dryness was higher in wine deriving from the mixed starter including *M. pulcherrima*, while the association *T. delbrueckii/S. cerevisiae* proved the most effective in reducing this sensation in the mouth.

4. Conclusions

The utilization of wine starters containing non-*Saccharomyces* yeasts in association with *S. cerevisiae* represents a valid tool for the achievement of different oenological objectives. Non-*Saccharomyces* yeasts modify the chemical-analytical profile of wines and through their impact on taste descriptors they may be utilized to modulate wine sensory properties. Accordingly, wines inoculated with non-*Saccharomyces/Saccharomyces* mixed starters presented a significant differentiation in the chemical-analytical composition, astringency, dryness perception and intensity of color. In particular, while the association *P. fermentans/S. cerevisiae* resulted in a more astringent wine, *T. delbrueckii/S. cerevisiae* emerged as the less astringent. Moreover, all associations exerted a positive effect on color intensity and wine produced by *S. bacillaris/S.cerevisiae* obtained the highest score. These last results also suggest the utilization of non-*Saccharomyces/S.cerevisiae* mixed starters for the management of Sangiovese color stability.

Author Contributions: Conceptualization, P.D., M.C., I.M., L.L.; methodology, P.D., M.C., I.M., L.L.; investigation, P.D., M.C., I.M., L.L., C.R., A.B.B.; formal analysis, L.L., C.R., A.B.B.; writing—original draft preparation, P.D., I.M.; writing—review and editing, all authors. All authors have read and agreed to the published version of the manuscript.

Funding: This study was part of a research project funded by the Italian Ministry of Economic Development and Consorzio Tuscania, Florence, Italy (http://www.ricercatuscania.it).

Conflicts of Interest: The authors declare no conflict of interest.

References

1. Bisson, L.F.; Joseph, C.L.; Domizio, P.Y. *Biology of Microorganisms on Grapes, in Must and in Wine*; Springer: Cham, Switzerland, 2017; pp. 65–101.
2. Bely, M.; Stoeckle, P.; Masneuf-Pomarède, I.; Dubourdieu, D. Impact of mixed *Torulaspora delbrueckii-Saccharomyces cerevisiae* culture on high-sugar fermentation. *Int. J. Food Microbiol.* **2008**, *122*, 312–320. [CrossRef]
3. Renault, P.; Miot-Sertier, C.; Marullo, P.; Lagarrigue, L.; Lonvaud-Funel, A.; Bely, M. Genetic characterization and phenotypic variability in *Torulaspora delbrueckii* species: Potential applications in the wine industry. *Int. J. Food Microbiol.* **2009**, *134*, 201–210. [CrossRef] [PubMed]
4. Comitini, F.; Gobbi, M.; Domizio, P.; Romani, C.; Lencioni, L.; Mannazzu, I.; Ciani, M. Selected non-*Saccharomyces* wine yeasts in controlled multistarter fermentations with *Saccharomyces cerevisiae*. *Food Microbiol.* **2011**, *28*, 873–882. [CrossRef] [PubMed]
5. Sadoudi, M.; Tourdot-Maréchal, R.; Rousseaux, S.; Steyer, D.; Gallardo-Chacón, J.-J.; Ballester, J.; Vichi, S.; Guérin-Schneider, R.; Caixach, J.; Alexandre, H. Yeast–yeast interactions revealed by aromatic profile analysis of Sauvignon Blanc wine fermented by single or co-culture of non-*Saccharomyces* and *Saccharomyces* yeasts. *Food Microbiol.* **2012**, *32*, 243–253. [CrossRef] [PubMed]
6. Van Breda, V.; Jolly, N.; Van Wyk, J. Characterisation of commercial and natural *Torulaspora delbrueckii* wine yeast strains. *Int. J. Food Microbiol.* **2013**, *163*, 80–88. [CrossRef]
7. Benito, S. The impact of *Torulaspora delbrueckii* yeast in winemaking. *Appl. Microbiol. Biotechnol.* **2018**, *102*, 3081–3094. [CrossRef]
8. Renault, P.; Coulon, J.; Moine, V.; Thibon, C.; Bely, M. Enhanced 3-sulfanylhexan-1-ol production in sequential mixed fermentation with *Torulaspora delbrueckii/Saccharomyces cerevisiae* reveals a situation of synergistic interaction between two industrial strains. *Front. Microbiol.* **2016**, *7*, 293–303. [CrossRef]
9. Belda, I.; Ruiz, J.; Beisert, B.; Navascués, E.; Marquina, D.; Calderón, F.; Rauhut, D.; Benito, S.; Santos, A. Influence of *Torulaspora delbrueckii* in varietal thiol (3-SH and 4-MSP) release in wine sequential fermentations. *Int. J. Food Microbiol.* **2017**, *257*, 183–191. [CrossRef]
10. Rantsiou, K.; Dolci, P.; Giacosa, S.; Torchio, F.; Tofalo, R.; Torriani, S.; Suzzi, G.; Rolle, L.; Cocolin, L. *Candida zemplinina* can reduce acetic acid produced by *Saccharomyces cerevisiae* in sweet wine fermentations. *Appl. Environ. Microbiol.* **2012**, *78*, 1987–1994. [CrossRef]

11. Englezos, V.; Rantsiou, K.; Torchio, F.; Rolle, L.; Gerbi, V.; Cocolin, L. Exploitation of the non-*Saccharomyces* yeast *Starmerella bacillaris* (synonym *Candida zemplinina*) in wine fermentation: Physiological and molecular characterizations. *Int. J. Food Microbiol.* **2015**, *199*, 33–40. [CrossRef]

12. Lencioni, L.; Taccari, M.; Ciani, M.; Domizio, P. *Zygotorulaspora florentina* and *Starmerella bacillaris* in multistarter fermentation with *Saccharomyces cerevisiae* to reduce volatile acidity of high sugar musts. *Aust. J. Grape Wine R.* **2018**, *24*, 368–372. [CrossRef]

13. Ciani, M.; Ferraro, L. Combined use of immobilized *Candida stellata* cells and *Saccharomyces cerevisiae* to improve the quality of wines. *J. Appl. Microbiol.* **1998**, *85*, 247–254. [CrossRef] [PubMed]

14. Soden, A.; Francis, I.L.; Oakey, H.; Henschke, P.A. Effects of co-fermentation with *Candida stellata* and *Saccharomyces cerevisiae* on the aroma and composition of Chardonnay wine. *Aust. J. Grape Wine Res.* **2000**, *6*, 21–30. [CrossRef]

15. Sipiczki, M. *Candida zemplinina* sp. nov., an osmotolerant and psychrotolerant yeast that ferments sweet botrytized wines. *Int. J. Syst. Evol. Microbiol.* **2003**, *53*, 2079–2083. [CrossRef]

16. Magyar, I.; Tóth, T. Comparative evaluation of some oenological properties in wine strains of *Candida stellata*, *Candida zemplinina*, *Saccharomyces uvarum* and *Saccharomyces cerevisiae*. *Food Microbiol.* **2011**, *28*, 94–100. [CrossRef]

17. Tofalo, R.; Schirone, M.; Torriani, S.; Rantsiou, K.; Cocolin, L.; Perpetuini, G.; Suzzi, G. Diversity of *Candida zemplinina* strains from grapes and Italian wines. *Food Microbiol.* **2012**, *29*, 18–26. [CrossRef]

18. Di Maio, S.; Genna, G.; Gandolfo, V.; Amore, G.; Ciaccio, M.; Oliva, D. Presence of *Candida zemplinina* in Sicilian musts and selection of a strain for wine mixed fermentations. *S. Afr. J. Enol. Vitic.* **2012**, *33*, 80–87. [CrossRef]

19. Canonico, L.; Comitini, F.; Oro, L.; Ciani, M. Sequential fermentation with selected immobilized non-*Saccharomyces* yeast for reduction of ethanol content in wine. *Front. Microbiol.* **2016**, *7*, 278. [CrossRef]

20. Englezos, V.; Rantsiou, K.; Cravero, F.; Torchio, F.; Ortiz-Julien, A.; Gerbi, V.; Rolle, L.; Cocolin, L. *Starmerella bacillaris* and *Saccharomyces cerevisiae* mixed fermentations to reduce ethanol content in wine. *Appl. Microbiol. Biot.* **2016**, *100*, 5515–5526. [CrossRef]

21. Kapsopoulou, K.; Mourtzini, A.; Anthoulas, M.; Nerantzis, E. Biological acidification during grape must fermentation using mixed cultures of *Kluyveromyces thermotolerans* and *Saccharomyces cerevisiae*. *World J. Microbiol. Biotechnol.* **2007**, *23*, 735–739. [CrossRef]

22. Gobbi, M.; Comitini, F.; Domizio, P.; Romani, C.; Lencioni, L.; Mannazzu, I.; Ciani, M. *Lachancea thermotolerans* and *Saccharomyces cerevisiae* in simultaneous and sequential co-fermentation: A strategy to enhance acidity and improve the overall quality of wine. *Food Microbiol.* **2013**, *33*, 271–281. [CrossRef]

23. Benito, Á.; Calderón, F.; Palomero, F.; Benito, S. Quality and composition of airén wines fermented by sequential inoculation of *Lachancea thermotolerans* and *Saccharomyces cerevisiae*. *Food Technol. Biotechnol.* **2016**, *54*, 135–144. [CrossRef]

24. Vilela, A. *Lachancea thermotolerans*, the non-*Saccharomyces* yeast that reduces the volatile acidity of wines. *Fermentation* **2018**, *4*, 56. [CrossRef]

25. Ciani, M.; Comitini, F.; Mannazzu, I.; Domizio, P. Controlled mixed culture fermentation: A new perspective on the use of non-*Saccharomyces* yeasts in winemaking. *FEMS Yeast Res.* **2010**, *10*, 123–133. [CrossRef] [PubMed]

26. Rosi, I.; Vinella, M.; Domizio, P. Characterization of β-glucosidase activity in yeasts of oenological origin. *J. Appl. Microbiol.* **1994**, *77*, 519–527. [CrossRef] [PubMed]

27. Fernández, M.; Ubeda, J.F.; Briones, A.I. Typing of non-*Saccharomyces* yeasts with enzymatic activities of interest in wine-making. *Int. J. Food Microbiol.* **2000**, *59*, 29–36. [CrossRef]

28. Barbosa, C.; Lage, P.; Esteves, M.; Chambel, L.; Mendes-Faia, A.; Mendes-Ferreira, A. Molecular and Phenotypic Characterization of *Metschnikowia pulcherrima* Strains from Douro Wine Region. *Fermentation* **2018**, *4*, 8. [CrossRef]

29. Zott, K.; Thibon, C.; Bely, M.; Lonvaud-Funel, A.; Dubourdieu, D.; Masneuf-Pomarede, I. The grape must non-*Saccharomyces* microbial community: Impact on volatile thiol release. *Int. J. Food Microbiol.* **2011**, *151*, 210–215. [CrossRef]

30. Belda, I.; Ruiz, J.; Navascue's, E.; Marquina, D.; Santos, A. Improvement of aromatic thiol release through the selection of yeasts with increased β-lyase activity. *Int. J. Food Microbiol.* **2016**, *225*, 1–8. [CrossRef]

31. Benito, S.; Palomero, F.; Morata, A.; Calderón, F.; Suárez-Lepe, J.A. New applications for *Schizosaccharomyces pombe* in the alcoholic fermentation of red wines. *Int. J. Food Sci. Technol.* **2012**, *47*, 2101–2108. [CrossRef]

32. Benito, S.; Palomero, F.; Calderón, F.; Palmero, D.; Suárez-Lepe, J.A. Selection of appropriate *Schizosaccharomyces* strains for winemaking. *Food Microbiol.* **2014**, *42*, 218–224. [CrossRef]

33. Romani, C.; Domizio, P.; Lencioni, L.; Gobbi, M.; Comitini, F.; Ciani, M.; Mannazzu, I. Polysaccharides and glycerol production by non-*Saccharomyces* wine yeasts in mixed fermentation. *Quad. Vitic. Enol. Univ. Torino* **2010**, *31*, 185–189.

34. Domizio, P.; Liu, Y.; Bisson, L.F.; Barile, D. Cell wall polysaccharides released during the alcoholic fermentation by *Schizosaccharomyces pombe* and *S. japonicus*: Quantification and characterization. *Food Microbiol.* **2017**, *61*, 136–149. [CrossRef] [PubMed]

35. Domizio, P.; Lencioni, L.; Calamai, L.; Portaro, L.; Bisson, L.F. Evaluation of the yeast *Schizosaccharomyces japonicus* for use in wine production. *Am. J. Enol. Vitic.* **2018**, *69*, 266–277. [CrossRef]

36. Romani, C.; Lencioni, L.; Gobbi, M.; Mannazzu, I.; Ciani, M.; Domizio, P. *Schizosaccharomyces japonicus*: A polysaccharide-overproducing yeast to be used in winemaking. *Fermentation* **2018**, *4*, 14. [CrossRef]

37. Domizio, P.; Romani, C.; Comitini, F.; Gobbi, M.; Ciani, M.; Lencioni, L.; Mannazzu, I. Potential spoilage non-*Saccharomyces* yeasts in mixed cultures with *Saccharomyces cerevisiae*. *Ann. Microbiol.* **2011**, *61*, 137–144. [CrossRef]

38. Domizio, P.; Romani, C.; Lencioni, L.; Comitini, F.; Gobbi, M.; Mannazzu, I.; Ciani, M. Outlining a future for non-*Saccharomyces* yeasts: Selection of putative spoilage wine strains to be used in association with *Saccharomyces cerevisiae* for grape juice fermentation. *Int. J. Food Microbiol.* **2011**, *147*, 170–180. [CrossRef] [PubMed]

39. Lencioni, L.; Romani, C.; Gobbi, M.; Comitini, F.; Ciani, M.; Domizio, P. Controlled mixed fermentation at winery scale using *Zygotorulaspora florentina* and *Saccharomyces cerevisiae*. *Int. J. Food Microbiol.* **2016**, *234*, 36–44. [CrossRef]

40. Pallmann, C.L.; Brown, J.A.; Olineka, T.L.; Cocolin, L.; Mills, D.A.; Bisson, L.F. Use of WL medium to profile native flora fermentations. *Am. J. Enol. Vitic.* **2001**, *52*, 198–203.

41. Di Stefano, R.; Guidoni, S. La determinazione dei polifenoli totali nei mosti e nei vini. *Vignevini* **1989**, *1*, 47–52.

42. Di Stefano, R.; Cravero, M.C.; Gentilini, N. Metodi per lo studio dei polifenoli dei vini. *L'Enotecnico* **1989**, *5*, 43–89.

43. Glories, Y. *Recherches Sur la Matiere Colorante des Vins Rouges*; l'Université de Bordeaux II: Bordeaux, France, 1978.

44. Delteil, D. Evaluation sensorielle du profil gustatif des vins. *Rev. Oenol.* **2000**, *94*, 21–23.

45. Granes, D.; Pic-Blateyron, L.; Negrel, J.; Bonnefond, C. L'analyse sensorielle descriptive quantifiée (ASDQ): Une methode pour un langage commun. *Rev. Franc. Oenol.* **2009**, *238*, 16–21.

46. Izquierdo-Cañas, P.M.; Palacios-Garcia, A.T.; Garcia-Romero, E. Enhancement of flavour properties in wines using sequential inoculations of non-*Saccharomyces* (*Hansenula* and *Torulaspora*) and *Saccharomyces* yeast starter. *Vitis* **2011**, *50*, 177–182.

47. Tronchoni, J.; Curiel, J.A.; Sáenz-Navajas, M.P.; Morales, P.; De-La-Fuente-Blanco, A.; Fernández-Zurbano, P.; Ferreira, V.; Gonzalez, R. Aroma profiling of an aerated fermentation of natural grape must with selected yeast strains at pilot scale. *Food Microbiol.* **2018**, *70*, 214–223. [CrossRef] [PubMed]

48. Contreras, A.; Hidalgo, C.; Henschke, P.A.; Chambers, P.J.; Curtin, C.; Varela, C. Evaluation of non-*Saccharomyces* yeasts for the reduction of alcohol content in wine. *Appl. Environ. Microbiol.* **2014**, *80*, 1670–1678. [CrossRef]

49. Gobbi, M.; de Vero, L.; Solieri, L.; Comitini, F.; Oro, L.; Giudici, P. Fermentative aptitude of non-*Saccharomyces* wine yeast for reduction in the ethanol content in wine. *Eur. Food Res. Technol.* **2014**, *239*, 41–48. [CrossRef]

50. Quirós, M.; Rojas, V.; Gonzalez, R.; Morales, P. Selection of non-*Saccharomyces* yeast strains for reducing alcohol levels in wine by sugar respiration. *Int. J. Food microbiol.* **2014**, *181*, 85–91. [CrossRef]

51. Contreras, A.; Curtin, C.; Varela, C. Yeast population dynamics reveal a potential collaboration between *Metschnikowia pulcherrima* and *Saccharomyces uvarum* for the production of reduced alcohol wines during Shiraz fermentation. *Appl. Microbiol. Biotechnol.* **2015**, *99*, 1885–1895. [CrossRef]

52. Contreras, A.; Hidalgo, C.; Schmidt, S.; Henschke, P.A.; Curtin, C.; Varela, C. The application of non-*Saccharomyces* yeast in fermentations with limited aeration as a strategy for the production of wine with reduced alcohol content. *Int. J. Food Microbiol.* **2015**, *205*, 7–15. [CrossRef]

53. Canonico, L.; Solomon, M.; Comitini, F.; Ciani, M.; Varela, C. Volatile profile of reduced alcohol wines fermented with selected non-*Saccharomyces* yeasts under different aeration conditions. *Food Microbiol.* **2019**, *84*, 103247. [CrossRef] [PubMed]

54. Varela, J.; Varela, C. Microbiological strategies to produce beer and wine with reduced ethanol concentration. *Curr. Opin. Biotechnol.* **2019**, *56*, 88–96. [CrossRef] [PubMed]

55. Whitener, M.E.B.; Stanstrup, J.; Panzeri, V.; Carlin, S.; Divol, B.; Du Toit, M.; Vrhovsek, U. Untangling the wine metabolome by combining untargeted SPME–GCxGC-TOF-MS and sensory analysis to profile Sauvignon blanc co-fermented with seven different yeasts. *Metabolomics* **2016**, *12*, 53. [CrossRef]

56. Nisiotou, A.; Sgouros, G.; Mallouchos, A.; Nisiotis, C.S.; Michaelidis, C.; Tassou, C.; Banilas, G. The use of indigenous *Saccharomyces cerevisiae* and *Starmerella bacillaris* strains as a tool to create chemical complexity in local wines. *Food Res. Int.* **2018**, *111*, 498–508. [CrossRef]

57. Benito, S. The impacts of *Lachancea thermotolerans* yeast strains on winemaking. *Appl. Microbiol. Biotechnol.* **2018**, *102*, 6775–6790. [CrossRef]

58. Chen, K.; Escott, C.; Loira, I.; del Fresno, J.M.; Morata, A.; Tesfaye, W.; Calderon, F.; Suárez-Lepe, J.A.; Han, S.; Benito, S. Use of non-*Saccharomyces* yeasts and oenological tannin in red winemaking: Influence on colour, aroma and sensorial properties of young wines. *Food Microbiol.* **2018**, *69*, 51–63. [CrossRef]

59. Medina, K.; Boido, E.; Dellacassa, E.; Carrau, F. Effects of non-*Saccharomyces* yeasts on color, anthocyanin and anthocyanin-derived pigments of Tannat grapes during fermentation. *Am. J. Enol. Vitic.* **2018**, *69*, 148–156. [CrossRef]

60. Medina, K.; Boido, E.; Fariña, L.; Dellacassa, E.; Carrau, F. Non-*Saccharomyces* and *Saccharomyces* strains co-fermentation increases acetaldehyde accumulation: Effect on anthocyanin-derived pigments in Tannat red wines. *Yeast* **2016**, *33*, 339–343. [CrossRef]

61. Vidal, S.; Francis, L.; Williams, P.; Kwiatkowski, M.; Gawel, R.; Cheynier, V.; Waters, E. The mouth-feel properties of polysaccharides and anthocyanins in a wine like medium. *Food. Chem.* **2004**, *85*, 519–525. [CrossRef]

62. Carvalho, E.; Mateus, N.; Plet, B.; Pianet, I.; Dufourc, E.; De Freitas, V. Influence of wine pectic polysaccharides on the interactions between condensed tannins and salivary proteins. *J. Food. Chem.* **2006**, *54*, 8936–8944. [CrossRef]

63. Chalier, P.; Angot, B.; Delteil, D.; Doco, T.; Gunata, Z. Interactions between aroma compounds and whole mannoprotein isolated from *Saccharomyces cerevisiae* strains. *Food Chem.* **2007**, *100*, 22–30. [CrossRef]

64. Juega, M.; Nunez, Y.P.; Carrascosa, A.V.; Martinez-Rodriguez, A.J. Influence of yeast mannoproteins in the aroma improvement of white wines. *J. Food Sci.* **2012**, *77*, M499–M504. [CrossRef]

65. Noble, A.C.; Bursick, G.F. The contribution of glycerol to perceived viscosity and sweetness in white wine. *Am. J. Enol. Vitic.* **1984**, *35*, 110–112.

66. Hufnagel, J.C.; Hofmann, T. Quantitative reconstruction of the nonvolatile sensometabolome of a red wine. *J. Agric. Food Chem.* **2008**, *56*, 9190–9199. [CrossRef] [PubMed]

67. Giaramida, P.; Ponticello, G.; Di Maio, S.; Squadrito, M.; Genna, G.; Barone, E.; Scacco, A.; Corona, O.; Amore, G.; Di Stefano, R.; et al. *Candida zemplinina* for production of wines with less alcohol and more glycerol. *S. Afr. J. Enol. Vitic.* **2013**, *34*, 204–211. [CrossRef]

68. Zara, G.; Mannazzu, I.; Del Caro, A.; Budroni, M.; Pinna, M.B.; Murru, M.; Farris, G.A.; Zara, S. Wine quality improvement through the combined utilisation of yeast hulls and *Candida zemplinina/Saccharomyces cerevisiae* mixed starter cultures. *Aust. J. Grape Wine Res.* **2014**, *20*, 199–207. [CrossRef]

69. Romano, P.; Suzzi, G. Higher alcohol and acetoin production by *Zygosaccharomyces* wine yeasts. *J. Appl. Bacteriol.* **1993**, *75*, 541–545. [CrossRef]

70. Rojas, V.; Gil, J.V.; Pinaga, F.; Manzanares, P. Acetate ester formation in wine by mixed cultures in laboratory fermentations. *Int. J. Food. Microbiol.* **2003**, *86*, 181–188. [CrossRef]

71. Moreira, N.; Mendes, F.; Guedes de Pinho, P.; Hogg, T.; Vasconcelos, I. Heavy sulphur compounds, higher alcohols and esters production profile of *Hanseniaspora uvarum* and *Hanseniaspora guilliermondii* grown as pure and mixed cultures in grape must. *Int. J. Food Microbiol.* **2008**, *124*, 231–238. [CrossRef]

72. Viana, F.; Gil, J.V.; Genove's, S.; Valle's, S.; Manzanares, P. Rational selection of non-*Saccharomyces* wine yeasts for mixed starters based on ester formation and enological traits. *Food Microbiol.* **2008**, *25*, 778–785. [CrossRef]

73. Herraiz, T.; Reglero, G.; Herraiz, M.; Martin-Alvarez, P.J.; Cabezudo, M.D. The influence of the yeast and type of culture on the volatile composition of wines fermented without sulfur dioxide. *Am. J. Enol. Vitic.* **1990**, *41*, 313–318.

74. Clemente-Jimenez, J.M.; Mingorance-Carzola, L.; Martinez-Rodriguez, S.; Las Heras-Vazquez, F.J.; Rodriguez-Vico, F. Molecular characterization and oenological properties of wine yeasts isolated during spontaneous fermentation of six varieties of grape must. *Food Microbiol.* **2004**, *21*, 149–155. [CrossRef]

75. Andorra, I.; Berradre, M.; Roze's, N.; Mas, A.; Guillamon, J.M.; Esteve-Zarzoso, B. Effect of pure and mixed cultures of the main wine yeast species on grape must fermentations. *Eur. Food Res. Technol.* **2010**, *231*, 215–224. [CrossRef]

76. Rojas, V.; Gil, J.V.; Pinaga, F.; Manzanares, P. Studies on acetate ester production by non-*Saccharomyces* wine yeasts. *Int. J. Food Microbiol.* **2001**, *70*, 283–289. [CrossRef]

77. Benito, S.; Hofmann, T.; Laier, M.; Lochbühler, B.; Schüttler, A.; Ebert, K.; Fritsch, S.; Röcker, J.; Rauhut, D. Effect on quality and composition of Riesling wines fermented by sequential inoculation with non-*Saccharomyces* and *Saccharomyces cerevisiae*. *Eur. Food Res. Technol.* **2015**, *241*, 707–717. [CrossRef]

78. Hranilovic, A.; Gambetta, J.M.; Schmidtke, L.; Boss, P.K.; Grbin, P.R.; Masneuf-Pomarede, I.; Bely, M.; Albertin, W.; Jiranek, V. Oenological traits of *Lachancea thermotolerans* show signs of domestication and allopatric di_erentiation. *Sci. Rep.* **2018**, *8*, 14812–14825. [CrossRef] [PubMed]

79. Escott, C.; Morata, A.; Loira, I.; Tesfaye, W.; Suarez-Lepe, J.A. Characterization of polymeric pigments and pyranoanthocyanins formed in microfermentations of non-*Saccharomyces* yeasts. *J. Appl. Microbiol.* **2016**, *121*, 1346–1356. [CrossRef]

80. Escott, C.; del Fresno, J.M.; Loira, I.; Morata, A.; Tesfaye, W.; González, M.C.; Suárez-Lepe, J.A. Formation of polymeric pigments in red wines through sequential fermentation of flavanol-enriched musts with non-*Saccharomyces* yeasts. *Food Chem.* **2018**, *239*, 975–983. [CrossRef]

81. Morata, A.; Loira, I.; Suárez Lepe, J.A. Influence of yeasts in wine colour. In *Grape and Wine Biotechnology*; Morata, A., Loira, I., Eds.; IntechOpen: London, UK, 2016; pp. 285–305.

82. Morata, A.; Loira, I.; Tesfaye, W.; Bañuelos, M.A.; González, C.; Suárez Lepe, J.A. *Lachancea thermotolerans* applications in wine technology. *Fermentation* **2018**, *4*, 53. [CrossRef]

83. Canuti, V.; Puccioni, S.; Storchi, P.; Zanoni, B.; Picchi, M.; Bertuccioli, M. Enological eligibility of grape clones based on the SIMCA method: The case of the Sangiovese cultivar from Tuscany. *Ital. J. Food Sci.* **2018**, *30*, 184–199.

Article

Sequential Non-*Saccharomyces* and *Saccharomyces cerevisiae* Fermentations to Reduce the Alcohol Content in Wine

Margarita García [1], Braulio Esteve-Zarzoso [2], Juan Mariano Cabellos [1] and Teresa Arroyo [1,*]

[1] Department of Food and Agricultural Science, IMIDRA, 28800 Alcalá de Henares, Spain; margarita_garcia_garcia@madrid.org (M.G.); juan.cabellos@madrid.org (J.M.C.)

[2] Department of Chemistry and Biotechnology, Rovira i Virgili University, 43007 Tarragona, Spain; braulio.esteve@urv.cat

* Correspondence: teresa.arroyo@madrid.org; Tel.: +34-918-879-486

Received: 21 May 2020; Accepted: 7 June 2020; Published: 10 June 2020

Abstract: Over the last decades, the average alcohol content of wine has increased due to climate change and consumer preferences for particular wine styles that resulted in increased grape sugar levels at harvest. Therefore, alcohol reduction is a current challenge in the winemaking industry. Among several strategies under study, the use of non-conventional yeasts in combination with *Saccharomyces cerevisiae* plays an important role for lowering ethanol production in wines nowadays. In the present work, 33 native non-*Saccharomyces* strains were assayed in sequential culture with a *S. cerevisiae* wine strain to determine their potential for reducing the alcohol content in Malvar white wines. Four of the non-*Saccharomyces* strains (*Wickerhamomyces anomalus* 21A-5C, *Meyerozyma guilliermondii* CLI 1217, and two *Metschnikowia pulcherrima* (CLI 68 and CLI 460)) studied in sequential combination with *S. cerevisiae* CLI 889 were best able to produce dry wines with decreased alcohol proportion in comparison with one that was inoculated only with *S. cerevisiae*. These sequential fermentations produced wines with between 0.8% (*v/v*) and 1.3% (*v/v*) lower ethanol concentrations in Malvar wines, showing significant differences compared with the control. In addition, these combinations provided favorable oenological characteristics to wines such as high glycerol proportion, volatile higher alcohols, and esters with fruity and sweet character.

Keywords: alcohol reduction; native yeast; non-*Saccharomyces*; sequential fermentation; wine

1. Introduction

At present, the increasing alcohol content in wines is closely related to climate change and consumer choice for full-bodied, rich, and ripe fruit flavor profiles, which often involve increased grape maturity [1–3]. In recent years, the worldwide trend towards more frequent warm periods during the grapevine growing season has increased sugar content in grapes and therefore the alcohol concentration in wines [4]. Thus, the average alcohol level has risen about 2% (*v/v*) over the past few decades in warm areas, and it is not uncommon to find wines with an alcohol content higher than 16% (*v/v*) [5]. Excessive alcohol concentration in wines can alter the sensory profile of wines, increasing bitterness, astringency, and hotness perception and masking some volatile compounds [6,7]. Additionally, wines with elevated alcohol content can lead to harmful health effects [8] and also increase costs in markets where taxes are linked to the ethanol level in many countries [9].

Among the various methodologies aimed at the reduction of alcohol content in wines, microbiological approaches may be promising to preserve organoleptic characteristics and quality in wines. In addition, they are profitable and easy to implement strategies that do not require the need for specialized equipment [10,11]. *Saccharomyces cerevisiae* is the principal microorganism selected for winemaking. This species completes fermentation of sugars due to its ability to produce and tolerate high concentrations of alcohol [12,13]. Unlike *S. cerevisiae*, non-*Saccharomyces* yeasts are not generally able to complete the fermentation process; thus, mixed or sequential inoculations with *S. cerevisiae* are required for this purpose [14–17]. Research efforts have therefore focused on developing new *S. cerevisiae* strains that produce less ethanol in wine [18] and on using non-*Saccharomyces* yeasts that metabolize sugar without producing ethanol or that do so with less efficiency [19].

Several investigations have employed non-*Saccharomyces* co-cultures as a tool for reducing the ethanol concentration in wine [19–29]. Here, the early inoculation of non-*Saccharomyces* strain transforms sugar to produce biomass and by-products, decreasing ethanol formation before addition of *S. cerevisiae* [2,30]. This action plan is particularly adequate to winemaking in warm regions, as in the case of the Madrid winegrowing region (Spain) under study in the present work. The climate in the Denomination of Origin (D.O.) "Vinos de Madrid" presents temperatures ranging from −8 °C in winter to 41 °C in summer, and rainfall ranges between 461 and 658 mm [31]. Winemakers in this region are working hard in order to elaborate new styles of wine that are more competitive in the market [32]. The knowledge and selection of native yeasts is a very important achievement to confer typicity and originality to the wine [33,34], and its use is also considered a reactive adaptation practice to climate change [35].

In this work, 33 native non-*Saccharomyces* strains from 13 different wine yeast species were tested with the aim of identifying yeasts that, in sequential fermentation with *S. cerevisiae*, could be used for reducing alcohol content in Malvar white wines, and additionally evaluating their positive impact on the quality of these wines. Moreover, no previous investigations have been carried out to select non-*Saccharomyces*/*S. cerevisiae* combinations with native yeasts from D.O. "Vinos de Madrid" (Madrid, Spain) directed towards ethanol reduction in wines.

2. Materials and Methods

2.1. Yeast Strains: Purity and Identity Control

A total of 33 non-*Saccharomyces* strains from the IMIDRA collection belonging to 10 different genera were used in this study (Table 1). All non-*Saccharomyces* strains were native from D.O. "Vinos de Madrid" vineyards and cellars [31,36]. The well-studied native strain, *S. cerevisiae* CLI 889, was employed as a control [31,34,37]. Cryogenically preserved (−80 °C) strains in 30% glycerol were subsequently seeded on YPD liquid medium (1% yeast extract, 1% meat peptone, and 2% glucose (Conda Laboratories, Madrid, Spain), w/v) and incubated for 24–48 h at 28 °C. Later, all strains were maintained at 4 °C on YPD plates.

To confirm yeast strain identifications, DNA extraction and rDNA 5.8S–ITS region PCR-RFLP analysis [38] were employed as described previously by Cordero-Bueso et al. [39]. Some of these strains were also sequenced [31,40], and the D1/D2 domain of the 26S rDNA gene was amplified using primers NL-1 and NL-4 [41].

Table 1. Yeast strains used in this study.

Species Name	Strain Code	Year of Isolation	Origin [1]	References [2]
Wickerhamomyces anomalus	CLI 1218	2007	Malvar [a]	[31,36]
	31-1C	2006	Garnacha [c]	This study
	21A-5C	2007	Garnacha [c]	[36]
	23A-6C	2007	Garnacha [c]	[36]
	5B-1C	2008	Garnacha [c]	This study
Candida stellata	6-5A	2006	Shiraz [c]	[36]
	2A-1B	2007	Shiraz [c]	This study

Table 1. *Cont.*

Species Name	Strain Code	Year of Isolation	Origin [1]	References [2]
Hanseniaspora valbyensis	CLI 194	1993	Garnacha [a]	[36]
	CLI 190	1993	Garnacha [a]	[36]
Hanseniaspora guilliermondii	CLI 417	1995	Malvar [a]	This study
	7A-3A	2007	Garnacha [c]	This study
	8A-8B	2007	Garnacha [c]	This study
	CLI 225	1994	Tempranillo [a]	[36,42]
	CLI 72	1993	Garnacha [a]	[36]
Hanseniaspora uvarum	CLI 903	1993	Airén [b]	[36,42]
Hanseniaspora vineae	CLI 3	1993	Tempranillo [a]	[36]
Torulaspora delbrueckii	LS1 FF2 3A	2009	Garnacha [a]	[33]
	LS2 FF2 1A	2009	Garnacha [a]	[33]
	CLI 918	2006	Malvar [a]	[16,40,42,43]
Metschnikowia pulcherrima	CLI 68	1993	Garnacha [a]	[36]
	CLI 457	1995	Malvar [a]	[16,36,40,42]
	CLI 463	1995	Malvar [a]	This study
	CLI 219	1994	Malvar [a]	[36,42]
	CLI 460	1995	Malvar [a]	[36,42]
	CLI 461	1995	Malvar [a]	This study
Lachancea thermotolerans	AMB FF4 10A	2009	Malvar [a]	[33]
	3-4A	2006	Shiraz [c]	[36]
	9-6C	2006	Malvar [a]	[16,40,42]
	CLI 1219	2007	Malvar [a]	[31,42]
Pichia membranifaciens	CLI 679	2006	Malvar [a]	[31,42]
Meyerozyma guilliermondii	CLI 1217	2006	Malvar [a]	[31,42]
Priceomyces carsonii	CLI 1221	2006	Malvar [a]	[31,42]
Zygosaccharomyces bailii	CLI 622	2009	Malvar [a]	[31,42]
Saccharomyces cerevisiae	CLI 889	2000	Airén [a]	[16,34,37,40,42,43]

[1] a, spontaneous fermentation; b, must; c, grape; [2] publications in which strains have been investigated.

2.2. Laboratory-Scale Fermentations

Bunches from healthy grapes of white Malvar (*Vitis vinifera* L. cv.) variety were collected from a vineyard in the Madrid winegrowing region of Spain (40°31′ N, 3°17′ W and 610 m altitude). The must was clarified by pectolytic enzymes (Enozym Altair, Agrovin, Spain) (0.01 g/L) at 4 °C and stored frozen until use. The main characteristics of Malvar must were pH 3.3; 23.3 °Brix, equivalent to about 230 g/L of reducing sugars; probable alcohol content, 13.5% (*v/v*); and yeast assimilable nitrogen (YAN), 170 mg/L.

The grape must was inoculated with a final concentration of 10^6 cells/mL from 48 h pre-cultures of each yeast strain (33 non-*Saccharomyces* and 1 *S. cerevisiae* as control strain). The fermentations were carried out in quadruplicate in 50 mL Falcon tubes containing 30 mL of sterile Malvar must. The trials were divided into two sections: Section I (pure culture), where strain growth was performed in aerobic conditions at 20 °C with continuous orbital shaking (130 rpm). The fermentation kinetic was controlled daily by weight loss. At 96 h, one duplicate of each trial was used to the study of dry weight, residual sugars (glucose + fructose), and volatile acidity (as g/L of acetic acid); and Section II (sequential culture), the other duplicate from Section I, was inoculated with 10^6 cells/mL *S. cerevisiae* CLI 889. In this case, Falcon tubes hermetically sealed and fitted with air locks ensured anaerobic conditions. The fermentation process was conducted at 20 °C with shaking at 130 rpm and was monitored daily until constant weight. Then, wine analyses were carried out.

Dry cell weight measurements were performed on samples from sections I and II. The wine samples were centrifuged (10,000 rpm, 5 min) and the pellets were washed with deionized water twice. Finally, dry weight was determined by filtering through a 0.45 μm pore size membrane filter (Millipore). Filters were heat-dried at 105 °C until constant weight was obtained.

2.3. Analitycal Determination of Wines

The concentration of glucose, fructose, glycerol, ethanol, and organic acids (malic, lactic, and acetic acids) was determined using a Waters 600E HPLC system (Waters, Milford, MA) at the end of fermentation. The HPLC was equipped with a Waters 2414 refractive index (RI) and Waters 2996 photodiode array detector (PDA) on a Rezex RHM−Monosaccharide H+ (8%) column (300 × 7.8 mm, Phenomenex, Torrance, CA, USA). The column was maintained at 65 °C, and 5 mM H_2SO_4 was used as mobile phase at a flow rate of 0.6 mL/min. In wine samples at 96 h, only residual sugars (glucose + fructose) and volatile acidity (as g/L of acetic acid) were measured with a multi analyzer LISA 200 (TDI, Barcelona, Spain), using enzymatic kits (TDI, Barcelona, Spain).

Quantification of major volatile compounds of wines was achieved using the gas chromatography coupled to flame ionization detector (GC−FID) technique. The GC system employed was an Agilent 6850 with a FID detector equipped with a column DB-Wax (60 m × 0.32 mm × 0.5 μm film thickness) from J&W Scientific (Folsom, CA, USA). The extraction and analysis methodologies of volatile analytes were performed following the procedure described by Ortega et al. [44]. Identification and quantification of the 32 individual major volatiles was performed using commercial pure standards. Calibration curves were drawn for each standard at 6 different concentration levels. Each standard was prepared in a synthetic wine solution (5 g/L of tartaric acid, dissolved in 13% of ethanol solution (*v/v*), at pH 3.4 adjusted with NaOH). The obtained coefficients of regression (R^2) were > 0.990 [32,45].

2.4. Statistical Treatment of Data

The data were analyzed with SPSS Statistics 25 software (SPSS Inc., Chicago, IL, USA). Analysis of variance was carried out by ANOVA Tukey's test to examine significant differences between samples. Thus, a principal component analysis (PCA) was used to study the contribution of oenological and aromatic variables to the differences between Malvar wines.

3. Results

The genetic identification of the 33 non-*Saccharomyces* strains from vineyards and cellars of D.O. "Vinos de Madrid" allowed them to be classified into 13 species belonging to 10 different genera. The initial strain selection was designed to include species frequently isolated from the winemaking environment. Moreover, our strategy for ethanol reduction was the use of one non-*Saccharomyces* strain that exhibited a low ethanol yield but consumed enough sugars to affect the ethanol concentration (Section I) and be compatible with *S. cerevisiae* in order to ensure the completion of fermentation (Section II).

3.1. Section I: Pure Culture of Non-Saccharomyces Strains

In this section of the work, we studied fermentative kinetics of non-*Saccharomyces* strains and the control strain (*S. cerevisiae* CLI 889) in pure cultures. The fermentative profiles during the first 96 h permitted the division of the strains into three different groups: A, B, and C (Figure 1). Group A was represented by four non-*Saccharomyces* species: *Wickerhamomyces anomalus* (two strains), *Candida stellata* (one strain), *Lachancea thermotolerans* (one strain), and *Hanseniaspora guilliermondii* (two strains), which showed similar CO_2 released to *S. cerevisiae* CLI 889 control strain. Group B included all *Torulaspora delbrueckii* and *Hanseniaspora valbyensis* strains studied in this work together with the other three *L. thermotolerans* strains. These strains showed less fermentative capacity than the control, with a CO_2 loss between 2.2−1.2 g against above 3 g liberated by *S. cerevisiae* CLI 889. Nine different non-*Saccharomyces* species were represented within group C. This latter group had 19 of

33 non-*Saccharomyces* strains studied, wherein their fermentation kinetics presented the lowest CO_2 liberation observed during the first 96 h.

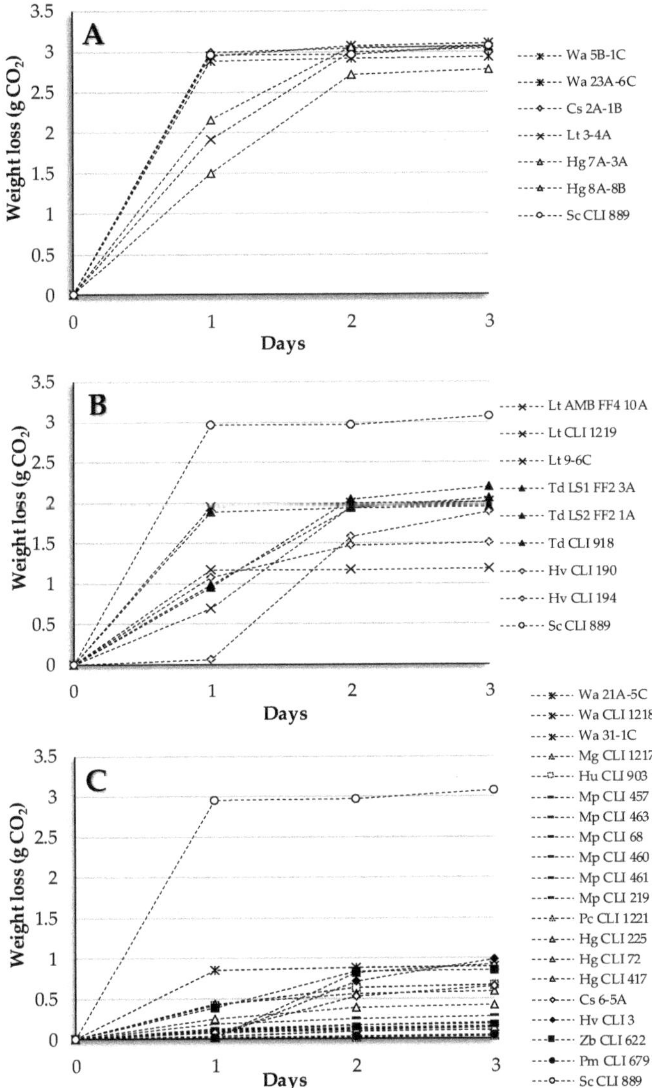

Figure 1. Release of CO_2 in aerobic conditions. (**A**) Six non-*Saccharomyces* strains showed similar CO_2 released to Sc CLI 889 control strain (white circles). (**B**) Eight non-*Saccharomyces* strains showed less CO_2 released than the control (white circles). (**C**) Nineteen non-*Saccharomyces* strains showed values below 1 g of CO_2 liberated from the control (white circles). Wa, *W. anomalus*; Cs, *C. stellata*; Lt, *L. thermotolerans*; Hg, *H. guilliermondii*; Sc, *S. cerevisiae*; Td, *T. delbrueckii*; Hv, *H. valbyensis*; Mg, *M. guilliermondii*; Hu, *H. uvarum*; Mp, *M. pulcherrima*; Pc, *P. carsonii*; Hv, *H. vineae*; Zb, *Z. bailii*; Pm, *P. membranifaciens*.

3.2. Section II: Sequential Culture of Non-Saccharomyces/S. cerevisiae Strains

After 96 h, *S. cerevisiae* CLI 889 was sequentially inoculated into all fermentations in Section I. A total of 8 days were needed by yeast strains to complete the fermentation process (Sections I and II).

Yeast isolates with fermentation behavior showing in group A (Section I) did not exhibit an increase on the CO_2 release, producing similar amounts of ethanol at the end of fermentation—all of these wines were about 13% (*v/v*).

When sequential fermentations finished, some strains combinations produced wines with ethanol concentration similar to the control (13%, *v/v*), and thus they were discarded as low-ethanol cultures. These sequential combinations that were not selected included the strains CLI 679, CLI 1218, 31-1C, CLI 457, CLI 72, CLI 219, CLI 461, CLI 463, CLI 1221, and CLI 903; all of them were classified into group C (Section I). Another group of non-*Saccharomyces*/*S. cerevisiae* fermentations, including CLI 918, CLI 194, CLI 1219, AMB FF4 10A, CLI 190, and LS1 FF2 3A strains from group B (Section I), and CLI 225, CLI 622, CLI 417, 6-5A, and CLI 3 strains from group C (Section I), increased by between 7% and 10% in terms of ethanol concentration, but high amounts of residual sugars were not fermented, and thus these combinations were not selected either; most of them belonged to group B (Section I) in which CO_2 liberated was lower than the control with values between 1.18 and 2.19 g. Finally, four non-*Saccharomyces*/*S. cerevisiae* sequential inoculations produced wines with decreased ethanol proportions compared with the control, and the residual sugars values were suitable for dry wines (<5 g/L residual sugar) (Table 2).

Table 2. Oenological parameters and cell dry weight for the best non-*Saccharomyces*/*S. cerevisiae* sequential combinations to reduce ethanol concentration in wines.

Parameters	Yeast Culture				
	Wa 21A-5C(S)	Mp CLI 68(S)	Mg CLI 1217(S)	Mp CLI 460(S)	Sc CLI 889(P)
Malic acid (g/L)	0.66 ± 0.12 [a]	0.60 ± 0.02 [a]	0.64 ± 0.02 [a]	0.45 ± 0.09 [a]	0.55 ± 0.02 [a]
Lactic acid (g/L)	2.10 ± 0.26 [a]	2.48 ± 0.30 [a]	2.50 ± 0.23 [a]	2.22 ± 0.53 [a]	2.31 ± 0.02 [a]
Acetic acid (g/L)	0.77 ± 0.08 [a]	0.86 ± 0.01 [ab]	0.78 ± 0.01 [ab]	0.71 ± 0.16 [a]	0.43 ± 0.00 [ac]
Glucose (g/L)	2.45 ± 0.30 [a]	2.80 ± 0.54 [a]	2.04 ± 0.39 [a]	2.97 ± 0.53 [a]	2.70 ± 0.01 [a]
Fructose (g/L)	1.06 ± 0.10 [a]	2.72 ± 0.60 [b]	0.69 ± 0.04 [ac]	1.93 ± 0.29 [ab]	1.01 ± 0.02 [a]
Glycerol (g/L)	7.83 ± 0.31 [a]	8.32 ± 0.10 [a]	7.06 ± 0.61 [ab]	9.30 ± 0.90 [ac]	7.60 ± 0.02 [a]
Alcohol degree (%)	12.05 ± 0.12 [a]	11.75 ± 0.05 [a]	11.77 ± 0.32 [a]	12.16 ± 0.26 [a]	13.00 ± 0.01 [b]
Dry weight (mg)	4.35 ± 0.07 [a]	3.73 ± 0.08 [a]	4.77 ± 0.57 [ab]	3.28 ± 0.61 [a]	2.95 ± 0.01 [ac]

Data are means ± standard deviation (*n* = 2). Data with different letters [a,b,c] within each row are significantly different (Tukey test; *p* < 0.05). (S), sequential culture; (P), pure culture.

3.3. Yeast Strain Sequential Combinations Selected as Low-Ethanol Producers

In order to reduce the ethanol content in wines, the selected non-*Saccharomyces* yeast strains were *W. anomalus* 21A-5C, *Metschnikowia pulcherrima* CLI 68, *Meyerozyma guilliermondii* CLI 1217, and *M. pulcherrima* CLI 460 used in sequential combination with *S. cerevisiae* CLI 889. These co-cultures produced wines with between 0.8% (*v/v*) and 1.3% (*v/v*) lower ethanol concentrations in Malvar wines, showing significant differences from the control (Table 2).

Sequential cultures inoculated with 21A-5C, CLI 68, and CLI 460 produced more glycerol than the control, highlighting *M. pulcherrima* CLI 460 strain with values significantly higher than the control (Table 2). There were no significantly differences in malic and lactic acid content, and fermentations with sequential combinations generated more acetic acid than the amount produced by the *S. cerevisiae* control (Table 2). Regarding dry weight, all sequential fermentations presented greater values compared with the control; in particular, sequential culture of *M. guilliermondii* CLI 1217 was 1.6-fold higher, showing significant differences (Table 2).

To find the aromatic composition of these wines, we studied 32 volatile compounds classified in alcohols, esters, acids, and aldehydes/ketones (Table 3). Sequential inoculation produced Malvar wines with greater total concentration of higher alcohols. The amounts of isoamyl alcohol (harsh, bitter) and β-phenylethyl alcohol (flowery, roses) were significantly higher in wines produced in sequential culture, increasing the total concentration of alcohols. The ethyl isovalerate and isoamyl acetate ester concentration responsible for fruity and sweet aromas were significantly different in wines generated from sequential inoculations. Regarding volatile acids, isobutyric acid and hexanoic acid were the main compounds responsible for the total concentration of volatile acids in all wines. The sequential culture *W. anomalus* 21A-5C/*S. cerevisiae* CLI 889 produced the highest concentration of the ketone acetoin. Finally, sequential cultures with *M. pulcherrima* strains (CLI 68 and CLI 460) and the control showed higher amounts of γ-butyrolactone, related to sweet aroma in wines.

Table 3. Major volatile compounds (mg/L) of wines produced in the Section II (sequential culture of non-*Saccharomyces* strains + *S. cerevisiae* CLI 889 and a control, *S. cerevisiae* CLI 889 pure culture).

Compound	Wa 21A-5C(S)	Mp CLI 68(S)	Mg CLI 1217(S)	Mp CLI 460(S)	Sc CLI 889(P)
1-Propanol	n.q.	n.q.	n.q.	n.q.	3.69 ± 0.13 [a]
1-Butanol	1.81 ± 0.14 [a]	0.50 ± 0.05 [b]	0.46 ± 0.05 [b]	0.48 ± 0.12 [b]	0.40 ± 0.10 [b]
Isobutanol	31.96 ± 1.31 [a]	33.51 ± 4.11 [a]	30.51 ± 1.28 [a]	49.46 ± 0.24 [b]	26.30 ± 0.95 [a]
Isoamyl alcohol	118.13 ± 1.88 [a]	114.56 ± 5.92 [ab]	122.03 ± 0.33 [a]	106.47 ± 0.53 [b]	91.37 ± 3.14 [c]
(Z)-3-Hexen-1-ol	0.12 ± 0.00 [a]	0.04 ± 0.00 [b]	0.05 ± 0.00 [c]	0.04 ± 0.00 [b]	0.20 ± 0.10 [d]
1-Hexanol	0.49 ± 0.00 [a]	0.26 ± 0.00 [b]	0.23 ± 0.03 [b]	0.22 ± 0.00 [b]	0.88 ± 0.05 [c]
Metionol	0.09 ± 0.00 [a]	0.37 ± 0.00 [ab]	0.32 ± 0.00 [ab]	0.53 ± 0.20 [b]	0.61 ± 0.10 [b]
Benzyl alcohol	0.17 ± 0.00 [a]	0.19 ± 0.00 [b]	0.27 ± 0.00 [c]	0.13 ± 0.00 [d]	0.15 ± 0.06 [e]
β-Phenylethyl alcohol	27.55 ± 0.05 [a]	21.27 ± 2.02 [b]	18.93 ± 1.04 [b]	23.03 ± 1.40 [ab]	10.53 ± 0.29 [c]
Σ Alcohols	**181.48 ± 3.37** [a]	**171.86 ± 11.99** [a]	**173.94 ± 0.01** [a]	**181.51 ± 1.83** [a]	**134.13 ± 4.92** [b]
Ethyl butyrate	0.21 ± 0.03 [a]	0.30 ± 0.01 [a]	0.30 ± 0.04 [a]	0.29 ± 0.03 [a]	0.31 ± 0.05 [a]
Ethyl isovalerate	1.35 ± 0.07 [a]	0.81 ± 0.01 [h]	0.98 ± 0.10 [h]	0.90 ± 0.06 [h]	0.28 ± 0.05 [c]
Ethyl isobutyrate	n.q.	n.q.	n.q.	n.q.	2.60 ± 0.37 [a]
Isoamyl acetate	2.07 ± 0.02 [a]	1.97 ± 0.04 [a]	2.80 ± 0.00 [b]	1.96 ± 0.17 [a]	0.99 ± 0.05 [c]
Ethyl hexanoate	0.03 ± 0.00 [a]	0.21 ± 0.00 [b]	0.13 ± 0.03 [ab]	0.20 ± 0.05 [b]	0.70 ± 0.06 [c]
Ethyl-3-hydroxybutyrate	0.16 ± 0.00 [a]	0.57 ± 0.00 [b]	0.68 ± 0.00 [c]	0.47 ± 0.00 [d]	0.32 ± 0.06 [e]
Hexyl acetate	0.05 ± 0.00 [a]	0.05 ± 0.00 [a]	0.06 ± 0.00 [a]	0.07 ± 0.00 [a]	0.07 ± 0.05 [a]
2-Phenylethyl acetate	0.31 ± 0.01 [a]	0.28 ± 0.01 [a]	0.24 ± 0.09 [a]	0.39 ± 0.06 [a]	0.76 ± 0.09 [b]
Diethyl succinate	n.q.	0.09 ± 0.00 [a]	0.05 ± 0.00 [a]	0.26 ± 0.08 [b]	6.57 ± 0.13 [c]
Ethyl octanoate	0.06 ± 0.00 [a]	0.17 ± 0.00 [b]	0.21 ± 0.00 [b]	0.18 ± 0.04 [b]	0.51 ± 0.06 [c]
Ethyl lactate	1.71 ± 0.56 [a]	8.23 ± 1.24 [b]	1.73 ± 0.39 [a]	5.93 ± 1.27 [bc]	3.32 ± 0.11 [ac]
Σ Esters	**5.98 ± 0.63** [a]	**12.68 ± 1.19** [b]	**7.19 ± 0.45** [a]	**10.63 ± 1.14** [b]	**16.43 ± 1.08** [c]
Isobutyric acid	4.86 ± 0.05 [a]	4.99 ± 0.34 [a]	3.25 ± 0.07 [b]	4.62 ± 0.04 [a]	2.89 ± 0.05 [b]
Butyric acid	0.22 ± 0.00 [a]	0.29 ± 0.01 [a]	0.25 ± 0.00 [a]	0.40 ± 0.14 [a]	0.23 ± 0.06 [a]
Isovaleric acid	1.82 ± 0.03 [a]	1.25 ± 0.04 [b]	1.29 ± 0.10 [b]	0.76 ± 0.01 [c]	0.74 ± 0.06 [c]
Hexanoic acid	0.90 ± 0.01 [a]	2.97 ± 0.65 [abc]	2.03 ± 0.60 [ac]	5.01 ± 0.76 [b]	3.11 ± 0.34 [bc]
Octanoic acid	0.41 ± 0.02 [a]	1.88 ± 0.09 [b]	1.26 ± 0.16 [c]	1.45 ± 0.15 [c]	2.18 ± 0.05 [b]
Decanoic acid	0.09 ± 0.00 [a]	0.18 ± 0.06 [a]	0.07 ± 0.00 [a]	0.14 ± 0.05 [a]	0.73 ± 0.09 [b]
Σ Acids	**8.30 ± 0.11** [a]	**11.57 ± 0.13** [bc]	**8.15 ± 0.60** [a]	**12.37 ± 1.14** [c]	**9.88 ± 0.65** [ab]
Diacetyle	0.58 ± 0.01 [a]	0.46 ± 0.17 [a]	0.51 ± 0.02 [a]	0.48 ± 0.04 [a]	0.63 ± 0.09 [a]
Furfural	n.q.	0.07 ± 0.00 [a]	0.04 ± 0.00 [a]	0.08 ± 0.00 [a]	0.07 ± 0.05 [a]
Benzaldehyde	0.01 ± 0.00 [a]	0.04 ± 0.00 [ab]	0.06 ± 0.00 [b]	0.05 ± 0.01 [b]	0.09 ± 0.05 [c]
Phenylacetaldehyde	n.q.	n.q.	n.q.	n.q.	0.52 ± 0.05 [a]
Acetoin	5.19 ± 0.46 [a]	1.48 ± 0.01 [b]	3.71 ± 0.00 [c]	2.30 ± 0.43 [b]	0.20 ± 0.10 [d]
Σ Aldehydes/Ketones	**5.78 ± 0.47** [a]	**2.04 ± 0.15** [b]	**4.32 ± 0.02** [c]	**2.91 ± 0.45** [bd]	**1.51 ± 0.34** [be]
γ-Butyrolactone	0.98 ± 0.00 [a]	6.78 ± 0.13 [b]	1.64 ± 0.00 [c]	5.41 ± 0.07 [d]	9.40 ± 0.10 [e]

Data are means ± standard deviation ($n = 2$). Data with different letters [a,b,c,d,e] within each row are significantly different (Tukey test; $p < 0.05$). n.q., not quantifiable. (S), sequential culture; (P), pure culture.

A PCA analysis was performed to cluster wines from sequential combinations and the control according to their oenological and aromatic composition. In the score plot for the first two principal components, PC1 and PC2 explain 75.9% of the total variance (Figure 2). PC1 was mainly determined by ethyl hexanoate (0.986), ethyl octanoate (0.950), decanoic acid (0.944), 2-phenylethyl acetate (0.916), total esters (0.916), 1-propanol (0.916), γ-butyrolactone (0.903), and alcohol degree (0.831); this component allowed us to differentiate the control *S. cerevisiae* fermentation from those fermentations conducted by sequential non-*Saccharomyces*/*S. cerevisiae* combinations. The principal constituents for PC2 were total volatile acids (0.903), butyric acid (0.840), hexanoic acid (0.810), glycerol (0.780), fructose (0.773), and isobutanol (0.740).

PC2 differentiated the sequential fermentations among them, showing clearly two groups. One group formed by sequential inoculations with *W. anomalus* 21A-5C/*S. cerevisiae* CLI 889 and *M. guilliermondii* CLI 1217/*S. cerevisiae* CLI 889, mostly related to dry weight, isoamyl alcohol, acetoin, total aldehydes/ketones, isoamyl acetate, ethyl isovalerate, and β-phenylethyl alcohol in the loadings plot (Figure 2B). Another group contained the sequential cultures with *M. pulcherrima* strains (CLI 68 and CLI 460), mainly classified by total acids, butyric acid, ethyl lactate, hexanoic acid, glycerol, fructose, and isobutanol.

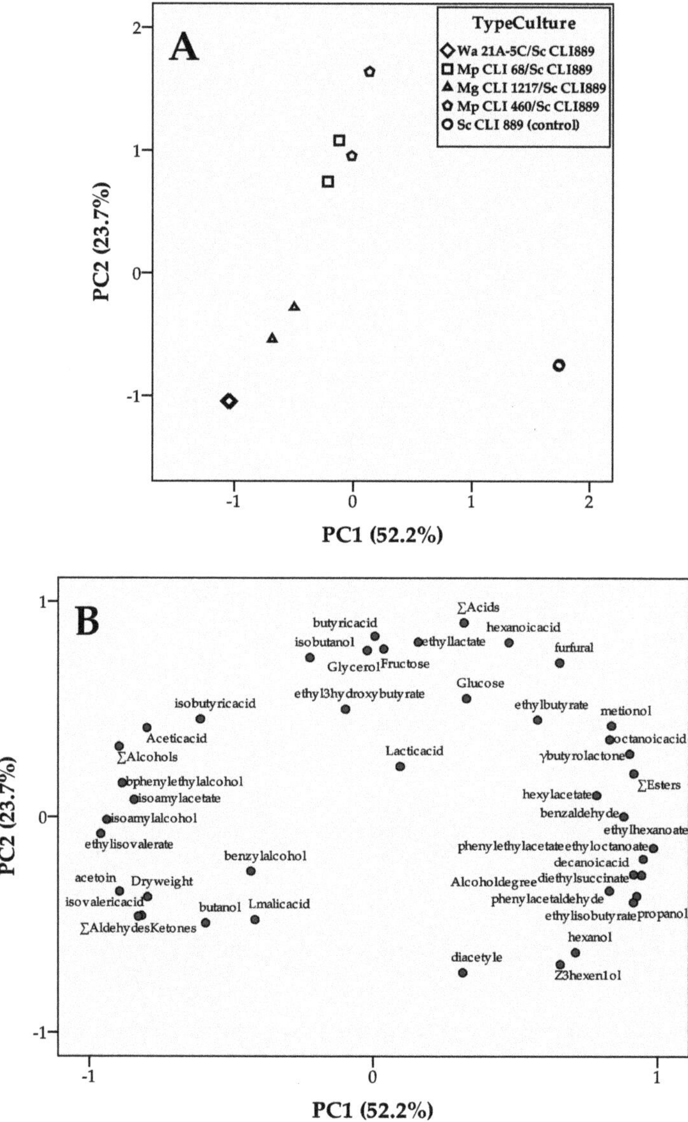

Figure 2. Principal component analysis (PCA) score plot (**A**) and loadings plot (**B**) using main fermentation parameters and 32 volatile compounds.

4. Discussion

A combination of quality, health, and economic reasons will force wine producers to find efficient strategies that enable the production of wines with lower ethanol content without detriment on sensory properties. In this work, the strategy employed for this purpose was the use of sequential combinations between non-*Saccharomyces* and *S. cerevisiae* yeast strains. It is well documented that non-*Saccharomyces* strains are often unable to consume all sugar present in a grape must [13,46]. Hence, sequential culture application would allow the completion of fermentation using one *S. cerevisiae* strain in a second instance [40,47–49]. The successful trials will be carried out by non-*Saccharomyces* strains with a low ethanol yield or those that are able to aerobically metabolize sugars without the simultaneous production of ethanol, prior to *S. cerevisiae* inoculation [2,50]. Regarding aeration regimen, some authors have suggested the use of aerobic yeasts in order to oxide sugars at early stages of winemaking and therefore decrease ethanol production [30,51,52]. After *S. cerevisiae* inoculation, researchers have favored anaerobic conditions to increase the ethanol yield of *Saccharomyces* strain and to avoid excessive oxidation of wine. The fermentation procedure programmed in this work was found to have positive results with other authors [50,51,53].

Several studies have evaluated the action of non-*Saccharomyces*/*S. cerevisiae* combinations in the reduction of ethanol content in wines [19–22,25–29,50,54]. In some cases, the lower ethanol yields resulted from high residual sugar at the end of fermentation [20,21,50]. By contrast, other research works have reported wines with a significant reduction in ethanol yield (0.6–1.7%, *v/v*) when using non-*Saccharomyces* and *S. cerevisiae* strains in mixed or sequential cultures. Contreras et al. [22] found that sequential inoculation of a selected *M. pulcherrima* strain (AWRI1149) with *S. cerevisiae* wine strain was the best combination for reducing the ethanol content in Chardonnay (0.9%, *v/v* lower than control) and Shiraz (1.6% *v/v* lower than control) wines. In the same way, Varela et al. [26] obtained Merlot wines fermented with *M. pulcherrima* with 1.0% less ethanol than *S. cerevisiae*-fermented wines at pilot scale. Further studies also showed ethanol reduction using immobilized selected strains of non-*Saccharomyces* yeasts followed by inoculation of free *S. cerevisiae* cells [25,47]. The sequential cultures of *M. pulcherrima* and *Starmerella bombicola* immobilized cells and *S. cerevisiae* free cells were the best for ethanol reduction with values 1.4% and 1.6% v/v, respectively [25]. In addition, ethanol lowering has been recorded in wines obtained by different *Saccharomyces* species. Using sterile Shiraz must, sequential inoculation of *M. pulcherrima* (AWRI1149) and *S. uvarum* (AWRI2846) with *S. cerevisiae* produced wines with 0.9% *v/v* less ethanol than *S. cerevisiae* alone [55]. Puškaš et al. [29] also observed that sequential cultures with *M. pulcherrima*, *S. bayanus*, and *S. cerevisiae* generated wines with 0.9% *v/v* lower ethanol than control. In the present work, the application of sequential cultures of native non-*Saccharomyces* strains (*W. anomalus* 21A-5C, *M. guilliermondii* CLI 1217, and *M. pulcherrima* CLI 68 and CLI 460) and *S. cerevisiae* CLI 889 generated a reduction of alcohol content between 0.8%–1.3% *v/v* in Malvar wines, where *M. pulcherrima* CLI 68/*S. cerevisiae* CLI 889 sequential inoculation produced the highest decrease in alcohol degree. On the other hand, *W. anomalus* has been described as low fermentative species in pure culture compared to *S. cerevisiae* [28,31]. This statement is consistent with our results where *W. anomalus* 21A-5C presented 121 g/L of residual sugars after the first 96 h (Table S1). In sequential culture with *S. cerevisiae*, previous works denoted that the presence of *W. anomalus* does not affect final alcohol contents [28]. Instead, the strain studied in this work (*W. anomalus* 21A-5C) produced wines with 0.9% *v/v* less ethanol than control, in agreement with Contreras et al. [19] who studied another strain of the *W. anomalus* species. Finally, the use of *M. guilliermondii* as a low-ethanol producer has not been well documented. Some research works have studied *M. guilliermondii* as a candidate for reducing ethanol content in wines, but none have considered its use for that purpose [19,56]. In contrast, *M. guilliermondii* CLI 1217 in sequential culture was the second-best option to decrease the ethanol concentration in Malvar wines (1.2% less ethanol than control).

Beyond ethanol, the growth of the four selected non-*Saccharomyces* affected glycerol and acetic acid concentrations in Malvar wines. Several studies have reported that the production of glycerol by yeasts leads to an increase in acetic acid concentration [57,58]. Wines produced with *M. pulcherrima* strains

CLI 68 and CLI 460 contained the greatest glycerol content (8.32 and 9.30 g/L, respectively) compared with other wines studied. The connection between *M. pulcherrima* and an increased glycerol production has been explained by the overexpression of the glycerol-3-phosphate dehydrogenase 1 (*GDP1*) gene in *S. cerevisiae* (associated with the conversion of dihydroxyacetone phosphate in glycerol-3-phosphate, an intermediate for glycerol formation). This gene is overinduced when *S. cerevisiae* coexists with *M. pulcherrima* in must fermentations [59]. Moreover, glycerol formation has been demonstrated as the best strategy, followed by yeasts for producing wines with lower ethanol content [60]. This compound is present in semi-sweet and dry wines ranging from 5 to 14 g/L, although glycerol imparts sweetness at a threshold of about 5.2 g/L in dry white wines [61]. Unlike glycerol, acetic acid imparts an objectionable character to wine at elevated concentrations. This volatile acid becomes undesirable at concentrations over 0.7–1.1 g/L, depending on the style of wine; its optimal concentration is 0.2–0.7 g/L [61]. One reason for elevated acetic acid levels is usually related to aeration, which could lead to elevated oxygen levels during fermentation [23,30,52]. However, more acetic acid was produced in Malvar wines during the anaerobic period than during aerobic fermentation in the current work, in agreement with results observed by Röcker et al. [24]. All sequential fermentation between four selected non-*Saccharomyces*/*S. cerevisiae* in this article produced wines with elevated volatile acidity (>0.7 g/L of acetic acid), significantly increased after *S. cerevisiae* inoculation (Table S1, see values of acetic acid caused by non-*Saccharomyces* fermentations). This noticeable increase could be caused by a lack of nutrition sources available for *S. cerevisiae* in the second part of fermentations [62,63]. Low YAN values (below 200 mg N/L, such as the Malvar must we studied) can also lead to elevated acetic acid concentration [64].

For selection of low-ethanol producing wine yeast, its impact on aroma profile is of great importance. Sequential cultures in this work had an important influence on higher alcohol proportions compared with the control. High levels of these volatile compounds (>300 mg/L) can have a detrimental effect on wine aroma, while concentrations below 300 mg/L can contribute positively to aroma complexity [65,66]. All wines produced using sequential inoculations presented values of higher alcohols below 300 mg/L.

It is worth noting that isoamyl alcohol (harsh, bitter) and β-phenylethyl alcohol (flowery, roses) are increased by sequential culture with *W. anomalus* 21A-5C and *M. guilliermondii* CLI 1217 strains. The ethyl isovalerate and isoamyl acetate esters, which impart fruity (banana) and sweet aromas, were also higher in these sequential cultures. In relation with *W. anomalus* species, these results agree with other publications [28,67–70]. Rojas et al. [67] indicated that one *W. anomalus* (*P. anomala*) strain produced the highest isoamyl acetate concentration in 48 h cultures in aerobiosis conditions; moreover, the increment in acetates was also observed in sequential cultures with *W. anomalus* and *S. cerevisiae* [28,70]. In addition to increasing alcohols, as well as ethyl and acetate esters [69], Airen white wines elaborated with *W. anomalus*/*S. cerevisiae* sequential cultures were judged to be better than *S. cerevisiae* monoculture due to their higher scores for descriptors as fruity and floral, and having an intense sweet smell and longer-lasting aftertaste [68]. Nevertheless, *M. guilliermondii* has been considered as a spoilage yeast in winemaking that is able to produce large amounts of volatile phenols [71], identified with horse, stable, leather, or medicinal notes [72]; in contrast, the *M. guilliermondii* CLI 1217 strain used in sequential culture in the present work has contributed to rising amounts of fusel alcohols and some esters related to fruity and floral character in Malvar white wines.

Apart from high levels of isoamyl alcohol and β-phenylethyl alcohol previously documented by authors [22,24,73–75], sequential cultures with *M. pulcherrima* strains (CLI 68 and CLI 460) also showed an elevated proportion of isobutanol (bitter, fusel, alcohol) compared to the wine fermented solely with *S. cerevisiae*. This high isobutanol content is in good agreement with the experimental data reported previously [24,28,76]. While some reports [22,73,77] have stated that wines inoculated with *M. pulcherrima*/*S. cerevisiae* contain higher concentration of esters, other studies [16,54,74,78,79] have noted that wines fermented with these yeast species in combination have lower concentrations, as in the case of this work. Moreover, it is worth mentioning that sequential culture with *M. pulcherrima* native strains presented higher concentration of esters with fruity aroma (ethyl isovalerate, isoamyl acetate and ethyl-3-hydroxybutyrate) than the control. On the other hand, Malvar wines elaborated with *M. pulcherrima* CLI 68 and CLI 460 strains are mostly related to volatile fatty acids. These compounds are generally associated with negative aromas in wine [80], although hexanoic, octanoic, and decanoic fatty acids impart mild and pleasant notes to wine at concentrations between 4 to 10 mg/L; however, their impact can be negative on wine at levels above 20 mg/L [81]. Thus, these fatty acids might have a positive effect on the aroma of *M. pulcherrima/S. cerevisiae* Malvar wines since their levels are below 20 mg/L.

Relative to wine fermented with the control, *S. cerevisiae* CLI 889, we found a higher total concentration of esters in Malvar wines using *S. cerevisiae* monoculture. This *S. cerevisiae* strain produced wines with a fruity and floral character due to the greater concentration of ethyl isobutyrate (pineapple), ethyl hexanoate (pineapple, apple), and 2-phenylethyl acetate (flowery, lilac) esters, being the perfect candidate to ferment Malvar musts, and improving the typicity of the wines produced in the area "Vinos de Madrid".

5. Conclusions

The present results indicated that sequential cultures of native non-*Saccharomyces* (*W. anomalus* 21A-5C, *M. guilliermondii* CLI 1217, and *M. pulcherrima* CLI 68 and CLI 460) with *S. cerevisiae* CLI 889 can be used as a strategy to reduce the ethanol levels in wines, whilst keeping the wine typicity of the area. These combinations could have a positive impact on glycerol content and the volatile profile of these wines, showing *W. anomalus* 21A-5C and *M. guilliermondii* CLI 1217 combinations with *S. cerevisiae* being mostly related to fruity and floral aroma compounds when compared with *M. pulcherrima* usage. However, further optimization will be required to control the acetic acid production in all sequential fermentations. Future work will focus on fermentations at a pilot scale through using the selected strains and having a second inoculation at different times, which will allow for the evaluation of the sensorial profile of the resulting wines.

Supplementary Materials: The following are available online at http://www.mdpi.com/2311-5637/6/2/60/s1, Table S1. Volatile acidity (as g/L of acetic acid), reducing sugars (glucose + fructose, g/L), and dry weight (mg) of wines after Section I (pure culture of non-*Saccharomyces* strains and the control under aerobic conditions).

Author Contributions: Investigation, M.G. and T.A.; methodology, M.G. and T.A.; formal analysis, M.G., B.E.-Z., J.M.C., and T.A.; writing—original draft preparation, M.G. and T.A.; writing—review and editing, M.G., B.E.-Z., J.M.C., and T.A. All authors have read and agreed to the published version of the manuscript.

Funding: This work was financially supported by the project RM2010-00009-C03-01 funded by the Instituto Nacional de Investigación y Tecnología Agraria y Alimentaria (INIA).

Conflicts of Interest: The authors declare no conflict of interest. The funders had no role in the design of the study; in the collection, analyses, or interpretation of data; in the writing of the manuscript; and in the decision to publish the results.

References

1. Mira de Orduña, R. Climate change associated effects on grape and wine quality and production. *Food Res. Int.* **2010**, *43*, 1844–1855. [CrossRef]
2. Gonzalez, R.; Quirós, M.; Morales, P. Yeast respiration of sugars by non-*Saccharomyces* yeast species: A promising and barely explored approach to lowering alcohol content of wines. *Trends Food Sci. Technol.* **2013**, *29*, 55–61. [CrossRef]
3. Liguori, L.; Russo, P.; Albanese, D.; Di Matteo, M. Production of low-alcohol beverages: Current status and perspectives. In *Food Processing for Increased Quality and Consumption*; Grumezescu, A., Holban, A.M., Eds.; Elsevier Inc.: Amsterdam, The Netherlands, 2018; pp. 347–382. ISBN 9780128114476.
4. Stockley, C.S.; Varela, C.; Coulter, A.; Dry, P.R.; Francis, I.L.; Muhlack, R.; Pretorius, I.S. *Controlling the Highs and the Lows of Alcohol in Wine*; Nova Science Publishers, Inc.: New York, NY, USA, 2012; ISBN 9781614706359.
5. Varela, C.; Dry, P.R.; Kutyna, D.R.; Francis, I.L.; Henschke, P.A.; Curtin, C.D.; Chambers, P.J. Strategies for reducing alcohol concentration in wine. *Aust. J. Grape Wine Res.* **2015**, *21*, 670–679. [CrossRef]

6. Fischer, U.; Noble, A. The effect of ethanol, catechin concentration, and pH on sourness and bitterness of wine. *Am. J. Enol. Vitic.* **1994**, *45*, 6–10.

7. Wilkinson, K.L.; Jiranck, V. Wine of reduced alcohol content: Consumer and society demand vs. industry willingness and ability to deliver. In *1st International Symposium Alcohol Level Reduction in Wine*; Institut des Sciences de la Vigne et du Vin: Villenave d'Ornon CEDEX, France, 2013; pp. 98–104.

8. Stockwell, T.; Zhao, J.; Panwar, S.; Roemer, A.; Naimi, T.; Chikritzhs, T. Do "moderate" drinkers have reduced mortality risk? A systematic review and meta-analysis of alcohol consumption and all-cause mortality. *J. Stud. Alcohol Drugs* **2016**, *77*, 185–198. [CrossRef]

9. Sharma, A.; Vandenberg, B.; Hollingsworth, B. Minimum pricing of alcohol versus volumetric taxation: Which policy will reduce heavy consumption without adversely affecting light and moderate consumers? *PLoS ONE* **2014**, *9*, e80936. [CrossRef]

10. Ciani, M.; Morales, P.; Comitini, F.; Tronchoni, J.; Canonico, L.; Curiel, J.A.; Oro, L.; Rodrigues, A.J.; Gonzalez, R. Non-conventional yeast species for lowering ethanol content of wines. *Front. Microbiol.* **2016**, *7*, 642. [CrossRef]

11. Rolle, L.; Englezos, V.; Torchio, F.; Cravero, F.; Río Segade, S.; Rantsiou, K.; Giacosa, S.; Gambuti, A.; Gerbi, V.; Cocolin, L. Alcohol reduction in red wines by technological and microbiological approaches: A comparative study. *Aust. J. Grape Wine Res.* **2017**, *24*, 62–74. [CrossRef]

12. Tilloy, V.; Ortiz-Julien, A.; Dequin, S. Reduction of ethanol yield and improvement of glycerol formation by adaptive evolution of the wine yeast *Saccharomyces cerevisiae* under hyperosmotic conditions. *Appl. Environ. Microbiol.* **2014**, *80*, 2623–2632. [CrossRef]

13. Fleet, G.H.; Heard, G.M. Yeast growth during fermentation. In *Wine Microbiology and Biotechnology*; Fleet, G.H., Ed.; Harwood Academic Publishers: Chur, Switzerland, 1993; pp. 27–54.

14. Andorrà, I.; Berradre, M.; Rozès, N.; Mas, A.; Guillamón, J.M.; Esteve-Zarzoso, B. Effect of pure and mixed cultures of the main wine yeast species on grape must fermentations. *Eur. Food Res. Technol.* **2010**, *231*, 215–224. [CrossRef]

15. Cominiti, F.; Gobbi, M.; Domizio, P.; Romani, C.; Lencioni, L.; Mannazzu, I.; Ciani, M. Selected non-*Saccharomyces* wine yeasts in controlled multistarter fermentations with *Saccharomyces cerevisiae*. *Food Microbiol.* **2011**, *28*, 873–882. [CrossRef]

16. García, M.; Arroyo, T.; Crespo, J.; Cabellos, J.M.; Esteve-Zarzoso, B. Use of native non-*Saccharomyces* strain: A new strategy in D.O. "Vinos de Madrid" (Spain) wines elaboration. *Eur. J. Food Sci. Technol.* **2017**, *5*, 1–31.

17. Dutraive, O.; Benito, S.; Fritsch, S.; Beisert, B.; Patz, C.; Rauhut, D. Effect of sequential inoculation with non-*Saccharomyces* and *Saccharomyces* yeasts on Riesling wine chemical composition. *Fermentation* **2019**, *5*, 79. [CrossRef]

18. Cuello, R.A.; Flores Montero, K.J.; Mercado, L.A.; Combina, M.; Ciklic, I.F. Construction of low-ethanol–wine yeasts through partial deletion of the *Saccharomyces cerevisiae* PDC2 gene. *AMB Express* **2017**, *7*, 67. [CrossRef]

19. Contreras, A.; Hidalgo, C.; Schmidt, S.; Henschke, P.A.; Curtin, C.; Varela, C. The application of non-*Saccharomyces* yeast in fermentations with limited aeration as a strategy for the production of wine with reduced alcohol content. *Int. J. Food Microbiol.* **2015**, *205*, 7–15. [CrossRef]

20. Ciani, M.; Beco, L.; Comitini, F. Fermentation behaviour and metabolic interactions of multistarter wine yeast fermentations. *Int. J. Food Microbiol.* **2006**, *108*, 239–245. [CrossRef]

21. Magyar, I.; Tóth, T. Comparative evaluation of some oenological properties in wine strains of *Candida stellata*, *Candida zemplinina*, *Saccharomyces uvarum* and *Saccharomyces cerevisiae*. *Food Microbiol.* **2011**, *28*, 94–100. [CrossRef]

22. Contreras, A.; Hidalgo, C.; Henschke, P.A.; Chambers, P.J.; Curtin, C.; Varela, C. Evaluation of non-*Saccharomyces* yeasts for the reduction of alcohol content in wine. *Appl. Environ. Microbiol.* **2014**, *80*, 1670–1678. [CrossRef]

23. Morales, P.; Rojas, V.; Quirós, M.; Gonzalez, R. The impact of oxygen on the final alcohol content of wine fermented by a mixed starter culture. *Appl. Microbiol. Biotechnol.* **2015**, *99*, 3993–4003. [CrossRef]

24. Röcker, J.; Strub, S.; Ebert, K.; Grossmann, M. Usage of different aerobic non-*Saccharomyces* yeasts and experimental conditions as a tool for reducing the potential ethanol content in wines. *Eur. Food Res. Technol.* **2016**, *242*, 2051–2070. [CrossRef]

25. Canonico, L.; Comitini, F.; Oro, L.; Ciani, M. Sequential fermentation with selected immobilized non-*Saccharomyces* yeast for reduction of ethanol content in wine. *Front. Microbiol.* **2016**, *7*, 278. [CrossRef]

26. Varela, C.; Barker, A.; Tran, T.; Borneman, A.; Curtin, C. Sensory profile and volatile aroma composition of reduced alcohol Merlot wines fermented with *Metschnikowia pulcherrima* and *Saccharomyces uvarum*. *Int. J. Food Microbiol.* **2017**, *252*, 1–9. [CrossRef]

27. Canonico, L.; Comitini, F.; Ciani, M. *Metschnikowia pulcherrima* selected strain for ethanol reduction in wine: Influence of cell immobilization and aeration condition. *Foods* **2019**, *8*, 378. [CrossRef]

28. Aplin, J.J.; White, K.P.; Edwards, C.G. Growth and metabolism of non-*Saccharomyces* yeasts isolated from Washington state vineyards in media and high sugar grape musts. *Food Microbiol.* **2019**, *77*, 158–165. [CrossRef]

29. Puškaš, V.S.; Miljić, U.D.; Djuran, J.J.; Vučurović, V.M. The aptitude of commercial yeast strains for lowering the ethanol content of wine. *Food Sci. Nutr.* **2020**, *8*, 278. [CrossRef]

30. Quirós, M.; Rojas, V.; Gonzalez, R.; Morales, P. Selection of non-*Saccharomyces* yeast strains for reducing alcohol levels in wine by sugar respiration. *Int. J. Food Microbiol.* **2014**, *181*, 85–91. [CrossRef]

31. Cordero-Bueso, G.; Esteve-Zarzoso, B.; Cabellos, J.M.; Gil-Díaz, M.; Arroyo, T. Biotechnological potential of non-*Saccharomyces* yeasts isolated during spontaneous fermentations of Malvar (*Vitis vinifera* cv. L.). *Eur. Food Res. Technol.* **2013**, *236*, 193–207. [CrossRef]

32. Gil, M.; Cabellos, J.M.; Arroyo, T.; Prodanov, M. Characterization of the volatile fraction of young wines from the Denomination of Origin "Vinos de Madrid" (Spain). *Anal. Chim. Acta* **2006**, *563*, 145–153. [CrossRef]

33. Tello, J.; Cordero-Bueso, G.; Aporta, I.; Cabellos, J.M.; Arroyo, T. Genetic diversity in commercial wineries: Effects of the farming system and vinification management on wine yeasts. *J. Appl. Microbiol.* **2012**, *112*, 302–315. [CrossRef]

34. Cordero-Bueso, G.; Esteve-Zarzoso, B.; Gil-Díaz, M.; García, M.; Cabellos, J.; Arroyo, T. Improvement of Malvar wine quality by use of locally-selected *Saccharomyces cerevisiae* strains. *Fermentation* **2016**, *2*, 7. [CrossRef]

35. Merloni, E.; Camanzi, L.; Mulazzani, L.; Malorgio, G. Adaptive capacity to climate change in the wine industry: A Bayesian Network approach. *Wine Econ. Policy* **2018**, *7*, 165–177. [CrossRef]

36. Arroyo, T.; Cordero, G.; Serrano, A.; Valero, E. β-Glucosidase production by non-*Saccharomyces* yeasts isolated from vineyard. In *Expression of Multidisciplinary Flavour Science*; Blank, I., Wüst, M., Yeretzian, C., Eds.; Institute of Chemistry and Biological Chemistry: Winterthur, Switzerland, 2010; pp. 359–362.

37. Arroyo, T. Estudio de la Influencia de Diferentes Tratamientos Enológicos en la Evolución de la Microbiota y en la Calidad de los vinos Elaborados con la Variedad "Airén", en la D.O. "Vinos de Madrid". Ph.D. Thesis, University of Alcalá, Madrid, Spain, 2000.

38. Esteve-Zarzoso, B.; Belloch, C.; Uruburu, F.; Querol, A. Identification of yeasts by RFLP analysis of the 5.8S rRNA gene and the two ribosomal internal transcribed spacers. *Int. J. Syst. Bacteriol.* **1999**, *49*, 329–337. [CrossRef] [PubMed]

39. Cordero-Bueso, G.; Arroyo, T.; Serrano, A.; Tello, J.; Aporta, I.; Vélez, M.D.; Valero, E. Influence of the farming system and vine variety on yeast communities associated with grape berries. *Int. J. Food Microbiol.* **2011**, *145*, 132–139. [CrossRef] [PubMed]

40. García, M.; Esteve-Zarzoso, B.; Crespo, J.; Cabellos, J.M.; Arroyo, T. Yeast monitoring of wine mixed or sequential fermentations made by native strains from D.O. "Vinos de Madrid" using real-time quantitative PCR. *Front. Microbiol.* **2017**, *8*, 2520. [CrossRef] [PubMed]

41. Kurtzman, C.P.; Robnett, C.J. Identification and phylogeny of ascomycetous yeasts from analysis of nuclear large subunit (26S) ribosomal DNA partial sequences. *Antonie Van Leeuwenhoek* **1998**, *73*, 331–371. [CrossRef]

42. García, M.; Greetham, D.; Wimalasena, T.T.; Phister, T.G.; Cabellos, J.M.; Arroyo, T. The phenotypic characterization of yeast strains to stresses inherent to wine fermentation in warm climates. *J. Appl. Microbiol.* **2016**, *121*, 215–233. [CrossRef] [PubMed]

43. García, M.; Apolinar-Valiente, R.; Williams, P.; Esteve-Zarzoso, B.; Arroyo, T.; Crespo, J.; Doco, T. Polysaccharides and oligosaccharides produced on Malvar wines elaborated with *Torulaspora delbrueckii* CLI 918 and *Saccharomyces cerevisiae* CLI 889 native yeasts from D.O. "Vinos de Madrid.". *J. Agric. Food Chem.* **2017**, *65*, 6656–6664. [CrossRef]

44. Ortega, C.; López, R.; Cacho, J.; Ferreira, V. Fast analysis of important wine volatile compounds-Development and validation of a new method based on gas chromatographic-flame ionisation detection analysis of dichloromethane microextracts. *J. Chromatogr. A* **2001**, *923*, 205–214. [CrossRef]

45. Balboa-Lagunero, T.; Arroyo, T.; Cabellos, J.M.; Aznar, M. Yeast selection as a tool for reducing key oxidation notes in organic wines. *Food Res. Int.* **2013**, *53*, 252–259. [CrossRef]
46. Fleet, G.H. Wine. In *Food Microbiology: Fundamentals and Frontiers*; Doyle, M.P., Beuchat, L.R., Eds.; ASM Press: Washington, DC, USA, 2007; pp. 863–890.
47. Ciani, M.; Canonico, L.; Oro, L.; Comitini, F. Sequential fermentation using non-*Saccharomyces* yeasts for the reduction of alcohol content in wine. *BIOWeb Conf.* **2014**, *3*, 02015. [CrossRef]
48. Loira, I.; Vejarano, R.; Bañuelos, M.A.; Morata, A.; Tesfaye, W.; Uthurry, C.; Villa, A.; Cintora, I.; Suárez-Lepe, J.A. Influence of sequential fermentation with *Torulaspora delbrueckii* and *Saccharomyces cerevisiae* on wine quality. *LWT Food Sci. Technol.* **2014**, *59*, 915–922. [CrossRef]
49. Vilela, A. Modulating wine pleasantness throughout wine-yeast co-inoculation or sequential inoculation. *Fermentation* **2020**, *6*, 22. [CrossRef]
50. Ciani, M.; Ferraro, L. Enhanced glycerol content in wines made with immobilized *Candida stellata* cells. *Appl. Environ. Microbiol.* **1996**, *62*, 128–132. [CrossRef] [PubMed]
51. Erten, H.; Campbell, I. The production of low-alcohol wines by aerobic yeasts. *J. Inst. Brew.* **2001**, *107*, 207–215. [CrossRef]
52. Rodrigues, A.J.; Raimbourg, T.; Gonzalez, R.; Morales, P. Environmental factors influencing the efficacy of different yeast strains for alcohol level reduction in wine by respiration. *LWT Food Sci. Technol.* **2016**, *65*, 1038–1043. [CrossRef]
53. Ferraro, L.; Fatichenti, F.; Ciani, M. Pilot scale vinification process using immobilized *Candida stellata* cells and *Saccharomyces cerevisiae*. *Process Biochem.* **2000**, *35*, 1125–1129. [CrossRef]
54. Canonico, L.; Solomon, M.; Comitini, F.; Ciani, M.; Varela, C. Volatile profile of reduced alcohol wines fermented with selected non-*Saccharomyces* yeasts under different aeration conditions. *Food Microbiol.* **2019**, *84*, 103247. [CrossRef]
55. Contreras, A.; Curtin, C.; Varela, C. Yeast population dynamics reveal a potential "collaboration" between *Metschnikowia pulcherrima* and *Saccharomyces uvarum* for the production of reduced alcohol wines during Shiraz fermentation. *Appl. Microbiol. Biotechnol.* **2014**, *99*, 1885–1895. [CrossRef]
56. Mestre Furlani, M.V.; Maturano, Y.P.; Combina, M.; Mercado, L.A.; Toro, M.E.; Vazquez, F. Selection of non-*Saccharomyces* yeasts to be used in grape musts with high alcoholic potential: A strategy to obtain wines with reduced ethanol content. *FEMS Yeast Res.* **2017**, *17*, 1–10. [CrossRef] [PubMed]
57. Remize, F.; Roustan, J.L.; Sablayrolles, J.M.; Barre, P.; Dequin, S. Glycerol overproduction by engineered *Saccharomyces cerevisiae* wine yeast strains leads to substantial changes in by-product formation and to a stimulation of fermentation rate in stationary phase. *Appl. Environ. Microbiol.* **1999**, *65*, 143–149. [CrossRef]
58. Eglinton, J.M.; Heinrich, A.J.; Pollnitz, A.P.; Langridge, P.; Henschke, P.A.; De Barros Lopes, M. Decreasing acetic acid accumulation by a glycerol overproducing strain of *Saccharomyces cerevisiae* by deleting the ALD6 aldehyde dehydrogenase gene. *Yeast* **2002**, *19*, 295–301. [CrossRef] [PubMed]
59. Sadoudi, M.; Rousseaux, S.; David, V.; Alexandre, H.; Tourdot-Maréchal, R. *Metschnikowia pulcherrima* influences the expression of genes involved in PDH bypass and glyceropyruvic fermentation in *Saccharomyces cerevisiae*. *Front. Microbiol.* **2017**, *8*, 1137. [CrossRef]
60. Varela, C.; Kutyna, D.R.; Solomon, M.R.; Black, C.A.; Borneman, A.; Henschke, P.A.; Pretorius, I.S.; Chambers, P.J. Evaluation of gene modification strategies for the development of low-alcohol-wine yeasts. *Appl. Environ. Microbiol.* **2012**, *78*, 6068–6077. [CrossRef] [PubMed]
61. Swiegers, J.H.; Bartowsky, E.J.; Henschke, P.A.; Pretorius, I.S. Yeast and bacterial modulation of wine aroma and flavour. *Aust. J. Grape Wine Res.* **2005**, *11*, 139–173. [CrossRef]
62. Beltran, G.; Esteve-Zarzoso, B.; Rozès, N.; Mas, A.; Guillamón, J.M. Influence of the timing of nitrogen additions during synthetic grape must fermentations on fermentation kinetics and nitrogen consumption. *J. Agric. Food Chem.* **2005**, *53*, 996–1002. [CrossRef] [PubMed]
63. Medina, K.; Boido, E.; Dellacassa, E.; Carrau, F. Growth of non-*Saccharomyces* yeasts affects nutrient availability for *Saccharomyces cerevisiae* during wine fermentation. *Int. J. Food Microbiol.* **2012**, *157*, 245–250. [CrossRef]
64. Vilanova, M.; Ugliano, M.; Varela, C.; Siebert, T.; Pretorius, I.S.; Henschke, P.A. Assimilable nitrogen utilisation and production of volatile and non-volatile compounds in chemically defined medium by *Saccharomyces cerevisiae* wine yeasts. *Appl. Microbiol. Biotechnol.* **2007**, *77*, 145–157. [CrossRef]
65. Styger, G.; Prior, B.; Bauer, F.F. Wine flavor and aroma. *J. Ind. Microbiol. Biotechnol.* **2011**, *38*, 1145–1159. [CrossRef]

66. Belda, I.; Ruiz, J.; Esteban-Fernández, A.; Navascués, E.; Marquina, D.; Santos, A.; Moreno-Arribas, M.V. Microbial contribution to wine aroma and its intended use for wine quality improvement. *Molecules* **2017**, *22*, 189. [CrossRef]

67. Rojas, V.; Gil, J.V.; Piñaga, F.; Manzanares, P. Studies on acetate ester production by non-*Saccharomyces* wine yeasts. *Int. J. Food Microbiol.* **2001**, *70*, 283–289. [CrossRef]

68. Izquierdo Cañas, P.M.; Palacios García, A.T.; García Romero, E. Enhancement of flavour properties in wines using sequential inoculations of non-*Saccharomyces* (*Hansenula* and *Torulaspora*) and *Saccharomyces* yeast starter. *Vitis* **2011**, *50*, 177–182.

69. Domizio, P.; Romani, C.; Lencioni, L.; Comitini, F.; Gobbi, M.; Mannazzu, I.; Ciani, M. Outlining a future for non-*Saccharomyces* yeasts: Selection of putative spoilage wine strains to be used in association with *Saccharomyces cerevisiae* for grape juice fermentation. *Int. J. Food Microbiol.* **2011**, *147*, 170–180. [CrossRef] [PubMed]

70. Izquierdo Cañas, P.M.; García-Romero, E.; Heras Manso, J.M.; Fernández-González, M. Influence of sequential inoculation of *Wickerhamomyces anomalus* and *Saccharomyces cerevisiae* in the quality of red wines. *Eur. Food Res. Technol.* **2014**, *239*, 279–286. [CrossRef]

71. Sáez, J.S.; Lopes, C.A.; Kirs, V.C.; Sangorrín, M.P. Enhanced volatile phenols in wine fermented with *Saccharomyces cerevisiae* and spoiled with *Pichia guilliermondii* and *Dekkera bruxellensis*. *Lett. Appl. Microbiol.* **2010**, *51*, 170–176. [CrossRef]

72. Suárez, R.; Suárez-Lepe, J.A.; Morata, A.; Calderón, F. The production of ethylphenols in wine by yeasts of the genera *Brettanomyces* and *Dekkera*: A review. *Food Chem.* **2007**, *102*, 10–21. [CrossRef]

73. Sadoudi, M.; Tourdot-Maréchal, R.; Rousseaux, S.; Steyer, D.; Gallardo-Chacón, J.J.; Ballester, J.; Vichi, S.; Guérin-Schneider, R.; Caixach, J.; Alexandre, H. Yeast-yeast interactions revealed by aromatic profile analysis of Sauvignon Blanc wine fermented by single or co-culture of non-*Saccharomyces* and *Saccharomyces* yeasts. *Food Microbiol.* **2012**, *32*, 243–253. [CrossRef]

74. González-Royo, E.; Pascual, O.; Kontoudakis, N.; Esteruelas, M.; Esteve-Zarzoso, B.; Mas, A.; Canals, J.M.; Zamora, F. Oenological consequences of sequential inoculation with non-*Saccharomyces* yeasts (*Torulaspora delbrueckii* or *Metschnikowia pulcherrima*) and *Saccharomyces cerevisiae* in base wine for sparkling wine production. *Eur. Food Res. Technol.* **2015**, *240*, 999–1012. [CrossRef]

75. Gobert, A.; Tourdot-Maréchal, R.; Morge, C.; Sparrow, C.; Liu, Y.; Quintanilla-Casas, B.; Vichi, S.; Alexandre, H. Non-*Saccharomyces* yeasts nitrogen source preferences: Impact on sequential fermentation and wine volatile compounds profile. *Front. Microbiol.* **2017**, *8*, 2175. [CrossRef]

76. Seguinot, P.; Ortiz-Julien, A.; Camarasa, C. Impact of nutrient availability on the fermentation and production of aroma compounds under sequential inoculation with *M. pulcherrima* and *S. cerevisiae*. *Front. Microbiol.* **2020**, *11*, 305. [CrossRef]

77. Rodríguez, M.E.; Lopes, C.A.; Barbagelata, R.J.; Barda, N.B.; Caballero, A.C. Influence of *Candida pulcherrima* Patagonian strain on alcoholic fermentation behaviour and wine aroma. *Int. J. Food Microbiol.* **2010**, *138*, 19–25. [CrossRef]

78. Varela, C.; Sengler, F.; Solomon, M.; Curtin, C. Volatile flavour profile of reduced alcohol wines fermented with the non-conventional yeast species *Metschnikowia pulcherrima* and *Saccharomyces uvarum*. *Food Chem.* **2016**, *209*, 57–64. [CrossRef]

79. Ruiz, J.; Belda, I.; Beisert, B.; Navascués, E.; Marquina, D.; Calderón, F.; Rauhut, D.; Santos, A.; Benito, S. Analytical impact of *Metschnikowia pulcherrima* in the volatile profile of Verdejo white wines. *Appl. Microbiol. Biotechnol.* **2018**, *102*, 8501–8509. [CrossRef]

80. Lambrechts, M.G.; Pretorius, I.S. Yeast and its importance to wine aroma-A review. *South African J. Enol. Vitic.* **2000**, *21*, 97–129. [CrossRef]

81. Jiang, B.; Zhang, Z. Volatile compounds of young wines from Cabernet Sauvignon, Cabernet Gernischet and Chardonnay varieties grown in the Loess Plateau region of China. *Molecules* **2010**, *15*, 9184–9196. [CrossRef]

 *fermentation*

Review

The Effect of Non-*Saccharomyces* and *Saccharomyces* Non-*Cerevisiae* Yeasts on Ethanol and Glycerol Levels in Wine

Nedret Neslihan Ivit [1,2]**, Rocco Longo** [3] **and Belinda Kemp** [4,5,*]

1   Perennia Food and Agriculture Inc., 32 Main Street, Kentville, NS B4N 1J5, Canada; nivit@perennia.ca
2   Office of Industry and Community Engagement, Acadia University, 210 Horton Hall, 18 University Ave, Wolfville, NS B4P 2R6, Canada
3   Tasmanian Institute of Agriculture, University of Tasmania, Prospect, TAS 7249, Australia; rocco.longo@utas.edu.au
4   Cool Climate Oenology and Viticulture Institute (CCOVI), Brock University, 1812 Sir Isaac Brock Way, St. Catharines, ON L2S 3A1, Canada
5   Department of Biological Science, Faculty of Maths and Science, Brock University, 1812 Sir Isaac Brock Way, St. Catharines, ON L2S 3A1, Canada
*   Correspondence: bkemp@brocku.ca

Received: 21 May 2020; Accepted: 28 July 2020; Published: 30 July 2020

Abstract: Non-*Saccharomyces* and *Saccharomyces* non-*cerevisiae* studies have increased in recent years due to an interest in uninoculated fermentations, consumer preferences, wine technology, and the effect of climate change on the chemical composition of grapes, juice, and wine. The use of these yeasts to reduce alcohol levels in wines has garnered the attention of researchers and winemakers alike. This review critically analyses recent studies concerning the impact of non-*Saccharomyces* and *Saccharomyces* non-*cerevisiae* on two important parameters in wine: ethanol and glycerol. The influence they have in sequential, co-fermentations, and solo fermentations on ethanol and glycerol content is examined. This review highlights the need for further studies concerning inoculum rates, aeration techniques (amount and flow rate), and the length of time before *Saccharomyces cerevisiae* sequential inoculation occurs. Challenges include the application of such sequential inoculations in commercial wineries during harvest time.

Keywords: non-*Saccharomyces*; *Saccharomyces* non-*cerevisiae*; yeast; wine; ethanol; glycerol

---

1. Introduction

After carbon dioxide (CO_2), ethanol and glycerol are the most abundant compounds produced during alcoholic fermentation. The levels of ethanol and glycerol in wine depend upon many factors, such as seasonal events affecting the concentration of grape sugar, and winemaking decisions, including fermentation conditions and fermenting yeasts [1].

At a commercial scale, inoculations with *Saccharomyces cerevisiae* strains are often preferred over those with non-*Saccharomyces* or *S.* non-*cerevisiae* yeasts, because the latter is considered responsible for incomplete fermentations (and consequently high levels of residual sugar in wine), and they produce high concentrations of acetic acid and ethyl acetate [2,3]. Nevertheless, non-*Saccharomyces* or *S.* non-*cerevisiae* yeasts are important to winemakers, particularly to those who target wines with unique sensory characters that are popularly recognised as typical of their geographical origin or variety [4–7]. These yeasts are also popular among winemakers who choose to produce less alcoholic wines [8].

Although non-*Saccharomyces* and *S.* non-*cerevisiae* yeasts are sought after for their specific oenological characteristics, it is a challenge for some of these yeasts to conduct a complete fermentation to a desired level of dryness. This is very important to winemakers, in part because finished wines with higher levels of residual sugars above 0.5 g/L require high doses of sulfur dioxide (SO_2) to ensure their microbial stability to prevent wine spoilage. Therefore, inoculations of non-*Saccharomyces/S. non-cerevisiae* in mixed cultures with *S. cerevisiae* strains, which have higher fermentation rates, have been studied to ensure complete fermentation.

The strategy of using selected mixed cultures for alcoholic fermentation is believed to be the key to produce wines with desirable characteristics that meet changing market demands with less ethanol but still with flavours comparable to standard wines [9]. This strategy is carried out by two different methods of inoculation: (1) co-inoculation, which involves concurrent inoculations of non-*Saccharomyces/S. non-cerevisiae* yeasts at high cell concentration (e.g., 10^7 cell/mL) with *S. cerevisiae*; and/or (2) sequential inoculation, which involves inoculating non-*Saccharomyces/S. non-cerevisiae* yeasts to start the fermentation and continue for a determined amount of time alone, and inoculating *S. cerevisiae* to take over and complete the fermentation [9,10]. The time period before carrying out the sequential *S. cerevisiae* inoculation and the *Saccharomyces*/non-*Saccharomyces* or *Saccharomyces* non-*cerevisiae* inoculum ratio are both important parameters that affect the fermentation kinetics and oenological outcomes, and the former generally varies between 1 and 3 days [11–14].

Many reviews have studied different perspectives of non-*Saccharomyces/S. non-cerevisiae* yeasts for modern winemaking practices, including their influence on different wine quality parameters with an emphasis on traits such as the primary (or varietal) and secondary (or fermentative) aromas of wines, acidity, freshness, as well as specific styles of wines (such as traditional method sparkling wines and red table wines) [3,9,15–21].

The aim of this review is to highlight those studies that have shown a direct link between the use of non-*Saccharomyces* or *S. non-cerevisiae* and the concentration of ethanol and glycerol in wine or synthetic media. The first part of this review provides an overview of ethanol and glycerol as contributors to wine sensory characteristics, and a general overview of non-*S. cerevisiae* or *S. non-cerevisiae* yeasts. The second part of this review provides more specific details of individual non-*S. cerevisiae* or *S. non-cerevisiae* species that are relevant to the wine industry. We conclude this review by suggesting what additional research might help winemakers have greater control over wine quality outcomes.

2. Ethanol Reduction

Ethanol is produced by yeast during the alcoholic fermentation and is generally found in the range of 11.5–15% *v/v* in wines. It is an important wine component that directly effects organoleptic properties, aging, and wine stability [22]. The impact of ethanol on the sensory profile of wines and other alcoholic beverages has been recently reviewed [23]. Ethanol influences taste and mouthfeel sensations, alters the sensation of sweetness, increases bitterness, decreases sourness, and contributes to the hotness sensation and body of the wine [24–27]. Ethanol can also decrease the volatility of aroma compounds by increasing their solubility in the wine [28], making small compounds such as fruity-driven ethyl esters and acetates less recognisable by human senses.

According to the International Organisation of Vine and Wine (OIV), the alcohol strength of wines must be a minimum of 8.5% *v/v*, although in cool climate wine regions, this value can be lowered to 7% *v/v* [29]. Over the past two decades, ethanol content in wines has been noticeably increasing in some regions by 0.1–1% per year [30,31]. Apart from hotter climates leading to higher sugar berry levels at harvest and therefore, higher alcohol contents in wine [32], one of the main reasons behind this progressive increase is consumer demand for specific wine styles, which are described as rich, well-structured, with a flavour profile dominated by dark, ripe fruits [33]. This style requires optimal grape maturity and higher sugar content of 240 g/L or more [34].

Nonetheless, an increasing trend for reduced alcohol in beverages (broadly defined as containing 9% to 13% *v/v* ethanol), and low-alcohol (0.5–2% *v/v*) wines by consumers has been recently observed [35,36]. Increasing health and safety consciousness and global initiatives towards moderating alcohol consumption are reasons for producing lower alcohol wines that appeal to wine drinkers [37].

Wines may also be subjected to higher taxation depending on their alcohol content, which increases the final cost of wines to the consumer [38]. Since ethanol is the main source of caloric content in wine, there is also a risk of a negative impact on wine export to countries where health labeling of foods and beverages served at restaurants is voluntary or mandatory [39].

During winemaking, high sugar and therefore ethanol can cause sluggish and stuck alcoholic fermentations and can be challenging for successful malolactic fermentations [40,41]. As a means to address these issues, methods have been studied that include lowering the final ethanol content of wine using a wide selection of interventions. These can be grouped into (1) pre- (e.g., viticultural, juice dilution, and fermentation of early harvest fruit); (2) concurrent (e.g., non-*Saccharomyces* yeasts, modified yeasts, and arrested fermentation) [42,43]; and (3) post-fermentation (e.g., non-membrane and membrane ethanol removal) [43–45]. The use of microbiological approaches such as inoculation with non-*Saccharomyces* yeasts for producing wines with less ethanol is a promising alternative to the removal of ethanol by membrane based-approaches [11,46–48]. The advantages associated with the use of low-ethanol/high-glycerol yielding yeasts include their relatively easy application and lower costs when compared to more expensive and less eco-friendly approaches, such as membrane contactors, nanofiltration, or the spinning cone column [49,50]. However, it is important to acknowledge that low-ethanol/high-glycerol yielding yeasts are much less effective than the membrane-based processes in terms of ethanol reductions [44], and they are perhaps only suitable when winemakers want to achieve a reduction in ethanol content by up to 3.0% *v/v* [49].

Inoculations with non-*Saccharomyces* yeasts can be used as a strategy to produce lower alcohol wines due to the yeasts' different sugar utilisation pathways, including respiration, alcoholic fermentation, and glycerol–pyruvic metabolisms, and different regulatory mechanisms, in comparison to *S. cerevisiae* [51]. While the theoretical sugar-to-ethanol yield for a complete fermentation by *S. cerevisiae* generally ranges from 90% to 95%, the residual sugar is consumed via alternative metabolic pathways and biomass biosynthesis [52]. On the other hand, ethanol yield and the by-products formed vary immensely amongst non-*Saccharomyces* yeasts [23]. For example, due to the Crabtree effect, *S. cerevisiae* prefers fermentation metabolism rather than respiration when the sugar amount exceeds 10 g/L [40]. In contrast, among non-*Saccharomyces* yeasts, there are strains and species that can consume sugar with aerobic respiration regardless of sugar concentration [53,54] without contributing significantly to the final ethanol level of the wine. Therefore, non-*Saccharomyces* yeasts have been studied under partial and controlled aeration strategies during fermentation to achieve lower ethanol by allowing part of the sugar to be consumed via respiration rather than alcoholic fermentation [47,55]. However, an increase in undesirable volatile compounds, such as acetic acid and ethyl acetate, are the main limiting factors of the aeration strategies that require the application of a proper aeration regime [55–57].

As a response to interest in reduced alcohol levels in wines, researchers studied non-*S. cerevisiae* and *S.* non-*cerevisiae* yeast species [49,56,58–60] (Table 1). Several non-*Saccharomyces* yeast strains were identified, including *Metschnikowia pulcherrima*, and two species of *Kluyveromyces*, which have the capacity to decrease ethanol yields by respiration [59]. *M. pulcherrima* AWRI 1149 was identified as a potential yeast to produce wine with a reduced ethanol concentration, having been identified following the evaluation of 50 non-*Saccharomyces* isolates under limited aeration conditions, and in sequential inoculations with *S. cerevisiae* [49]. A similar study with 48 non-*Saccharomyces* yeast isolates identified *Torulaspora delbrueckii* AWRI 1152 and *Zygosaccharomyces bailii* AWRI 1578 yeasts as suitable for reducing ethanol [56]. More recently, the respiratory, fermentative, and physiological characteristics of 114 non-*Saccharomyces* yeasts were evaluated [60]. Taking into account their ability to reduce ethanol content *Hanseniaspora uvarum* BHu9 and BHu11, *Hanseniaspora osmophila* BHo51, *Starmerella bacillaris* (synonym. *Candida zemplinina*) BSb55, and *Candida membranaefaciens* BCm71 were selected as candidates for co-fermentations.

3. Glycerol

Glycerol is the most abundant yeast metabolism by-product after ethanol and CO_2. It is produced from dihydrodroxyacetone phosphate, which is first reduced to glycerol-3-phosphate via glycerol-3-phosphate dehydrogenase (GPDH), and then converted into glycerol by a specific phosphatase. This is a non-volatile 3-hydroxy alcohol, which is a polyol also known as a sugar alcohol. Glycerol is a viscous liquid at room temperature and appears to contribute to mouthfeel and viscosity at, or above 28 g/L and sweetness in the range of 5–12 g/L [42,61–63]. However, Nieuwoudt, et al. [64] did not find any link between wine quality (expressed as the number of medals received at a wine competition) and concentrations of glycerol in wine, and Goold et al. [42] concluded in their review that glycerol had only a minimal influence on the viscosity of wine.

The synthesis of glycerol and acetic acid, in addition to ethanol, are both linked to redox balance [22,42]. The significance of glycerol synthesis to redox balance has been suggested to be due to the inability of mutants (unable to synthesise glycerol) to grow in anaerobic conditions [22,65]. During the stationary phase of yeast during fermentation, glycerol synthesis has been found to be associated with redox balance by removing excess reducing power [22,66].

Many factors can influence the production of glycerol, which is in general more abundant in wines fermented with non-*Saccharomyces* than those fermented with *S. cerevisiae* [67–69], and in red wine (approximately 10.5 g/L) compared to white wines (approximately 7 g/L) [64]. Glycerol is generally more abundant in red wines in part because red juice typically ferments at higher temperatures (20–25 °C) than white wines (<20 °C). Yet, fermenting temperatures positively influence the production of glycerol by yeasts, and non-*Saccharomyces* and *S. non-cerevisiae* are not an exception. For example, increasing the fermentation temperature from 16 to 20 °C increased the glycerol content from 1.69 to 3.04 g/L in co-fermentations of *Candida stellata* and *S. cerevisiae* [52]. A significant increase in the glycerol content of a grape juice was also observed after increasing the temperature from 12 to 25 °C, with fermentations carried out by pure *Saccharomyces paradoxus* reporting an increment of approximately 2.5 g/L, for example [12]. The sugar level of grapes at harvest (and therefore in the juice) also influences the production of glycerol, because this compound is accumulated by yeast to combat dehydration by balancing the intracellular osmolarity with that of the medium [1]. This effect has become even more evident in recent times due to hotter seasonal temperatures compressing the ripening windows of different grape varieties in warm climates. This has meant that winemakers delay the harvest date because of wineries operating at full capacity, thereby causing a part of the crop to overripe in the vineyard [70]. Hranilovic, et al. [71] reported that the glycerol content of a Shiraz wine from early harvest grapes (approximately 265 g/L TSS) was much higher than those from the late harvest (approximately 325 g/L TSS), with a *M. pulcherrima* strain (followed by a *S. cerevisiae* inoculum), producing early and late harvest wines with 10.51 and 12.59 g/L glycerol, respectively. Juice with high sugar concentration also leads to an excess in acetic acid, which can be explained by yeasts trying to maintain redox balances by using surplus $NAD(P)^+$ accumulated during the synthesis of fermentation metabolites [62,72–74].

The growth of non-*Saccharomyces* yeast species such as *Lachancea thermotolerans*, *T. delbrueckii*, and *M. pulcherrima* strictly depends upon oxygen availability [57,75,76]. When the effect of oxygen availability on glycerol production by non-*Saccharomyces* was evaluated, oxygenation at three dissolved oxygen levels of 0.08, 0.41, and 1.71 mg/L resulted in glycerol reduction [76]. This was evident for a *T. delbrueckii* strain in co-fermentation with *S. cerevisiae*, with the glycerol content decreasing from 6.79 g/L in the *T. delbrueckii* anaerobic treatment up to 1.09 g/L in the co-inoculation treatment with 1.71 mg/L of dissolved oxygen. Different results were reported by Morales et al. [57], who observed increased glycerol yields for *M. pulcherrima/S. cerevisiae* mixed cultures under controlled oxygenation conditions (sparged with pure air, nitrogen, or mixtures of both) during the first 48 h of fermentation, and anaerobic conditions thereafter.

The non-*Saccharomyces* yeasts, which have the capacity to redirect the sugar consumption for the production of alternative compounds, rather than ethanol, have been studied in wines with reduced ethanol content. These alternative compounds could be glycerol and pyruvic acid produced via glycerol–pyruvic metabolisms. Alternatively, before sugar is utilised during alcoholic fermentation, sugars can be consumed via respiratory metabolism [40], which is the case with various non-*Saccharomyces* yeasts with lower Crabtree effect.

Other factors can have an impact on the formation of glycerol by non-*Saccharomyces*. These include the concentration of nitrogen and sulfites [8]. Limited nitrogen concentrations in the must (in the form of amino acids and ammonium) can lead to a significant increase of glycerol production. By contrast, higher levels of sulfur dioxide lead to higher levels of glycerol [30]. Increased glycerol production has been found to be linked to increased acetic acid, which is easily detected due to its vinegar smell [3,18,77].

Upon a thorough literature review, we identified five non-*Saccharomyces* yeasts that have been widely studied due to their different oenological traits; *Schizosaccharomyces pombe, Metschnikowia pulcherrima, Lachancea thermotolerans, Candida stellata,* and *Torulaspora delbrueckii*. These five non-*Saccharomyces* yeasts have different pathways that result in ethanol reduction and glycerol production while influencing different parameters of the chemical composition of wines. The following sections of this review present current knowledge regarding the effect of alternative yeasts that influence ethanol and glycerol concentrations while highlighting gaps in our knowledge that require further research.

4. Schizosaccharomyces Pombe

Schizosaccharomyces pombe is a widely studied yeast due to its particular ability to moderate wine acidity via malic acid degradation [78]. Among its other promising traits is the ability to enhance the color of red wines and reduce Ochratoxin A, biogenic amines, and ethyl carbamate [79–81]. The most detrimental metabolites produced by spoilage yeasts in pure culture or spoiled juices have been found to decease in mixed fermentations carried out at the laboratory scale [2,82]. The main characteristics of *S. pombe* and its application in winemaking were reviewed recently [83,84]. *S. pombe* is commercially available as an alternative method to de-acidity wine [19].

Unlike some non-*Saccharomyces* yeast species, *S. pombe* is capable of fermenting wines up to comparable concentrations of *Saccharomyces,* in the range of 10–15% *v/v* ethanol, depending on the strain and presence of aeration [85]. Malo-alcoholic fermentation [86] and the glycerol–pyruvic pathway [87] observed in *S. pombe* inoculations have different impacts on the final ethanol content of the wines. Therefore, while some studies reported an ethanol reduction with inoculations involving *S. pombe* [87], others have reported no difference or even an increase [88–91].

The use of *S. pombe* (strain 938) for white winemaking was investigated by Benito et al. [87], with sole, mixed, and sequential fermentations in conjunction with *S. cerevisiae (Cru Blanc)*. All the strains in the study were able to ferment the wines to dryness. The sole fermentation of *S. pombe* showed 0.65% *v/v* lower ethanol compared to *S. cerevisiae* alone, values being 13.18% *v/v* and 12.53% *v/v*, respectively. Similar results were obtained with the sequential inoculation, with 0.4% *v/v* lower ethanol value in comparison to the control treatment [87]. Loira et al. [88] studied the effect on Syrah wine sensory quality of *S. pombe* strains (938, V1 and 4.2) in mixed and sequential fermentations with *S. cerevisiae* strain 7VA. *S. pombe* was not used as a sole inoculum in this study. Ethanol reduction did not occur in the mixed or sequential fermentations with *S. pombe* and *S. cerevisiae* compared to treatments with a sole inoculum of *S. cerevisiae*. In this study, slightly higher ethanol values that were not statistically significant were reported for mixed and sequential fermentations with *S. pombe* strains (13.2% to 13.5% *v/v*) compared to *S. cerevisiae* (13.2% *v/v*) [88]. Although the same strains of *S. pombe* (938) were used in these studies, different delay times were applied for the sequential inoculation of *S. cerevisiae* (48 h versus 7 days), and different strains of *S. cerevisiae* were used for the comparison.

Increased ethanol levels have been reported in studies that used *S. pombe* where juices had high malic acid content. *S. pombe* (Y0119) was used in a sequential inoculation with *S. cerevisiae* (NT116) to ferment Kei-apple (*Dovyalis caffra L.*) juice, which contained high malic acid (45 g/L). An increase in ethanol level (6.08% *v/v*) was reported compared to sole inoculation with *S. cerevisiae* (4.67% *v/v*) [91]. Similarly, slightly higher ethanol values (approximately 0.2% to 0.5% *v/v*) were reported in a study where *S. pombe* was used in sequential inoculation with *S. cerevisiae* for fermenting plum juice (*Prunus domestica* L.), compared to *S. cerevisiae* as a control [90]. Studies that included *S. pombe* that measured glycerol reported that the yeast produced lower levels of glycerol than the other yeasts in the study (Table 2) [20].

5. Metschnikowia Pulcherrima

Metschnikowia pulcherrima is a non-*Saccharomyces* yeast that is commercially available from many manufacturers [10,19]. Its ability to enhance varietal aroma compounds [50,92] and reduce the ethanol content of wines has raised interest in its commercial use [93,94]. Recently, the impact of *M. pulcherrima* in winemaking has been reviewed [95].

The ability of *M. pulcherrima* for withstanding ethanol concentrations of up to 7% *v/v* has been reported by Combina et al. [96]. Recently, four strains of *M. pulcherrima* were identified as being able to actively grow at a higher ethanol concentration (9% *v/v*) [97]. Sixty-two of the 65 strains isolated from the Douro region of Portugal in this study were able to tolerate a 6% *v/v* ethanol level. On the other hand, its ability to ferment was reported up to levels of 4% *v/v* in micro-fermentations conducted in pasteurised grape must [13]. Consequently, different co-inoculation and sequential inoculation strategies have been studied [14,57,93].

Contreras et al. [49] identified a *M. pulcherrima* strain that can reduce the ethanol content of wine through part of a sequential inoculation with *S. cerevisiae*. Lower ethanol levels of 0.9% and 1.6% *v/v* were achieved in Chardonnay and Shiraz wines, respectively, compared to a sole inoculum of *S. cerevisiae*, which found ethanol levels of 15.1% *v/v* in Chardonnay and 13.8% *v/v* in Shiraz wines. A similar result of 0.9% *v/v* ethanol reduction was reported in a study carried out with the sequential inoculation of *M. pulcherrima* with *S. cerevisiae* compared to a sole inoculum of *S. cerevisiae* that produced ethanol levels of 13.2% *v/v* [93]. Furthermore, a mixed inoculum of *M. pulcherrima* and *S. uvarum* with sequential inoculation of *S. cerevisiae* was conducted. Two different *M. pulcherrima/S. uvarum* inoculum ratios were applied (1×10^6 cells/mL to 1×10^5 cells/mL, and 1×10^6 cells/mL to 1×10^4 cells/mL). Both resulted in an ethanol reduction of 1.7% *v/v*, along with higher concentrations of succinic acid and glycerol, compared to wine fermented with *S. cerevisiae* [93]. Varela et al. [50] studied *M. pulcherrima* and *S. uvarum*, both of which were found to be able to produce wines with reduced ethanol content. Along with the ethanol reduction, the sensory profile and volatile aromatic composition of Merlot wines were studied. Fermentation was conducted via co-inoculation using *M. pulcherrima* (1×10^6 cells/mL) and *S. cerevisiae* (1×10^5 cells/mL). An ethanol reduction of 1.0% *v/v*, along with higher concentrations of ethyl acetate, total esters, total higher alcohols, and total sulfur compounds were detected in wines fermented with a co-inoculation of *M. pulcherrima* and *S. cerevisiae*, compared to wines fermented with *S. cerevisiae*. Although a higher production of ethyl acetate and total sulfur-containing compounds was detected, the sensory panel did not detect associated negative attributes in the wines [50].

The effect of different aeration regimes and immobilisation on the ethanol reduction with selected strains of *M. pulcherrima* has been studied [98]. An ethanol reduction of 1.38% *v/v* was achieved in first 72 h of fermentation of Verdicchio must with *M. pulcherrima* under an aeration flow of 20 mL/L/min, compared to the control that used *S. cerevisiae* [98]. A blend of Malvasia and Viura (Macabeo) must was fermented in a study by Morales et al. [57] using a mixed culture of *M. pulcherrima* and *S. cerevisiae*, and using different aeration regimes (sparged with air or nitrogen). The lowest ethanol values were reported for the treatments sparged with air and fermented with the mixed culture of *M. pulcherrima* and *S. cerevisiae*, regardless of the inoculum level of *S. cerevisiae* (1% or 10%). In these treatments, 11% *v/v* ethanol was obtained, compared to 14.7% *v/v* and 12.9% *v/v* ethanol in the treatments fermented with *S. cerevisiae* sparged with nitrogen and air, respectively. However, high acetic acid values (higher than 0.65 g/L) obtained under air-sparged treatment produced wines that would have been unacceptable for consumers and do not meet market regulations were reported. In contrast, treatments sparged with nitrogen and fermented with the mixed culture of *M. pulcherrima* and *S. cerevisiae* had an ethanol reduction of 0.8% *v/v* compared to those fermented with *S. cerevisiae*, with acceptable levels of acetic acid (lower than 0.1 g/L) [57]. The same level of ethanol reduction was achieved (0.8% *v/v*) using a sequential fermentation with Viura/Macabeo and Malvasia varieties using *M. pulcherrima* with *S. cerevisiae* compared to the control (*S. cerevisiae*) under non-aerated conditions [99].

The ability of selected immobilised non-*Saccharomyces* yeasts (*Starmerella bombicola* (formerly named *Candida stellata*), *Metschnikowia pulcherrima*, *Hanseniaspora osmophila*, and *Hanseniaspora uvarum*) to reduce the ethanol content in wine via sequential fermentation has been also studied [14]. In synthetic grape juice, the sequential inoculation of *M. pulcherrima* had 1.14% to 1.35% *v/v*, with 48 and 72 h delays in inoculation of *S. cerevisiae*, respectively. Ethanol concentration was reduced by 1.10% to 1.46% *v/v* in natural grape juice in the same study [14]. Similarly, Röcker et al. [47] studied five non-*Saccharomyces* yeast strains including *M. pulcherrima* for a sequential inoculation with *S. cerevisiae* var. *bayanus* strain under three different aeration conditions (aeration with sterile pressured air during 15 and 5 days, and under regulated oxygen content of 20% dissolved oxygen), to ferment Riesling must. Ethanol concentration was reduced by up to 3.8% *v/v* within 3 days of aeration, although the authors reported an increase in vinegar (associated with acetic acid) and oxidation sensory attributes [47].

Three non-*Saccharomyces* yeast strains (*M. pulcherrima*, *T. delbrueckii*, and *Zygosaccharomyces bailii*) fermented Chardonnay using sequential inoculation with *S. cerevisiae* under three different aeration conditions (no air addition, 5 mL/min aeration (0.025 VVM), 10 mL/min aeration (0.05 VVM)). The authors reported that the sequential inoculation with *M. pulcherrima* reduced alcohol by 1.6% *v/v*, which was the highest ethanol reduction among all the non-*Saccharomyces* yeast strains in the study, compared to *S. cerevisiae* [43]. In this study, applied aeration conditions did not cause an increase in the acetic acid production. However, in the wines produced with a sequential inoculation of *M. pulcherrima* and *S. cerevisiae*, an over-production of ethyl acetate (280 mg/L) was detected under the aeration regime of 0.05 VVM. This indicates an oxygen threshold for the over-production of this compound [43], which causes unpleasant odors such as nail polish remover and vinegar [100].

The ability of *M. pulcherrima* to reduce the final ethanol content via its respiratory characteristic has been shown with various studies (Table 1). Recently, *M. pulcherrima* has been reported to produce lower levels of glycerol under semi-anaerobic conditions than when the air flow into the fermentations was 1 mL/L/min and 20 mL/L/min (Table 2). Therefore, taken together in respect to ethanol and glycerol content, future research with *M. pulcherrima* (solo, sequential, or co-fermentations) should include a range of aeration strategies, a range of inoculum rates, and an investigation into the length of time before inoculation of *S. cerevisiae* yeast, to render its application feasible at a commercial winemaking scale to achieve ethanol reduction without compromising or enhancing sensory characteristics [14,43,47,59,93,97].

6. Lachancea Thermotolerans

Lachancea thermotolerans (previously *Kluyveromyces thermotolerans*) is available commercially from yeast manufacturers [19,101]. This yeast has specific oenological traits including a positive influence on wines' sensory profile [102] and total acidity [13,103]. The main characteristics of *L. thermotolerans* and its effects on winemaking were recently reviewed [101,104–106].

L. thermotolerans in pure culture was able to reach 10.46% *v/v* ethanol levels in micro-fermentations of pasteurised commercial white wine must, conducted at 25 °C, by leaving more than 50 g/L residual sugar. The control wine inoculated with a pure culture of *S. cerevisiae* EC-1118 fermented the must to dryness (less than 6 g/L residual sugar), reaching an ethanol level of 13.04% *v/v* [102]. The study continued at industrial scale, and an ethanol reduction of 0.7% *v/v* was achieved with a Sangiovese must [102]. In the study by Del Fresno et al. [89], a higher ethanol reduction of 1.2% *v/v* was achieved. This was compared to a different *S. cerevisiae* strain (7VA), which was sequentially inoculated later than the previous study (6 days), with a higher inoculum amount (both at 10^8 cell/mL). The laboratory-scale fermentation using Tempranillo must conducted at 25 °C and 12.66% *v/v* ethanol level used sequential inoculation, while sole inoculation of *S. cerevisiae* 7VA was 13.84% *v/v* [89].

As well as its ability to decrease the pH of wine, the metabolic pathway of *L. thermotolerans* that can turn sugars into lactic acid is also described as a way to reduce the level of alcohol in wines [18]. A study conducted using micro-vinifications of Airén must conducted at 25 °C [80] with *L. thermotolerans* 617 (100 mL containing 2.27×10^7 CFU/mL) followed by the sequential inoculation of *S. cerevisiae* 87 (10^7 CFU/mL of) with 96 h of delay, an ethanol reduction of 0.4% *v/v* was achieved, compared to the sole inoculation of *S. cerevisiae* 87. This sequential inoculation also showed a higher lactic acid production (3.18 g/L) and lower pH (3.52, compared to 3.74 in control wine). The sensory panel perceived this treatment to have better sensorial properties, with higher scores for sweetness, despite similar levels of residual sugar compared to control. This was explained by the possible effect of higher L-lactic acid concentration produced by the effect of *L. thermotolerans*.

Other beneficial compositional effects reported in *L. thermotolerans* treated wines include increases in glycerol at concentrations high enough (>5 g/L) that they could be of sensory relevance. As shown by Kapsopoulou et al. [107] for grape must containing 160 g/L sugar (with 7.4 g/L titratable acidity, at pH 3.5), glycerol formation was significantly higher (5.75 g/L) when *S. cerevisiae* TH941 (5×10^5 cell/mL) was inoculated 3 days after the inoculation of *L. thermotolerans* SCM952 (5×10^5 cell/mL). This was compared to sole *S. cerevisiae* fermentations (4.82 g/L) and other sequential inoculations (after 1 and 2 days). The authors attributed this to the increased survival of the *L. thermotolerans* strain in the co-fermentation treatment in comparison to the other treatments. Likewise, a very high glycerol level (11.22 g/L) was reported in Sangiovese red wines obtained at a commercial scale by the sequential inoculation of *S. cerevisiae* EC1118 (10^6 cell/mL) 2 days after that of *L. thermotolerans* 101 (10^7 cell/mL) [102]. This value was significantly higher compared to the control (9.02 g/L) and the co-inoculated *L. thermotolerans* and *S. cerevisiae* treatments (9.68 g/L). Comitini et al. [13] inoculated non-*Saccharomyces* yeasts in combination with the *S. cerevisiae* EC1118 strain as a starter. All non-*Saccharomyces* strains were inoculated at 10^7 cell/mL, while the EC1118 starter strain was inoculated at three different concentrations: $10^7, 10^5$, and 10^3 cell/mL. Of all the inoculations tested, those of *L. thermotolerans* with *S. cerevisiae* produced higher levels of glycerol (6.95–7.58 g/L) at all three ratios than when *S. cerevisiae* was inoculated on its own (6.23–6.56 g/L). Contrary to the *C. zemplinina* and *M. pulcherrima* inoculations, no significant differences were found between the three different inoculum ratios (1:1, 100:1, 10,000:1) of *L. thermotolerans* with *S. cerevisiae*. This suggests that glycerol production may not correlate with cell concentration and persistence of the *L. thermotolerans* yeast, at least according to this study. Where the effects of sequential and co-inoculations of *L. thermotolerans* with *S. cerevisiae* yeasts on glycerol concentration were evaluated, the glycerol content in sequentially inoculated fermentations (7.55 g/L) was higher than those observed in the co-inoculated treatments (7.18 g/L) [80].

These findings suggest that sequential fermentation with *L. thermotolerans* and *S. cerevisiae* remains a viable option for winemakers. Further investigations on the sensory quality of the wines could help identify if these variations in glycerol are discernable by consumers and preserve the wines from the formation of off-flavours.

7. Candida Stellata/Starmerella Bombicola

Candida stellata is a widely studied yeast due to its positive contributions to wine, including its capacity to produce desirable metabolites such as glycerol [72], and its ability to carry out enzymatic activities that have positive effects on wine sensory attributes. An extensive review of *C. stellata* was carried out by García et al. [108]. Recent studies uncovered the mistake of referring to *C. zemplinina* instead of *C. stellata*, which may explain the disputable characterisations of the oenological traits of these species [109]. For instance, recently, a commonly used strain of *C. stellata*, DBVPG 3827, was reclassified as *Starmella bombicola* [108].

Traditionally dominating overripe, infected, or botrytised grape berries, *C. stellata* is reported to tolerate at least 9% *v/v* ethanol concentration, while at 15 °C, its growth was recorded at 11% *v/v* ethanol concentration, with decreased tolerance at both 10 °C and 30 °C [22,108,110].

Soden et al. [72] studied the effect of inoculations with *C. stellata* and *S. cerevisiae* on Chardonnay juice. The treatments included sole yeast fermentations of *C. stellata* and *S. cerevisiae*, as well as co-inoculation and sequential inoculation conducted at a room temperature of 18 °C. *C. stellata* in sole fermentation was not able to consume all the sugar, reaching an ethanol concentration of only 5.8% *v/v*. Therefore, *S. cerevisiae* was used for the sequential inoculation, and added after the fermentation activity of *C. stellata* had ceased, with an inoculation density of 5×10^6 cells/mL for both yeasts. The resulting wine was dry and had a significantly lower ethanol concentration (11.8% *v/v*) in comparison to the control with a sole inoculation of *S. cerevisiae* (12.5% *v/v*) [72]. The same authors also highlighted in their study the increases in glycerol concentration in the wines fermented with *C. stellata* in comparison to sole *S. cerevisiae* fermentations. The glycerol levels varied from 5.2 g/L in the co-inoculation treatment up to 15.7 g/L in the sequential inoculation trial. Similar results for ethanol reduction were found by Ferraro et al. [111] but using immobilised cells of *C. stellata* and *S. cerevisiae* on Trebbiano Toscano grape must fermented at 20 °C. Inoculum of *S. cerevisiae* at 5×10^6 cells/mL was added after 3 days of fermentation. The sequential fermentation of immobilised cells of *C. stellata* and *S. cerevisiae* reached an ethanol level of 10.6% *v/v* compared to the control (11.24% *v/v*), which was inoculated with *S. cerevisiae* only [111]. The ethanol reduction was explained by the significant increase (approximately 70%) in glycerol as a consequence of low fermentation rate and reduced production of ethanol with immobilised cells of *C. stellata* [112].

Immobilised cells of *Starmerella bombicola* (formerly referred to as *Candida stellata*) were used in sequential inoculation with *S. cerevisiae* on Verdicchio grape must. The fermentation was conducted at 25 °C and the inoculation of *S. cerevisiae* (1×10^6 cell/mL) with 72 h of delay. The effect of the removal of the immobilised *S. bombicola* cells was investigated [14]. The 1.07% and 1.64% *v/v* less ethanol achieved with and without the removal of *S. bombicola* beads, respectively, was notable compared to inoculation with *S. cerevisiae*. A significant increase was seen in the concentration of by-products such as glycerol and succinic acid. The ethanol reduction was explained by the production of by-products from glycerol–pyruvic fermentation or other metabolic pathways [14].

Both growing in similar environmental conditions, including high sugar-containing musts, and possessing similar taxonomic and oenological profile characteristics, *C. zemplinina* (synonym *Starmerella bacillaris* [113]) and *C. stellata* have produced contrasting results in previous literature. Magyar and Tóth [114] evaluated the oenological characteristics of some yeast strains, including four strains of both *C. stellata* and *C. zemplinina*. While both *C. stellata* and *C. zemplinina* showed a strong fructophilic character, *C. stellata* showed higher ethanol and glycerol yield and the same level of volatile acidity compared to *C. zemplinina* [114].

The role of *Candida zemplinina* (synonym *Starmerella bacillaris*) as a tool to produce wines with less ethanol levels but higher glycerol concentrations has been extensively studied and recently reviewed [115]. *Candida* isolates were obtained from Sicilian musts by Di Maio et al. [116] and sequential inoculations with three different *C. zemplinina* strains and *S. cerevisiae* (NDA21) were conducted on Nero d'Avola must. The highest ethanol reduction was 0.3% *v/v* obtained with the sequential inoculation of *C. zemplinina* Cz3 strain, compared to a pure inoculation of *S. cerevisiae* (NDA21), along with higher glycerol content [116]. A similar level of ethanol reduction (0.3% *v/v*) and higher glycerol production was recorded in the study conducted by Rolle et al. [117] on Barbera must. In this study, two different *C. zemplinina* strains were used (FC54 and C.z03). The inoculation of *C. zemplinina* (10^6 cells/mL) was followed by the inoculation of *S. cerevisiae* Uvaferm BC (10^6 cells/mL) after 2 days. Fermentations of natural grape must by sequential inoculations of *C. zemplinina* with *S. cerevisiae* EC1118 increased the level of glycerol (5.45–6.30 g/L in the final wine), and remarkably produced less acetaldehyde and total SO_2 compared to the other yeasts [118].

8. Torulaspora Delbrueckii

Torulaspora delbrueckii was one of the first commercially available non-*Saccharomyces* yeasts and is currently available in dry or frozen form from many yeast manufacturers and suppliers [3,10,19]. Some of the oenological traits of *T. delbrueckii* reported in studies include optimising wine quality parameters i.e., enhancing aroma composition, and positively impacting the foam properties for traditional methods of sparkling wine [119–122]. Studies have been conducted using this yeast on different wine styles, and its effect on the winemaking practices was recently reviewed [123–125].

Significant ethanol reductions due to the utilisation of *T. delbrueckii* in sequential fermentations have been reported when compared to pure inoculations with *S. cerevisiae* (Table 1). Nevertheless, some studies reported slight ethanol reductions (lower than 0.2% *v/v*) [125] or none [88]. Additional, higher ethanol reduction levels were achieved when aeration was integrated during the fermentation process, which stimulates aerobic metabolism [56]. Contreras et al. [56] carried out a sequential inoculation with *T. delbrueckii* AWRI1152, followed by inoculation of *S. cerevisiae* AWRI1631 when 50% of sugar was consumed. Four different aeration regimes were applied (air or nitrogen), and fermentations were conducted at 22 °C with agitation (200 rpm) in a chemically defined grape juice medium. With the aeration at 5 mL/min (0.025 VVM) for the first 24 h of the fermentation, an ethanol reduction of 1.5% *v/v* was achieved, compared to the control of *S. cerevisiae* under anaerobic conditions. No increase in acetic acid levels occurred; however, the impact on the flavour profile of the wine requires further investigation [56]. To answer this question, Canonico et al. [43] studied three non-*Saccharomyces* yeast strains (*M. pulcherrima*, *T. delbrueckii*, and *Zygosaccharomyces bailii*) and their ability to reduce ethanol under limited aeration conditions. In this study, sequential inoculation with *T. delbrueckii* resulted in 0.9% to 1% *v/v* lower ethanol, depending on the aeration strategy (0.025 VVM and 0.05 VVM) compared to control of *S. cerevisiae* under anaerobic conditions. Furthermore, wines fermented with *T. delbrueckii* under aerobic conditions showed a favorable balance between ethanol reduction and volatile profile [43].

Recent studies showed that the nutrient supplementation has a positive correlation on the ethanol yield of *T. delbrueckii* [126,127]. Additionally, Mecca et al. [127] studied three commercially available *T. delbrueckii* strains and compared some oenological characteristics. Significant differences were reported in ethanol yields, as well as in the volatile aroma compounds [127]. Concerning the glycerol content of wines made using *T. delbrueckii*, a range of 4.1–8.9 g/L has been reported in wines (Table 2) [125,128,129]. It has been suggested that *T. delbrueckii* has a more developed glycerol–pyruvic pathway than other yeasts, although some studies have found no differences in glycerol production [91,124,130].

9. Other Non-Saccharomyces and Saccharomyces Non-cerevisiae Yeasts

In addition to the yeasts mentioned in the sections above, new yeasts are constantly being investigated for their ability to ferment wines and their role in uninoculated fermentations. Contreras et al. [56] studied *Zygosaccharomyces bailii* in sequential inoculation under different aeration conditions in chemically defined grape juice medium. With the aeration at 5 mL/min (0.025 VVM) throughout the fermentation, an ethanol reduction of 2% (*v/v*) was achieved, as well as a significantly lower acetic acid compared to the control of *S. cerevisiae* under anaerobic conditions. Using the same *Z. bailii* strain in sequential inoculation, in Chardonnay must, Canonico et al. [43] reported a significant ethanol reduction (1% *v/v*) under aeration (0.05 VVM). In this study, a lower ethanol reduction (0.8% *v/v*) was achieved with a lower aeration (0.025 VVM); however, the resultant wine had a promising volatile profile including individual esters, higher alcohols, and volatile acidity.

Various studies have reported ethanol reduction with non-*Saccharomyces* yeasts from different genus, including *Pichia* and *Hanseniaspora* (Table 1). Maturano et al. [11] studied two non-*Saccharomyces* yeasts, *H. uvarum* and *C. membranaefaciens*, with the sequential inoculation with *S. cerevisiae* on Malbec must. Taking a step further, three fermentation factors—inoculum size, time prior to inoculation of *Saccharomyces cerevisiae*, and fermentation temperature—were optimised using a Box–Behnken experimental design [11]. By applying the optimised factors, the highest ethanol reduction with *H. uvarum* was achieved with an inoculum size of 5×10^6 cells/mL and a delay of 48 h 37 min before *S. cerevisiae* inoculation with the fermentation temperature of 25 °C. However, for *C. membranaefaciens*, the optimised factors were different: an inoculum size of 2.72×10^6 cells/mL, delay of 24 h 15 min before *S. cerevisiae* inoculation, and fermentation temperature of 25 °C was used. The study showed that the time before the inoculation of *S. cerevisiae* affected the ethanol production of the non-*Saccharomyces* yeasts. The high sugar consumption ability of *H. uvarum* via oxidative metabolism was reported to be the reason for this [11].

The application of sequential inoculations with long delays prior to *S. cerevisiae* inoculation in winery environments could be challenging. Competitive native or wild *S. cerevisiae* species present in the winery environment can dominate the fermentation before achieving the expected effect from the inoculated non-*Saccharomyces* yeasts [14]. A recent study by Canonico et al. [43] reported the volatile profile of reduced ethanol wines. Finding an acceptable balance between ethanol reduction, volatile aroma profile, and sensory characteristics of the wines is crucial. Therefore, studies conducted on a pilot scale that includes sensory analysis of the reduced alcohol wines made from non-*Saccharomyces* yeasts is lacking. Nitrogen management is an important factor to achieve ethanol reduction, which has been highlighted by authors in previous studies [126,127,131]. The specific nutrient needs of non-*Saccharomyces* yeasts used for ethanol reduction purposes should be further studied.

Table 1. Ethanol reduction in wines produced from mixed fermentations with non-*Saccharomyces* and *Saccharomyces* non-*cerevisiae* with *S. cerevisiae* yeast.

Grape Variety	Wine Style	Ethanol Reduction % (v/v)	Inoculation	Reference
Schizosaccharomyces pombe				
Airén	White still	0.4	Sequential inoculation with *S. cerevisiae*	[87]
Airén	White still	0.65	Pure inoculation	[87]
Schizosaccharomyces japonicus				
Trebbiano	White still	2.4	Sequential inoculation *S. japonicus* (immobilised)+ *S. cerevisiae*	[132]
Trebbiano	White still	1.7	Co-inoculation *S. japonicus* (immobilised)+ *S. cerevisiae*	[132]
Metschnikowia pulcherrima				
Malvasia/Viura	White still	0.8	Sequential inoculation with *S. cerevisiae* (aeration)	[99]
Chardonnay	White still	0.9	Sequential inoculation with *S. cerevisiae*	[49]
Shiraz	Red still	0.9	Sequential inoculation with *S. cerevisiae*	[93]
Merlot	Red still	1	Co-inoculation with *S. cerevisiae*	[50]
Synthetic grape juice	–	1.1–1.3	Sequential inoculation *M. pulcherrima* (immobilised)+ *S. cerevisiae*	[14]
Verdicchio	White still	1.2–1.6	Sequential inoculation *M. pulcherrima* (immobilised)+ *S. cerevisiae*	[14]
Chardonnay	White still	0.7–1.6	Sequential inoculation (aeration)	[43]
Shiraz	Red still	1.6	Sequential inoculation with *S. cerevisiae*	[49]

Table 1. *Cont.*

Grape Variety	Wine Style	Ethanol Reduction % (*v/v*)	Inoculation	Reference
Malvasia/Viura	White still	3.7	Sequential inoculation with *S. cerevisiae* (aeration)	[57]
Riesling	White still	3.8	Sequential inoculation (aeration)	[47]
Lachancea thermotolerans				
Shiraz	Red still	0.4	Sequential inoculation with *S. cerevisiae*	[71]
Airén	White still	0.4	Sequential inoculation with *S. cerevisiae*	[80]
Sangiovese	Red still	0.7	Sequential inoculation with *S. cerevisiae*	[102]
Tempranillo	Red still	1.2	Sequential inoculation with *S. cerevisiae*	[89]
Candida stellata/Starmerella bombicola				
Trebbiano	White still	0.6	Sequential inoculation *C. stellata* (immobilised) + *S. cerevisiae*	[111]
Chardonnay	White still	0.7	Sequential inoculation with *S. cerevisiae*	[72]
Verdicchio	White still	1.6	Sequential inoculation with *S. bombicola* (immobilised) + *S. cerevisiae*	[14]
Candida zemplinina/Starmerella bacillaris				
Nero d'Avola	Rosé still	0.3	Sequential inoculation with *S. cerevisiae*	[116]
Barbera	Red still	0.3	Sequential inoculation with *S. cerevisiae*	[117]
Barbera	Red still	0.7	Sequential inoculation with *S. cerevisiae*	[133]
Riesling	White still	0.8	Sequential inoculation with *S. cerevisiae* (aeration)	[47]
Torulaspora delbrueckii				
Airén	White still	0.3	Sequential inoculation with *S. cerevisiae*	[134]
Corvina, Rondinella, Corvinone	Red still	0.45	Sequential inoculation with *S. cerevisiae* (aeration)	[135]
Tempranillo	Red still	0.5	Sequential inoculation with *S. cerevisiae*	[130]
Chardonnay	White still	0.9–1.0	Sequential inoculation with *S. cerevisiae* (aeration)	[43]
Chemically defined grape juice medium	–	1.5	Sequential inoculation with *S. cerevisiae* (aeration)	[56]
Malvasia/Viura	White still	0.5	Sequential inoculation with *S. cerevisiae* (aeration)	[99]
Zygosaccharomyces bailii				
Chardonnay	White still	1.0	Sequential inoculation with *S. cerevisiae* (aeration)	[43]
Chemically defined grape juice medium	–	2.0	Sequential inoculation with *S. cerevisiae* (aeration)	[56]
Pichia kluvyeri				
Riesling	White still	0.25	Sequential inoculation with *S. cerevisiae*	[136]
Riesling	White still	3.0	Sequential inoculation with *S. cerevisiae* (aeration)	[47]
Pichia guilliermodii				
Riesling	White still	2.0	Sequential inoculation with *S. cerevisiae* (aeration)	[47]
Hanseniaspora uvarum				
Pinotage	Red still	0.8	Sequential inoculation with *S. cerevisiae*	[137]

Table 1. *Cont.*

Grape Variety	Wine Style	Ethanol Reduction % (*v/v*)	Inoculation	Reference
Synthetic grape juice	–	0.8–1.0	Sequential inoculation *H. osmophila* (immobilised)+ *S. cerevisiae*	[14]
Verdicchio	White still	1.0–1.2	Sequential inoculation *H. osmophila* (immobilised)+ *S. cerevisiae*	[14]
Sauvignon blanc	White still	1.3	Sequential inoculation with *S. cerevisiae*	[137]
Hanseniaspora opuntiae				
Pinotage	Red still	0.6	Sequential inoculation with *S. cerevisiae*	[137]
Sauvignon blanc	White still	1.3	Sequential inoculation with *S. cerevisiae*	[137]
Hanseniaspora osmophila				
Synthetic grape juice	–	0.8–1.3	Sequential inoculation *H. osmophila* (immobilised)+ *S. cerevisiae*	[14]
Verdicchio	White still	1	Sequential inoculation *H. osmophila* (immobilised)+ *S. cerevisiae*	[14]
Saccharomyces uvarum				
Shiraz	Red still	0.8	Sequential inoculation with *S. cerevisiae*	[93]
Merlot	Red still	1.7	Pure inoculation	[50]

Recent studies were conducted with the aim of determining the combinations of two non-*Saccharomyces* yeast species (*S. pombe* and *L. thermotolerans*) [138]. The combination of these yeasts was studied at a laboratory scale using micro-vinifications of Tempranillo grape must [79,89,139,140]. Higher levels of glycerol concentrations were reported in mixed fermentations with *L. thermotolerans* (CONCERTO™) and *S. pombe* (inoculated simultaneously at 10^6 cfu/mL), compared to a control with *S. cerevisiae* [79,139,140]. Glycerol increase varied between 0.27 and 0.71 g/L, which may be related to the strain of *S. pombe* used (V2 and 4.5) and control *S. cerevisiae* (88 and CECT 87). A different strain of *S. pombe* (938) was studied in combination with *L. thermotolerans* (CONCERTO™) [89]. Two inoculum ratios of *S. pombe/L. thermotolerans* (1:1 and 1:3) were used for the laboratory-scale fermentations with Tempranillo must. The authors reported significantly lower glycerol values in mixed fermentation treatments (5.02 g/L and 6.78 g/L) compared to the control fermentation with *S. cerevisiae* 7VA (7.42 g/L) [89]. Therefore, more studies need to be conducted to understand the effect of the interaction between these two non-*Saccharomyces* yeasts.

Combinations of non-*Saccharomyces* and *Saccharomyces* non-*cerevisiae* yeasts for wine fermentations have been investigated, including *M. pulcherrima* and *S. uvarum* [93,141]. *M. pulcherrima* and *S. uvarum* were used in mixed inoculum in the study [93]. Glycerol levels of 12.30 and 12.48 g/L were reported from two different inoculum ratios of *M. pulcherrima/S. uvarum* (1×10^6 cells/mL to 1×10^5 cells/mL, and 1×10^6 cells/mL to 1×10^4 cells/mL, respectively) in laboratory-scale fermentations of Shiraz must. These values were significantly higher compared to the control wine fermented with *S. cerevisiae*, where 7.91 g/L glycerol concentration was reported. Although the highest level of glycerol (14.55 g/L) was reported in sequential inoculation with *S. uvarum* and *S. cerevisiae*, considering the values of ethanol reduction, a mixed inoculum of *M. pulcherrima* and *S. uvarum* could be a promising combination [93]. The same treatments were applied to Chardonnay must in the study by Varela et al. [141]. Similarly, significantly higher glycerol levels were reported in both inoculum ratios of mixed fermentation of *M. pulcherrima/S. uvarum* (11.90 g/L and 12.63 g/L) compared to control wine fermented with *S. cerevisiae* (8.20 g/L). Unlike the study of Contreras et al. [93], the reported values in the mixed fermentations were not significantly different from the sequential inoculation of *S. uvarum* and *S. cerevisiae*, in which a glycerol level of 11.19 g/L was reported [141]. Considering the ethanol decrease and glycerol levels together, the combined inoculation of *M. pulcherrima* and *S. uvarum* shows promising results.

Applying different non-*Saccharomyces* and *Saccharomyces* non-*cerevisiae* yeasts, a wide range of glycerol concentrations has been reported in studies (Table 2) of still and sparkling wines, as well as white and red wines (3.5 g/L to 15.9 g/L). Glycerol levels above 5.2 g/L may directly affect the style of the wine by influencing sensory characteristics including the sweetness, body, and structure of the wines [56]. Therefore, the selection criteria for the use of these yeasts should always consider the sensory characteristics of the final wines. In the case of sparkling wines, glycerol levels have an impact on the viscosity, volatile aroma compounds, and foaming [142]. High levels of glycerol produced by some *Saccharomyces* non-*cerevisiae* yeasts, such as *Saccharomyces kudriavzevii* [143] or *Schizosaccharomyces japonicus* [132] during base wine fermentation, may not make them suitable candidates for achieving a complete second alcoholic fermentation in sparkling winemaking [8]. Additionally, studies have reported an increase of acetic acid when glycerol is produced by some yeasts [8,73].

Table 2. Glycerol concentrations in wines from non-*Saccharomyces* and *Saccharomyces* non-*cerevisiae* yeast studies.

Grape Variety	Wine Style	Glycerol Concentration (g/L)	Method of Detection	Reference
Torulaspora delbrueckii				
Tempranillo	Red still	* 8.6–8.9	Enzymatically with MIURA One oenological analyser	[129]
Chemically defined grape juice medium	N/A	9.3	HPLC	[56]
Tempranillo	Red still	6.7	Y15 enzymatic analyser	[125]
Viura/Macabeo	White still	4.1	Enzymatically with MIURA One oenological analyser	[128]
Hanseniaspora uvarum				
Negromaro	Red still	5.5	HPLC	[144]
Chemically defined grape juice medium	N/A	3.5	HPLC	[56]
Metschnikowia pulcherrima				
Tempranillo	Red still	* 8.2–8.6	Enzymatically with MIURA One oenological analyser	[129]
Viura/Macabeo	White still	4.8	Enzymatically with MIURA One oenological analyser	[128]
Chardonnay	White still	5.5–7.8	HPLC	[43]
Schizosaccharomyces pombe ‡				
Airén	Sparkling wine	4.7	Y15 enzymatic analyser	[145]
Tempranillo	Red sparkling wine	5.0	Y15 enzymatic analyser	[145]
Schizosaccharomyces japonicus				
Trebbiano	White still	15.9	HPLC	[132]
Saccharomycodes ludwigii ‡				
Airén	Sparkling wine	5.0	Y15 enzymatic analyser	[145]
Tempranillo	Red sparkling wine	5.1	Y15 enzymatic analyser	[145]
Saccharomyces uvarum ‡				
Synthetic grape must	N/A	5.2	HPLC	[12]
Cabernet franc	Red wine	10–12	Enzymatically using Megazyme International assay kit	[146]
Lachancea thermotolerans				
Tempranillo	Red still	* 8.2–8.3	Enzymatically with MIURA One oenological analyser	[129]
Viura/Macabeo	White still	4.7	Enzymatically with MIURA One oenological analyser	[128]

Table 2. *Cont.*

Grape Variety	Wine Style	Glycerol Concentration (g/L)	Method of Detection	Reference
		Starmerella bacillaris		
Synthetic grape must	N/A	* 7.7–8.2	HPLC	[128]
		Williopsis pratensis		
Tempranillo	Red still	8.0	Enzymatically with MIURA One oenological analyser	[129]
		Zygosaccharomyces bailii		
Tempranillo	Red still	7.8	Enzymatically with MIURA One oenological analyser	[129]
Viura/Macabeo	White still	5.6	Enzymatically with MIURA One oenological analyser	[128]
		Candida vini		
Tempranillo	Red still	7.9	Enzymatically with MIURA One oenological analyser	[129]

* Two or more products of the same yeast strain used in the study. ‡ *Saccharomyces* non-*cerevisiae* yeast.

10. Conclusions

This review focused on studies concerning non-*Saccharomyces* and *Saccharomyces* non-*cerevisiae* yeasts, and their effect in solo, sequential, and co-fermentation with *S. cerevisiae* or other non-*Saccharomyces* yeasts, on the ethanol and glycerol content of wines. The application of large-scale sequential inoculations of yeast in a commercial scale winery could be challenging during a busy harvest period, especially if the length of time varies between yeast additions. There is a distinct lack of studies concerning the long-term effect of these yeasts on wine from aging on lees and in bottles, as well as the effect of high glycerol levels in sparkling wines on sensory characteristics. Further research should include the inoculum rate of the yeasts, aging ability of the wines on lees, aeration techniques including flow rate, and time before the sequential yeast is added.

Author Contributions: B.K. conceived the idea, supervised the writing, contributed to the glycerol section and produced Table 2. N.N.I. contributed to the introduction and yeast sections, and produced Table 1. R.L. wrote the alcohol section and contributed to other section within the paper. All authors have read and agreed to the published version of the manuscript.

Funding: This research received no external funding.

Acknowledgments: The authors wish to thank Hannah Charnock, Brock University, St. Catharines, Ontario, Canada for designing the graphic abstract that accompanies this article using Biorender Software, (Toronto, Canada).

Conflicts of Interest: The authors declare no conflict of interest.

References

1. Scanes, K.; Hohrnann, S.; Prior, B. Glycerol production by the yeast *Saccharomyces cerevisiae* and its relevance to wine: A review. *S. Afr. J. Enol. Vitic.* **1998**, *19*, 17–24. [CrossRef]
2. Domizio, P.; Romani, C.; Lencioni, L.; Comitini, F.; Gobbi, M.; Mannazzu, I.; Ciani, M. Outlining a future for non-*Saccharomyces* yeasts: Selection of putative spoilage wine strains to be used in association with *Saccharomyces cerevisiae* for grape juice fermentation. *Int. J. Food Microbiol.* **2011**, *147*, 170–180. [CrossRef] [PubMed]
3. Jolly, N.P.; Varela, C.; Pretorius, I.S. Not your ordinary yeast: Non-*Saccharomyces* yeasts in wine production uncovered. *FEMS Yeast Res.* **2014**, *14*, 215–237. [CrossRef] [PubMed]
4. Francesca, N.; Gaglio, R.; Alfonzo, A.; Settanni, L.; Corona, O.; Mazzei, P.; Romano, R.; Piccolo, A.; Moschetti, G. The wine: Typicality or mere diversity? The effect of spontaneous fermentations and biotic factors on the characteristics of wine. *Agric. Agric. Sci. Procedia* **2016**, *8*, 769–773. [CrossRef]

5. Vigentini, I.; Barrera Cardenas, S.; Valdetara, F.; Faccincani, M.; Panont, C.A.; Picozzi, C.; Foschino, R. Use of native yeast strains for in-bottle fermentation to face the uniformity in sparkling wine production. *Front. Microbiol.* **2017**, *8*, 1225. [CrossRef]

6. Ciani, M.; Canonico, L.; Oro, L.; Comitini, F. Footprint of Nonconventional Yeasts and Their Contribution in Alcoholic Fermentations. In *Biotechnological Progress and Beverage Consumption*; Academic Press: Cambridge, MA, USA, 2020; pp. 435–465.

7. Ciani, M.; Capece, A.; Comitini, F.; Canonico, L.; Siesto, G.; Romano, P. Yeast interactions in inoculated wine fermentation. *Front. Microbiol.* **2016**, *7*, 555. [CrossRef]

8. García, M.; Esteve-Zarzoso, B.; Cabellos, J.M.; Arroyo, T. Sequential non-*Saccharomyces* and *Saccharomyces cerevisiae* fermentations to reduce the alcohol content in wine. *Fermentation* **2020**, *6*, 60. [CrossRef]

9. Padilla, B.; Gil, J.V.; Manzanares, P. Past and future of non-*Saccharomyces* yeasts: From spoilage microorganisms to biotechnological tools for improving wine aroma complexity. *Front. Microbiol.* **2016**, *7*, 411. [CrossRef]

10. Petruzzi, L.; Capozzi, V.; Berbegal, C.; Corbo, M.R.; Bevilacqua, A.; Spano, G.; Sinigaglia, M. Microbial resources and enological significance: Opportunities and benefits. *Front. Microbiol.* **2017**, *8*, 995. [CrossRef]

11. Maturano, Y.P.; Mestre, M.V.; Kuchen, B.; Toro, M.E.; Mercado, L.A.; Vazquez, F.; Combina, M. Optimization of fermentation-relevant factors: A strategy to reduce ethanol in red wine by sequential culture of native yeasts. *Int. J. Food Microbiol.* **2019**, *289*, 40–48. [CrossRef]

12. Alonso-del-Real, J.; Lairón-Peris, M.; Barrio, E.; Querol, A. Effect of temperature on the prevalence of *Saccharomyces* non-*cerevisiae* species against a *S. cerevisiae* wine strain in wine fermentation: Competition, physiological fitness, and influence in final wine composition. *Front. Microbiol.* **2017**, *8*, 150. [CrossRef] [PubMed]

13. Comitini, F.; Gobbi, M.; Domizio, P.; Romani, C.; Lencioni, L.; Mannazzu, I.; Ciani, M. Selected non-*Saccharomyces* wine yeasts in controlled multistarter fermentations with *Saccharomyces cerevisiae*. *Food Microbiol.* **2011**, *28*, 873–882. [CrossRef]

14. Canonico, L.; Comitini, F.; Oro, L.; Ciani, M. Sequential fermentation with selected immobilized non-*Saccharomyces* yeast for reduction of ethanol content in wine. *Front. Microbiol.* **2016**, *7*, 278. [CrossRef]

15. Pretorius, I.S. Conducting wine symphonics with the aid of yeast genomics. *Beverages* **2016**, *2*, 36. [CrossRef]

16. Benito, Á.; Calderón, F.; Benito, S. The influence of non-*Saccharomyces* species on wine fermentation quality parameters. *Fermentation* **2019**, *5*, 54. [CrossRef]

17. Pérez-Torrado, R.; Barrio, E.; Querol, A. Alternative yeasts for winemaking: *Saccharomyces* non-*cerevisiae* and its hybrids. *Crit. Rev. Food Sci. Nutr.* **2018**, *58*, 1780–1790. [CrossRef] [PubMed]

18. Vilela, A. Use of nonconventional yeasts for modulating wine acidity. *Fermentation* **2019**, *5*, 27. [CrossRef]

19. Morata, A.; Escott, C.; Bañuelos, M.A.; Loira, I.; del Fresno, J.M.; González, C.; Suárez-Lepe, J.A. Contribution of non-*Saccharomyces* yeasts to wine freshness. A review. *Biomolecules* **2020**, *10*, 34. [CrossRef]

20. Ivit, N.N.; Kemp, B. The impact of non-*Saccharomyces* yeast on traditional method sparkling wine. *Fermentation* **2018**, *4*, 73. [CrossRef]

21. Morata, A.; Escott, C.; Loira, I.; Del Fresno, J.M.; González, C.; Suárez-Lepe, J.A. Influence of *Saccharomyces* and non-*Saccharomyces* yeasts in the formation of pyranoanthocyanins and polymeric pigments during red wine making. *Molecules* **2019**, *24*, 4490. [CrossRef]

22. Jackson, R. Fermentation. In *Wine Science: Principles and Applications (Food Science and Technology)*; Academic Press: Amsterdam, The Netherlands, 2014; Chapter 7, p. 493.

23. Ickes, C.M.; Cadwallader, K.R. Effects of ethanol on flavor perception in alcoholic beverages. *Chemosens. Percept.* **2017**, *10*, 119–134. [CrossRef]

24. King, E.S.; Dunn, R.L.; Heymann, H. The influence of alcohol on the sensory perception of red wines. *Food Qual. Pref.* **2013**, *28*, 235–243. [CrossRef]

25. Nurgel, C.; Pickering, G. Modeling of sweet, bitter and irritant sensations and their interactions elicited by model ice wines. *J. Sens. Stud.* **2006**, *21*, 505–519. [CrossRef]

26. Fischer, U.; Noble, C.A. The effect of ethanol, catechin concentration, and ph on sourness and bitterness of wine. *Am. J. Enol. Vitic.* **1994**, *45*, 6–10.

27. Gawel, R.; Sluyter, S.V.; Waters, E.J. The effects of ethanol and glycerol on the body and other sensory characteristics of Riesling wines. *Aust. J. Grape Wine Res.* **2007**, *13*, 38–45. [CrossRef]

28. Saha, B.; Longo, R.; Torley, P.; Saliba, A.; Schmidtke, L. SPME method optimized by Box-Behnken design for impact odorants in reduced alcohol wines. *Foods* **2018**, *7*, 127. [CrossRef]

29. OIV. *Oiv International Standard for the Labelling of Wines 2015*; OIV: Paris, France, 2015.

30. Godden, P.; Wilkes, E.; Johnson, D. Trends in the composition of Australian wine 1984–2014. *Aust. J. Grape Wine Res.* **2015**, *21*, 741–753. [CrossRef]

31. Alston, J.M.; Fuller, K.B.; Lapsley, J.T.; Soleas, G.; Tumber, K.P. Splendide mendax: False label claims about high and rising alcohol content of wine. *J. Wine Econ.* **2015**, *10*, 275–313. [CrossRef]

32. Van Leeuwen, C.; Darriet, P. The impact of climate change on viticulture and wine quality. *J. Wine Econ.* **2016**, *11*, 150–167. [CrossRef]

33. Williamson, P.O.; Robichaud, J.; Francis, I.L. Comparison of Chinese and Australian consumers' liking responses for red wines. *Aust. J. Grape Wine Res.* **2012**, *18*, 256–267. [CrossRef]

34. Bindon, K.; Varela, C.; Kennedy, J.; Holt, H.; Herderich, M. Relationships between harvest time and wine composition in *Vitis vinifera* L. cv. Cabernet sauvignon 1. Grape and wine chemistry. *Food Chem* **2013**, *138*, 1696–1705. [CrossRef] [PubMed]

35. Jordão, A.; Vilela, A.; Cosme, F. From sugar of grape to alcohol of wine: Sensorial impact of alcohol in wine. *Beverages* **2015**, *1*, 292–310. [CrossRef]

36. Bucher, T.; Deroover, K.; Stockley, C. Low-alcohol wine: A narrative review on consumer perception and behaviour. *Beverages* **2018**, *4*, 82. [CrossRef]

37. Saliba, A.; Ovington, L.; Moran, C. Consumer demand for low-alcohol wine in an Australian sample. *Int. J. Wine Res.* **2013**, *5*, 1–8. [CrossRef]

38. Sharma, A.; Sinha, K.; Vandenberg, B. Pricing as a means of controlling alcohol consumption. *Br. Med Bull.* **2017**, *123*, 149–158. [CrossRef]

39. Annunziata, A.; Pomarici, E.; Vecchio, R.; Mariani, A. Do consumers want more nutritional and health information on wine labels? Insights from the eu and USA. *Nutrients* **2016**, *8*, 416. [CrossRef]

40. Ribéreau-Gayon, P.; Dubourdieu, D.; Donèche, B.; Lonvaud, A. Conditions of Yeast Development. In *Handbook of Enology, Volume 1: The Microbiology of Wine and Vinifications*; John Wiley & Sons: Hoboken, NJ, USA, 2006; pp. 79–114.

41. Zapparoli, G.; Tosi, E.; Azzolini, M.; Vagnoli, P.; Krieger, S. Bacterial inoculation strategies for the achievement of malolactic fermentation in high-alcohol wines. *S. Afr. J. Enol. Vitic.* **2009**, *30*, 49–55. [CrossRef]

42. Goold, H.D.; Kroukamp, H.; Williams, T.C.; Paulsen, I.T.; Varela, C.; Pretorius, I.S. Yeast's balancing act between ethanol and glycerol production in low-alcohol wines. *Microb. Biotechnol.* **2017**, *10*, 264–278. [CrossRef]

43. Canonico, L.; Solomon, M.; Comitini, F.; Ciani, M.; Varela, C. Volatile profile of reduced alcohol wines fermented with selected non-*Saccharomyces* yeasts under different aeration conditions. *Food Microbiol.* **2019**, *84*, 103247. [CrossRef]

44. Longo, R.; Blackman, J.W.; Torley, P.J.; Rogiers, S.Y.; Schmidtke, L.M. Changes in volatile composition and sensory attributes of wines during alcohol content reduction. *J. Sci. Food Agric.* **2017**, *97*, 8–16. [CrossRef]

45. Longo, R.; Blackman, J.W.; Antalick, G.; Torley, P.J.; Rogiers, S.Y.; Schmidtke, L.M. A comparative study of partial dealcoholisation versus early harvest: Effects on wine volatile and sensory profiles. *Food Chem.* **2018**, *261*, 21–29. [CrossRef]

46. Varela, C.; Dry, P.; Kutyna, D.; Francis, I.; Henschke, P.; Curtin, C.; Chambers, P. Strategies for reducing alcohol concentration in wine. *Aust. J. Grape Wine Res.* **2015**, *21*, 670–679. [CrossRef]

47. Röcker, J.; Strub, S.; Ebert, K.; Grossmann, M. Usage of different aerobic non-*Saccharomyces* yeasts and experimental conditions as a tool for reducing the potential ethanol content in wines. *Eur. Food Res. Technol.* **2016**, *242*, 2051–2070. [CrossRef]

48. Ciani, M.; Morales, P.; Comitini, F.; Tronchoni, J.; Canonico, L.; Curiel, J.A.; Oro, L.; Rodrigues, A.J.; Gonzalez, R. Non-conventional yeast species for lowering ethanol content of wines. *Front. Microbiol.* **2016**, *7*, 642. [CrossRef] [PubMed]

49. Contreras, A.; Hidalgo, C.; Henschke, P.A.; Chambers, P.J.; Curtin, C.; Varela, C. Evaluation of non-*Saccharomyces* yeasts for the reduction of alcohol content in wine. *Appl. Environ. Microbiol.* **2014**, *80*, 1670–1678. [CrossRef] [PubMed]

50. Varela, C.; Barker, A.; Tran, T.; Borneman, A.; Curtin, C. Sensory profile and volatile aroma composition of reduced alcohol Merlot wines fermented with *Metschnikowia pulcherrima* and *Saccharomyces uvarum*. *Int. J. Food Microbiol.* **2017**, *252*, 1–9. [CrossRef] [PubMed]

51. Ciani, M.; Comitini, F. Use of non-*Saccharomyces* yeasts in red winemaking. In *Red Wine Technology*; Antonio, M., Ed.; Elsevier: Amsterdam, The Netherlands, 2019; pp. 51–68.

52. Ciani, M.; Comitini, F. Influence of temperature and oxygen concentration on the fermentation behaviour of *Candida stellata* in mixed fermentation with *Saccharomyces cerevisiae*. *World J. Microbiol. Biotechnol.* **2006**, *22*, 619–623. [CrossRef]

53. Alexander, M.; Jeffries, T. Respiratory efficiency and metabolite partitioning as regulatory phenomena in yeasts. *Enzym. Microb. Technol.* **1990**, *12*, 2–19. [CrossRef]

54. De Deken, R. The crabtree effect: A regulatory system in yeast. *Microbiology* **1966**, *44*, 149–156. [CrossRef] [PubMed]

55. Canonico, L.; Galli, E.; Ciani, E.; Comitini, F.; Ciani, M. Exploitation of three non-conventional yeast species in the brewing process. *Microorganisms* **2019**, *7*, 11. [CrossRef]

56. Contreras, A.; Hidalgo, C.; Schmidt, S.; Henschke, P.A.; Curtin, C.; Varela, C. The application of non-*Saccharomyces* yeast in fermentations with limited aeration as a strategy for the production of wine with reduced alcohol content. *Int. J. Food Microbiol.* **2015**, *205*, 7–15. [CrossRef] [PubMed]

57. Morales, P.; Rojas, V.; Quirós, M.; Gonzalez, R. The impact of oxygen on the final alcohol content of wine fermented by a mixed starter culture. *Appl. Microbiol. Biotechnol.* **2015**, *99*, 3993–4003. [CrossRef] [PubMed]

58. Gonzalez, R.; Quirós, M.; Morales, P. Yeast respiration of sugars by non-*Saccharomyces* yeast species: A promising and barely explored approach to lowering alcohol content of wines. *Trends Food Sci. Technol.* **2013**, *29*, 55–61. [CrossRef]

59. Quirós, M.; Rojas, V.; Gonzalez, R.; Morales, P. Selection of non-*Saccharomyces* yeast strains for reducing alcohol levels in wine by sugar respiration. *Int. J. Food Microbiol.* **2014**, *181*, 85–91. [CrossRef]

60. Mestre Furlani, M.V.; Maturano, Y.P.; Combina, M.; Mercado, L.A.; Toro, M.E.; Vazquez, F. Selection of non-*Saccharomyces* yeasts to be used in grape musts with high alcoholic potential: A strategy to obtain wines with reduced ethanol content. *FEMS Yeast Res.* **2017**, *17*. [CrossRef]

61. Ballester-Tomás, L.; Prieto, J.A.; Gil, J.V.; Baeza, M.; Randez-Gil, F. The Antarctic yeast *Candida sake*: Understanding cold metabolism impact on wine. *Int. J. Food Microbiol.* **2017**, *245*, 59–65. [CrossRef]

62. Mbuyane, L.L.; de Kock, M.; Bauer, F.F.; Divol, B. *Torulaspora delbrueckii* produces high levels of c5 and c6 polyols during wine fermentations. *FEMS Yeast Res.* **2018**, *18*, foy084. [CrossRef]

63. Puškaš, V.S.; Miljić, U.D.; Djuran, J.J.; Vučurović, V.M. The aptitude of commercial yeast strains for lowering the ethanol content of wine. *Food Sci. Nutr.* **2020**, *8*, 1489–1498. [CrossRef]

64. Nieuwoudt, H.; Prior, B.; Pretorius, L.; Bauer, F. Glycerol in South African table wines: An assessment of its relationship to wine quality. *S. Afr. J. Enol. Vitic.* **2002**, *23*, 22–30. [CrossRef]

65. Nissen, T.L.; Hamann, C.W.; Kielland-Brandt, M.C.; Nielsen, J.; Villadsen, J. Anaerobic and aerobic batch cultivations of *Saccharomyces cerevisiae* mutants impaired in glycerol synthesis. *Yeast* **2000**, *16*, 463–474. [CrossRef]

66. Roustan, J.; Sablayrolles, J. Impact of the addition of electron acceptors on the by-products of alcoholic fermentation. *Enzym. Microb. Technol.* **2002**, *31*, 142–152. [CrossRef]

67. Rantsiou, K.; Dolci, P.; Giacosa, S.; Torchio, F.; Tofalo, R.; Torriani, S.; Suzzi, G.; Rolle, L.; Cocolin, L. *Candida zemplinina* can reduce acetic acid produced by *Saccharomyces cerevisiae* in sweet wine fermentations. *Appl. Environ. Microbiol.* **2012**, *78*, 1987–1994. [CrossRef] [PubMed]

68. Domizio, P.; Liu, Y.; Bisson, L.; Barile, D. Cell wall polysaccharides released during the alcoholic fermentation by *Schizosaccharomyces pombe* and *S. Japonicus*: Quantification and characterization. *Food Microbiol.* **2017**, *61*, 136–149. [CrossRef] [PubMed]

69. Di Gianvito, P.; Perpetuini, G.; Tittarelli, F.; Schirone, M.; Arfelli, G.; Piva, A.; Patrignani, F.; Lanciotti, R.; Olivastri, L.; Suzzi, G. Impact of *Saccharomyces cerevisiae* strains on traditional sparkling wines production. *Food Res. Int.* **2018**, *109*, 552–560. [CrossRef]

70. Longo, R.; Blackman, J.W.; Antalick, G.; Torley, P.J.; Rogiers, S.Y.; Schmidtke, L.M. Volatile and sensory profiling of Shiraz wine in response to alcohol management: Comparison of harvest timing versus technological approaches. *Food Res. Int.* **2018**, *109*, 561–571. [CrossRef]

71. Hranilovic, A.; Li, S.; Boss, P.K.; Bindon, K.; Ristic, R.; Grbin, P.R.; Van der Westhuizen, T.; Jiranek, V. Chemical and sensory profiling of Shiraz wines co-fermented with commercial non-*Saccharomyces* inocula. *Aust. J. Grape Wine Res.* **2018**, *24*, 166–180. [CrossRef]

72. Soden, A.; Francis, I.; Oakey, H.; Henschke, P. Effects of co-fermentation with *Candida stellata* and *Saccharomyces cerevisiae* on the aroma and composition of chardonnay wine. *Aust. J. Grape Wine Res.* **2000**, *6*, 21–30. [CrossRef]

73. Eglinton, J.M.; Heinrich, A.J.; Pollnitz, A.P.; Langridge, P.; Henschke, P.A.; de Barros Lopes, M. Decreasing acetic acid accumulation by a glycerol overproducing strain of *Saccharomyces cerevisiae* by deleting the ald6 aldehyde dehydrogenase gene. *Yeast* **2002**, *19*, 295–301. [CrossRef]

74. Li, H.; Du, G.; Li, H.-L.; Wang, H.-L.; Yan, G.-L.; Zhan, J.-C.; Huang, W.-D. Physiological response of different wine yeasts to hyperosmotic stress. *Am. J. Enol. Vitic.* **2010**, *61*, 529–535. [CrossRef]

75. Pérez-Nevado, F.; Albergaria, H.; Hogg, T.; Girio, F. Cellular death of two non-*Saccharomyces* wine-related yeasts during mixed fermentations with *Saccharomyces cerevisiae*. *Int. J. Food Microbiol.* **2006**, *108*, 336–345.

76. Shekhawat, K.; Bauer, F.F.; Setati, M.E. Impact of oxygenation on the performance of three non-*Saccharomyces* yeasts in co-fermentation with *Saccharomyces cerevisiae*. *Appl. Microbiol. Biotechnol.* **2017**, *101*, 2479–2491. [CrossRef]

77. Prior, B.; Toh, T.; Jolly, N.; Baccari, C.; Mortimer, R. Impact of yeast breeding for elevated glycerol production on fermentative activity and metabolite formation in Chardonnay wine. *S. Afr. J. Enol. Vitic.* **2000**, *21*, 92–99. [CrossRef]

78. Silva, S.; Ramón-Portugal, F.; Andrade, P.; Abreu, S.; de Fatima Texeira, M.; Strehaiano, P. Malic acid consumption by dry immobilized cells of *Schizosaccharomyces pombe*. *Am. J. Enol. Vitic.* **2003**, *54*, 50–55.

79. Benito, Á.; Calderón, F.; Benito, S. The combined use of *Schizosaccharomyces pombe* and *Lachancea thermotolerans*—Effect on the anthocyanin wine composition. *Molecules* **2017**, *22*, 739. [CrossRef] [PubMed]

80. Benito, Á.; Calderón, F.; Palomero, F.; Benito, S. Quality and composition of Airén wines fermented by sequential inoculation of *Lachancea thermotolerans* and *Saccharomyces cerevisiae*. *Food Technol. Biotechnol.* **2016**, *54*, 135–144. [CrossRef] [PubMed]

81. Cecchini, F.; Morassut, M.; Moruno, E.G.; Di Stefano, R. Influence of yeast strain on Ochratoxin A content during fermentation of white and red must. *Food Microbiol.* **2006**, *23*, 411–417. [CrossRef]

82. Escott, C.; Del Fresno, J.M.; Loira, I.; Morata, A.; Suárez-Lepe, J.A. *Zygosaccharomyces rouxii*: Control strategies and applications in food and winemaking. *Fermentation* **2018**, *4*, 69. [CrossRef]

83. Loira, I.; Morata, A.; Palomero, F.; González, C.; Suárez-Lepe, J.A. *Schizosaccharomyces pombe*: A promising biotechnology for modulating wine composition. *Fermentation* **2018**, *4*, 70. [CrossRef]

84. Benito, S. The impacts of *Schizosaccharomyces* on winemaking. *Appl. Microbiol. Biotechnol.* **2019**, *103*, 4291–4312. [CrossRef]

85. Suárez-Lepe, J.; Palomero, F.; Benito, S.; Calderón, F.; Morata, A. Oenological versatility of *Schizosaccharomyces* spp. *Eur. Food Res. Technol.* **2012**, *235*, 375–383. [CrossRef]

86. Palomero, F.; Morata, A.; Benito, S.; Calderón, F.; Suárez-Lepe, J. New genera of yeasts for over-lees aging of red wine. *Food Chem.* **2009**, *112*, 432–441. [CrossRef]

87. Benito, S.; Palomero, F.; Morata, A.; Calderon, F.; Palmero, D.; Suárez-Lepe, J. Physiological features of *Schizosaccharomyces pombe* of interest in making of white wines. *Eur. Food Res. Technol.* **2013**, *236*, 29–36. [CrossRef]

88. Loira, I.; Morata, A.; Comuzzo, P.; Callejo, M.J.; González, C.; Calderón, F.; Suárez-Lepe, J.A. Use of *Schizosaccharomyces pombe* and *Torulaspora delbrueckii* strains in mixed and sequential fermentations to improve red wine sensory quality. *Food Res. Int.* **2015**, *76*, 325–333. [CrossRef]

89. Del Fresno, J.M.; Morata, A.; Loira, I.; Bañuelos, M.A.; Escott, C.; Benito, S.; Chamorro, C.G.; Suárez-Lepe, J.A. Use of non-*Saccharomyces* in single-culture, mixed and sequential fermentation to improve red wine quality. *Eur. Food Res. Technol.* **2017**, *243*, 2175–2185. [CrossRef]

90. Miljić, U.; Puškaš, V.; Vučurović, V.; Muzalevski, A. Fermentation characteristics and aromatic profile of plum wines produced with indigenous microbiota and pure cultures of selected yeast. *J. Food Sci.* **2017**, *82*, 1443–1450. [CrossRef]

91. Minnaar, P.; Jolly, N.; Paulsen, V.; Du Plessis, H.; Van Der Rijst, M. *Schizosaccharomyces pombe* and *Saccharomyces cerevisiae* yeasts in sequential fermentations: Effect on phenolic acids of fermented kei-apple (*Dovyalis caffra* L.) juice. *Int. J. Food Microbiol.* **2017**, *257*, 232–237. [CrossRef]

92. Sadoudi, M.; Tourdot-Maréchal, R.; Rousseaux, S.; Steyer, D.; Gallardo-Chacón, J.-J.; Ballester, J.; Vichi, S.; Guérin-Schneider, R.; Caixach, J.; Alexandre, H. Yeast–yeast interactions revealed by aromatic profile analysis of sauvignon blanc wine fermented by single or co-culture of non-*Saccharomyces* and *Saccharomyces* yeasts. *Food Microbiol.* **2012**, *32*, 243–253. [CrossRef]

93. Contreras, A.; Curtin, C.; Varela, C. Yeast population dynamics reveal a potential collaboration between *Metschnikowia pulcherrima* and *Saccharomyces uvarum* for the production of reduced alcohol wines during Shiraz fermentation. *Appl. Microbiol. Biotechnol.* **2015**, *99*, 1885–1895. [CrossRef]

94. Duarte, F.L.; Egipto, R.; Baleiras-Couto, M.M. Mixed fermentation with *Metschnikowia pulcherrima* using different grape varieties. *Fermentation* **2019**, *5*, 59. [CrossRef]

95. Morata, A.; Loira, I.; Escott, C.; del Fresno, J.M.; Bañuelos, M.A.; Suárez-Lepe, J.A. Applications of *Metschnikowia pulcherrima* in wine biotechnology. *Fermentation* **2019**, *5*, 63. [CrossRef]

96. Combina, M.; Elía, A.; Mercado, L.; Catania, C.; Ganga, A.; Martinez, C. Dynamics of indigenous yeast populations during spontaneous fermentation of wines from Mendoza, Argentina. *Int. J. Food Microbiol.* **2005**, *99*, 237–243. [CrossRef] [PubMed]

97. Barbosa, C.; Lage, P.; Esteves, M.; Chambel, L.; Mendes-Faia, A.; Mendes-Ferreira, A. Molecular and phenotypic characterization of *Metschnikowia pulcherrima* strains from douro wine region. *Fermentation* **2018**, *4*, 8. [CrossRef]

98. Canonico, L.; Comitini, F.; Ciani, M. *Metschnikowia pulcherrima* Selected Strain for Ethanol Reduction in Wine: Influence of Cell Immobilization and Aeration Condition. *Foods* **2019**, *8*, 378. [CrossRef]

99. Tronchoni, J.; Curiel, J.A.; Sáenz-Navajas, M.P.; Morales, P.; De-La-Fuente-Blanco, A.; Fernández-Zurbano, P.; Ferreira, V.; Gonzalez, R. Aroma profiling of an aerated fermentation of natural grape must with selected yeast strains at pilot scale. *Food Microbiol.* **2018**, *70*, 214–223. [CrossRef] [PubMed]

100. Ribéreau-Gayon, P.; Glories, Y.; Maujean, A.; Dubourdieu, D. Alcohols and Other Volatile Compounds. In *Handbook of Enology, Volume 2: The Chemistry of Wine Stabilization and Treatments*, 2nd ed.; John Wiley and Sons Ltd.: Hoboken, NJ, USA, 2006; pp. 51–64.

101. Benito, S. The impacts of *Lachancea thermotolerans* yeast strains on winemaking. *Appl. Microbiol. Biotechnol.* **2018**, *102*, 6775–6790. [CrossRef] [PubMed]

102. Gobbi, M.; Comitini, F.; Domizio, P.; Romani, C.; Lencioni, L.; Mannazzu, I.; Ciani, M. *Lachancea thermotolerans* and *Saccharomyces cerevisiae* in simultaneous and sequential co-fermentation: A strategy to enhance acidity and improve the overall quality of wine. *Food Microbiol.* **2013**, *33*, 271–281. [CrossRef] [PubMed]

103. Balikci, E.K.; Tanguler, H.; Jolly, N.P.; Erten, H. Influence of *Lachancea thermotolerans* on cv. Emir wine fermentation. *Yeast* **2016**, *33*, 313–321. [CrossRef]

104. Morata, A.; Loira, I.; Tesfaye, W.; Bañuelos, M.A.; González, C.; Suárez Lepe, J.A. *Lachancea thermotolerans* applications in wine technology. *Fermentation* **2018**, *4*, 53. [CrossRef]

105. Porter, T.J.; Divol, B.; Setati, M.E. *Lachancea* yeast species: Origin, biochemical characteristics and oenological significance. *Food Res. Int.* **2019**, *119*, 378–389. [CrossRef]

106. Vilela, A. *Lachancea thermotolerans*, the Non-*Saccharomyces* Yeast that Reduces the Volatile Acidity of Wines. *Fermentation* **2018**, *4*, 56. [CrossRef]

107. Kapsopoulou, K.; Mourtzini, A.; Anthoulas, M.; Nerantzis, E. Biological acidification during grape must fermentation using mixed cultures of *Kluyveromyces thermotolerans* and *Saccharomyces cerevisiae*. *World J. Microbiol. Biotechnol.* **2007**, *23*, 735–739. [CrossRef]

108. García, M.; Esteve-Zarzoso, B.; Cabellos, J.M.; Arroyo, T. Advances in the study of *Candida stellata*. *Fermentation* **2018**, *4*, 74. [CrossRef]

109. Csoma, H.; Sipiczki, M. Taxonomic reclassification of *Candida stellata* strains reveals frequent occurrence of *Candida zemplinina* in wine fermentation. *FEMS Yeast Res.* **2008**, *8*, 328–336. [CrossRef]

110. Gao, C.; Fleet, G. The effects of temperature and pH on the ethanol tolerance of the wine yeasts, *Saccharomyces cerevisiae, Candida stellata* and *Kloeckera apiculata*. *J. Appl. Bacteriol.* **1988**, *65*, 405–409. [CrossRef]

111. Ferraro, L.; Fatichenti, F.; Ciani, M. Pilot scale vinification process using immobilized *Candida stellata* cells and *Saccharomyces cerevisiae*. *Process Biochem.* **2000**, *35*, 1125–1129. [CrossRef]

112. Ciani, M.; Ferraro, L. Enhanced glycerol content in wines made with immobilized *Candida stellata* cells. *Appl. Environ. Microbiol.* **1996**, *62*, 128–132. [CrossRef]

113. Wang, C.; Esteve-Zarzoso, B.; Cocolin, L.; Mas, A.; Rantsiou, K. Viable and culturable populations of *Saccharomyces cerevisiae, Hanseniaspora uvarum* and *Starmerella bacillaris* (synonym *Candida zemplinina*) during Barbera must fermentation. *Food Res. Int.* **2015**, *78*, 195–200. [CrossRef]

114. Magyar, I.; Tóth, T. Comparative evaluation of some oenological properties in wine strains of *Candida stellata, Candida zemplinina, Saccharomyces uvarum* and *Saccharomyces cerevisiae*. *Food Microbiol.* **2011**, *28*, 94–100. [CrossRef]

115. Englezos, V.; Giacosa, S.; Rantsiou, K.; Rolle, L.; Cocolin, L. *Starmerella bacillaris* in winemaking: Opportunities and risks. *Curr. Opin. Food Sci.* **2017**, *17*, 30–35. [CrossRef]

116. Di Maio, S.; Genna, G.; Gandolfo, V.; Amore, G.; Ciaccio, M.; Oliva, D. Presence of *Candida zemplinina* in Sicilian musts and selection of a strain for wine mixed fermentations. *S. Afr. J. Enol. Vitic.* **2012**, *33*, 80–87. [CrossRef]

117. Rolle, L.; Englezos, V.; Torchio, F.; Cravero, F.; Río Segade, S.; Rantsiou, K.; Giacosa, S.; Gambuti, A.; Gerbi, V.; Cocolin, L. Alcohol reduction in red wines by technological and microbiological approaches: A comparative study. *Aust. J. Grape Wine Res.* **2017**, *24*, 62–74. [CrossRef]

118. Binati, R.L.; Lemos Junior, W.J.F.; Luzzini, G.; Slaghenaufi, D.; Ugliano, M.; Torriani, S. Contribution of non-*Saccharomyces* yeasts to wine volatile and sensory diversity: A study on *Lachancea thermotolerans, Metschnikowia spp.* and *Starmerella bacillaris* strains isolated in Italy. *Int. J. Food Microbiol.* **2020**, *318*, 108470. [CrossRef] [PubMed]

119. Loira, I.; Vejarano, R.; Bañuelos, M.; Morata, A.; Tesfaye, W.; Uthurry, C.; Villa, A.; Cintora, I.; Suárez-Lepe, J. Influence of sequential fermentation with *Torulaspora delbrueckii* and *Saccharomyces cerevisiae* on wine quality. *LWT-Food Sci. Technol.* **2014**, *59*, 915–922. [CrossRef]

120. Arslan, E.; Çelik, Z.D.; Cabaroğlu, T. Effects of pure and mixed autochthonous *Torulaspora delbrueckii* and *Saccharomyces cerevisiae* on fermentation and volatile compounds of narince wines. *Foods* **2018**, *7*, 147. [CrossRef] [PubMed]

121. González-Royo, E.; Pascual, O.; Kontoudakis, N.; Esteruelas, M.; Esteve-Zarzoso, B.; Mas, A.; Canals, J.M.; Zamora, F. Oenological consequences of sequential inoculation with non-*Saccharomyces* yeasts (*Torulaspora delbrueckii* or *Metschnikowia pulcherrima*) and *Saccharomyces cerevisiae* in base wine for sparkling wine production. *Eur. Food Res. Technol.* **2015**, *240*, 999–1012. [CrossRef]

122. Canonico, L.; Comitini, F.; Ciani, M. *Torulaspora delbrueckii* for secondary fermentation in sparkling wine production. *Food Microbiol.* **2018**, *74*, 100–106. [CrossRef]

123. Ramírez, M.; Velázquez, R. The yeast *Torulaspora delbrueckii*: An interesting but difficult-to-use tool for winemaking. *Fermentation* **2018**, *4*, 94. [CrossRef]

124. Benito, S. The impact of *Torulaspora delbrueckii* yeast in winemaking. *Appl. Microbiol. Biotechnol.* **2018**, *102*, 3081–3094. [CrossRef]

125. Belda, I.; Navascués, E.; Marquina, D.; Santos, A.; Calderon, F.; Benito, S. Dynamic analysis of physiological properties of *Torulaspora delbrueckii* in wine fermentations and its incidence on wine quality. *Appl. Microbiol. Biotechnol.* **2015**, *99*, 1911–1922. [CrossRef]

126. Taillandier, P.; Lai, Q.P.; Julien-Ortiz, A.; Brandam, C. Interactions between *Torulaspora delbrueckii* and *Saccharomyces cerevisiae* in wine fermentation: Influence of inoculation and nitrogen content. *World J. Microbiol. Biotechnol.* **2014**, *30*, 1959–1967. [CrossRef]

127. Mecca, D.; Benito, S.; Beisert, B.; Brezina, S.; Fritsch, S.; Semmler, H.; Rauhut, D. Influence of nutrient supplementation on *Torulaspora delbrueckii* wine fermentation aroma. *Fermentation* **2020**, *6*, 35. [CrossRef]

128. Escribano, R.; González-Arenzana, L.; Portu, J.; Garijo, P.; López-Alfaro, I.; López, R.; Santamaría, P.; Gutiérrez, A. Wine aromatic compound production and fermentative behaviour within different non-*Saccharomyces* species and clones. *J. Appl. Microbiol.* **2018**, *124*, 1521–1531. [CrossRef] [PubMed]

129. Escribano-Viana, R.; González-Arenzana, L.; Portu, J.; Garijo, P.; López-Alfaro, I.; López, R.; Santamaría, P.; Gutiérrez, A.R. Wine aroma evolution throughout alcoholic fermentation sequentially inoculated with non-*Saccharomyces/Saccharomyces* yeasts. *Food Res. Int.* **2018**, *112*, 17–24. [CrossRef] [PubMed]

130. Belda, I.; Ruiz, J.; Beisert, B.; Navascués, E.; Marquina, D.; Calderón, F.; Rauhut, D.; Benito, S.; Santos, A. Influence of *Torulaspora delbrueckii* in varietal thiol (3-SH and 4-MSP) release in wine sequential fermentations. *Int. J. Food Microbiol.* **2017**, *257*, 183–191. [CrossRef] [PubMed]

131. Rodrigues, A.J.; Raimbourg, T.; Gonzalez, R.; Morales, P. Environmental factors influencing the efficacy of different yeast strains for alcohol level reduction in wine by respiration. *LWT-Food Sci. Technol.* **2016**, *65*, 1038–1043. [CrossRef]

132. Domizio, P.; Lencioni, L.; Calamai, L.; Portaro, L.; Bisson, L.F. Evaluation of the yeast *Schizosaccharomyces japonicus* for use in wine production. *Am. J. Enol. Vitic.* **2018**, *69*, 266–277. [CrossRef]

133. Englezos, V.; Rantsiou, K.; Cravero, F.; Torchio, F.; Ortiz-Julien, A.; Gerbi, V.; Rolle, L.; Cocolin, L. *Starmerella bacillaris* and *Saccharomyces cerevisiae* mixed fermentations to reduce ethanol content in wine. *Appl. Microbiol. Biotechnol.* **2016**, *12*, 5515–5526. [CrossRef]

134. Izquierdo Canas, P.M.; Palacios Garcia, A.T.; Garcia Romero, E. Enhancement of flavour properties in wines using sequential inoculations of non-*Saccharomyces* (*Hansenula* and *Torulaspora*) and *Saccharomyces* yeast starter. *VITIS-J. Grapevine Res.* **2011**, *50*, 177–182.

135. Azzolini, M.; Fedrizzi, B.; Tosi, E.; Finato, F.; Vagnoli, P.; Scrinzi, C.; Zapparoli, G. Effects of *Torulaspora delbrueckii* and *Saccharomyces cerevisiae* mixed cultures on fermentation and aroma of Amarone wine. *Eur. Food Res. Technol.* **2012**, *235*, 303–313. [CrossRef]

136. Benito, A.; Hofmann, T.; Laier, M.; Lochbüler, B.; Schüttler, A.; Ebert, K.; Fritsch, S.; Röcker, J.; Rauhut, D. Effect on quality and composition of Riesling wines fermented by sequential inoculation with non-*Saccharomyces* and *Saccharomyces cerevisiae*. *Eur. Food Res. Technol.* **2015**, *241*, 707–717. [CrossRef]

137. Rossouw, D.; Bauer, F.F. Exploring the phenotypic space of non-*Saccharomyces* wine yeast biodiversity. *Food Microbiol.* **2016**, *55*, 32–46. [CrossRef]

138. Benito, S. Combined use of *Lachancea thermotolerans* and *Schizosaccharomyces pombe* in winemaking: A review. *Microorganisms* **2020**, *8*, 655. [CrossRef]

139. Benito, Á.; Calderón, F.; Benito, S. Combined use of *S. pombe* and *L. thermotolerans* in winemaking. Beneficial effects determined through the study of wines' analytical characteristics. *Molecules* **2016**, *21*, 1744. [CrossRef] [PubMed]

140. Benito, Á.; Calderón, F.; Benito, S. Mixed alcoholic fermentation of *Schizosaccharomyces pombe* and *Lachancea thermotolerans* and its influence on mannose-containing polysaccharides wine composition. *AMB Express* **2019**, *9*, 17. [CrossRef] [PubMed]

141. Varela, C.; Sengler, F.; Solomon, M.; Curtin, C. Volatile flavour profile of reduced alcohol wines fermented with the non-conventional yeast species *Metschnikowia pulcherrima* and *Saccharomyces uvarum*. *Food Chem.* **2016**, *209*, 57–64. [CrossRef] [PubMed]

142. Kemp, B.; Alexandre, H.; Robillard, B.; Marchal, R. Effect of production phase on bottle-fermented sparkling wine quality. *J. Agric. Food Chem.* **2015**, *63*, 19–38. [CrossRef] [PubMed]

143. Arroyo-López, F.N.; Pérez-Torrado, R.; Querol, A.; Barrio, E. Modulation of the glycerol and ethanol syntheses in the yeast *Saccharomyces kudriavzevii* differs from that exhibited by *Saccharomyces cerevisiae* and their hybrid. *Food Microbiol.* **2010**, *27*, 628–637. [CrossRef] [PubMed]

144. De Benedictis, M.; Bleve, G.; Grieco, F.; Tristezza, M.; Tufariello, M. An optimized procedure for the enological selection of non-*Saccharomyces* starter cultures. *Antonie Van Leeuwenhoek* **2011**, *99*, 189–200. [CrossRef]

145. Ivit, N.N.; Loira, I.; Morata, A.; Benito, S.; Palomero, F.; Suárez-Lepe, J.A. Making natural sparkling wines with non-*Saccharomyces* yeasts. *Eur. Food Res. Technol.* **2018**, *244*, 925–935. [CrossRef]

146. Kelly, J.M.; van Dyk, S.A.; Dowling, L.K.; Pickering, G.J.; Kemp, B.; Inglis, D.L. *Saccharomyces uvarum* yeast isolate consumes acetic acid during fermentation of high sugar juice and juice with high starting volatile acidity. *OENO ONE Int. J. Vine Wine Sci.* **2020**, *54*, 199–211.

 *fermentation*

Article

Effect of *Candida intermedia* LAMAP1790 Antimicrobial Peptides against Wine-Spoilage Yeasts *Brettanomyces bruxellensis* and *Pichia guilliermondii*

Rubén Peña, Jeniffer Vílches, Camila G.-Poblete and María Angélica Ganga *

Laboratorio de Biotecnología y Microbiología Aplicada, Departamento de Ciencia y Tecnología de los Alimentos, Universidad de Santiago de Chile, Santiago 917020, Chile; ruben.pena@usach.cl (R.P.); jeniffernatalia221@gmail.com (J.V.); camila.gonzalezpo@usach.cl (C.G.-P.)
* Correspondence: angelica.ganga@usach.cl; Tel.: +56-2-27184509

Received: 25 May 2020; Accepted: 19 June 2020; Published: 1 July 2020

Abstract: Wine spoilage yeasts are one of the main issues in the winemaking industry, and the control of the *Brettanomyces* and *Pichia* genus is an important goal to reduce economic loses from undesired aromatic profiles. Previous studies have demonstrated that *Candida intermedia* LAMAP1790 produces antimicrobial peptides of molecular mass under 10 kDa with fungicide activity against *Brettanomyces bruxellensis*, without affecting the yeast *Saccharomyces cerevisiae*. So far, it has not been determined whether these peptides show biocontroller effect in this yeast or other spoilage yeasts, such as *Pichia guilliermondii*. In this work, we determined that the exposure of *B. bruxellensis* to the low-mass peptides contained in the culture supernatant of *C. intermedia* LAMAP1790 produces a continuous rise of reactive oxygen species (ROS) in this yeast, without presenting a significant effect on membrane damage. These observations can give an approach to the antifungal mechanism. In addition, we described a fungicide activity of these peptides fraction against two strains of *P. guilliermondii* in a laboratory medium. However, carrying out assays on synthetic must, peptides must show an effect on the growth of *B. bruxellensis*. Moreover, these results can be considered as a start to develop new strategies for the biocontrol of spoilage yeast.

Keywords: antimicrobial peptides; *Brettanomyces bruxellensis*; *Candida intermedia*; *Pichia guilliermondii*; reactive oxygen species

---

1. Introduction

Alcoholic fermentation is the process of monosaccharide's conversion to ethanol and CO_2. Therefore, the anaerobic metabolism of *Saccharomyces cerevisiae* is the main cause of wine fermentation. However, in spontaneous wine production other yeasts participate such as *Hanseniaspora*, *Candida*, *Pichia*, and *Metschnikowia* genera, among others [1]. Nevertheless, these yeasts have lower fermentation capacity, and are not able to grow in high ethanol concentration conditions, given that the *Saccharomyces* genus is the predominant during the final stages of fermentation [1].

Among the unfavorable growth wine conditions, several yeasts are capable to proliferate and generate undesired characteristics in the final product. *Brettanomyces bruxellensis* has been described as the main spoilage yeast during the maturity stage of wine in barrels [2,3]. This yeast has the capacity of transforming hydroxycinnamic acids into vinyl and ethyl derivates, which produce off-flavors in wine [4,5]. Additionally, this aromatic defect can be produced in early stages of fermentation by other yeasts such as *Pichia guilliermondii* [2]. Among these, there are strains which can transform *p*-coumaric acid in 4-vinylphenol in similar proportions as described for *B. bruxellensis*, being that *P. guilliermondii* is a potential problem for winemaking [6]. Because of this, in this industry the use of sulfites is a widespread strategy to control growth of undesired microorganisms. Nevertheless, several strains are resistant and the use of sulfites in high quantities is potentially unsafe for human health [7]. As a result, alternative strategies such as antimicrobial peptides (AMPs) have been proposed to biocontrol spoilage microorganisms [8–12]. AMPs are low molecular mass peptides with amphipathic characteristic which can affect the growth of several microorganisms by permeabilization of plasmatic membranes and/or by increasing the reactive oxygen species [13–15]. Previously, Peña et al. [16,17] have described antimicrobial peptides production in *Candida intermedia*, which reduce the viability of different *B. bruxellensis* strains in a laboratory medium without affecting the growth of fermentative yeast *S. cerevisiae*. However, it has not yet been described how these peptides affect *B. bruxellensis* and if they are able to inhibit the growth of *P. guilliemondii*. Thus, the aim of this work was to explore the cellular damage produced by *C. intermedia* LAMAP1790 peptides above *B. bruxellensis* and determine if they can control the growth of yeast *B. bruxellensis* and *P. guilliermondii* using laboratory culture mediums and synthetic must. This knowledge will allow to determine the antimicrobial peptides produced for *C. intermedia* as a possible biocontroller in the wine industry.

2. Materials and Methods

2.1. Strains and Culture Media

The strains of *B. bruxellensis* LAMAP2480, *C. intermedia* LAMAP1790, *Pichia guilliermondii* LAMAP3202, LAMAP3203, *S. cerevisiae* BY4741, and EC1118 were obtained from the culture collection at the Laboratory of Biotechnology and Applied Microbiology, University of Santiago de Chile. *C. intermedia* LAMAP1790 was isolated in Chile from must in the early stages of fermentative process [18] and *B. bruxellensis* LAMAP2480 was isolated from Chilean wine [19]. Both strains of *P. guilliermondii* were isolated from Argentinian vineyards. The strain LAMAP3202 and LAMAP3203 was characterized by Sangorrín et al. (2013), labeled as P7 and P8 strains respectively [20]. All strains used in this work were grown on GYEB media (yeast extract 5 g/L and glucose 20 g/L, adjusted to pH 5.0 with 100 mmol/L phosphate-citrate buffer) [21].

2.2. Obtained Supernatant with Antifungal Activity of C. intermedia and Characterization of the Protein Nature of This Activity

To obtain the supernatant with antifungal activity from *C. intermedia* LAMAP1790, the yeast was inoculated in 100 mL GYEB medium during 48 h at 28 °C with orbital agitation at 120 rpm. Then, the culture was centrifuged during 10 min at 5900× *g* to obtain saturated culture supernatant. Afterward, a cut-off of total proteins present in the supernatant was done by means of ultrafiltration in devises *Amicon® Ultra-15* with 10 kDa cutoff (Merck-Millipore®, Darmstadt, Germany). In this work, the antifungal supernatant is defined as the fraction obtained from ultrafiltration which only contains proteins of molecular mass under 10 kDa. This antifungal supernatant was sterilized using disposable filters with 0.22 µm pore size (Membrane Solutions LLC®, Windham, NH, USA) and stored at −20 °C to be used later. To determine whether the antifungal activity is related with the presence of peptides with molecular mass under 10 kDa, the antifungal supernatant was treated with 2 mg/mL protease of *Streptomyces griseus* (Sigma-Aldrich®, St. Louis, MO, USA) during 4 h at 37 °C.

2.3. Determination of the Cellular Damage Produced on B. bruxellensis by Exposure to Antifungal Supernatant of C. intermedia

The obtained antifungal supernatant of *C. intermedia* LAMAP1790 was assessed to determine if it produces: (a) membrane permeability or (b) rise of the reactive oxygen species (ROS) on exposed *B. bruxellensis* cells during different periods, similar to described by [22] and [23]. Then, 3×10^5 *B. bruxellensis* LAMAP2480 cells were exposed individually to 1 mL of sterile antifungal supernatant and incubated during 12 h and 24 h at 28 °C. As positive control, a similar number of cells with 600 µg/mL zymolyase 100*T* (Amsbio®, Abingdon, OX, UK) was inoculated at 37 °C during 2 h and then exposed to 30% H_2O_2 during 30 min. As negative control, the same concentration of cells was used and at 28 °C in buffer HEPES saline 1× pH 7.0 (70 mM NaCl, 0.75 mM Na_2HPO_4, 25 mM HEPES) were incubated. After treatments, the cells were washed 3 times with buffer HEPES saline 1× pH 7.0. To facilitate the observation, yeast was stained with calcofluor white (Sigma®) in 1:1 proportion with KOH to 10% p/v. The membrane permeability was assessed by means of staining with 2 µM propidium iodide (Sigma®) and the accumulation of ROS was determined by means of staining with 10 µM 6-carboxy-2′,7′-dichlorodihydrofluorescein diacetate (C400; Thermo-Scientific®, Waltham, MA, USA). The fluorescent cells were observed using the epifluorescence microscope Moticam Pro BA410 (Motic®, Xiamen, China), with 40× fluorescence microscope objective lent.

2.4. Screening the Antifungal Activity of C. intermedia LAMAP1790 on B. bruxellensis, P. guilliermondii, and S. cerevisiae

The qualitative determination of the antifungal activity of *C. intermedia* LAMAP1790 on *B. bruxellensis* LAMAP2480, *P. guilliermondii* LAMAP3203, LAMAP3203 and *S. cerevisiae* EC1118 strains was carried out following the methodology used by [16]. For this, 1×10^5 cells from each strain were inoculated in 25 mL warm agar MBA (5 g/L yeast extract, 5 g/L peptone, 20 g/L glucose and 15 g/L agar, adjusted to pH 5.0 with 100 mmol/L phosphate-citrate buffer and supplemented with 0.03 g/L of methylene blue). Each inoculated media was plated into petri dishes. A surface inoculation was carried out using 10 µL of 1×10^8 cells/mL suspension from *C. intermedia* LAMAP1790. As control, the same surface inoculation of *S. cerevisiae* BY4741 was carried out. The plates were incubated for 7 days at 28 °C and every assay was evaluated six times. The qualitative determination was done by observation and measuring the inhibition halo present in the plates.

2.5. Antifungal Activity of Low Mass Peptide Fraction Obtained from C. intermedia Antifungal Supernatant against B. bruxellensis, S. cerevisiae and P. guilliermondii in Synthetic Must

The obtention of 100X concentrated low mass peptide fraction (under 10 kDa) was performed by lyophilization (IlShineBioBase® freeze dryer, Dongducheon-si, Gyeonggi-do, Korea) of 3 L to sterile antifungal supernatant derived from cultures of *C. intermedia* LAMAP1790 in GYEB medium. The total protein quantification in the fraction was done according to [24]. The evaluation of the antifungal activity was done using simultaneous inoculation of *S. cerevisiae* EC1118 and *B. bruxellensis* LAMAP2480 or *P. guilliermondii* LAMAP3202 in synthetic grape must, (100 g/L glucose, 100 g/L fructose, 5 g/L maleic acid, 0.5 g/L citric acid, 3 g/L tartaric acid, 0.75 g/L potassium phosphate, 0.5 g/L potassium sulfate, 0.155 g/L calcium chloride, 0.25 g/L magnesium sulfate, 0.2 g/L sodium chloride, 4 mg/L manganese sulfate, 1.5 mg/L calcium pantenoate, 2 mg/L nicotinic acid, 0.25 mg/L thiamine hydrochloride and 0.003 mg/L biotin; pH 3.5) [25], Previously, each strain was adapted to the media using a procedure described by [26]. To the antifungal assays, 5 mL synthetic must was inoculated with 1×10^2 cells of *S. cerevisiae* EC1118, *B. bruxellensis* LAMAP2480 or *P. guilliermondii* LAMAP3202 strains individually (determined by direct yeast count in Neubauer chamber), and supplemented with 1 µg of low mass peptide fraction. As a control, the same procedure was carried out, but the medium was supplemented with 1 µg of total proteins obtained from the concentrate sterile culture supernatant of *S. cerevisiae* BY4741 (*IlShineBioBase® freeze dryer*, Dongducheon-si, Gyeonggi-do, Korea). Each assay was incubated for 21 days, and every 3 days a cellular count of the cultures was carried out on YPD agar plates (5 g/L yeast extract, 5 g/L peptone, 20 g/L glucose and 20 g/L agar) incubated for 7 days at 28 °C. The count of spoilage yeast in the mixed culture was performed in YPD agar plates supplemented with 0.01% *v/v* of cycloheximide, according to [27].

2.6. Statistical Analysis

All the data was analyzed using the Kruskal–Wallis test, with an initial analysis of the distribution goodness of fit using the Kolmogorov–Smirnov test. All analysis was carried out with Statgraphics Centurion XVI Software (Statpoint Technologies Inc., Warrenton, VA, USA). The significant differences were validated with a probability < 0.05.

3. Results and Discussion

One of the most important aspects in the study of AMPs is to determine its antifungal action mechanism. In relation to antimicrobial peptides to biocontrol contaminant microorganisms in winemaking, Enrique et al. (2008) [9] studied the antifungal effect of the synthetic peptide LfcinB$_{17\text{-}31}$ on *B. bruxellensis*, determining that its action mechanism is related to the penetration of the peptides into the cell cytoplasm. Additionally, by fluorescence microscopy, Branco et al. (2017) [12] have described that saccharomycin (antifungal peptides produced by *S. cerevisiae* CCMI885 strain) produce cell membrane disruption and internalization of the peptides in *Hanseniaspora guilliermondii* and *B. bruxellensis*. Our previous results have demonstrated that *C. intermedia* LAMAP1790 releases peptides in the culture medium with masses under 4.6 kDa, which show selective antifungal activity on *B. bruxellensis* strains [16,17]. With the purpose of defining the cell damaged produced by the antifungal supernatant of *C. intermedia* LAMAP1790 (which contains these peptides) on *B. bruxellensis* LAMAP2480, different assays were carried out using calcofluor white (CW), propidium iodide (PI), or 6-carboxy-2′,7′-dichlorodihydrofluorescein diacetate (C400) [12,22,28] (Figure 1).

Under optimum growth condition, the yeast wall stains bright blue-white by the CW assembly [28], being impermeable to PI [12] and the C400 cannot be oxidized to its fluorescence form [22]. As a negative control, cells were inoculated in buffer HEPES saline pH 7.0 (Figure 1A–C), it can be observed that neither the wall cellular nor the impermeability of the membrane was affected. Only 29.91 ± 7.29% of the observed cells in the medium show green fluorescence (Figure 1C) and none show red color. This increase of green fluorescence would be related to the lack of nutrients that *B. bruxellensis* had during the 24 h trail, due has been reported that such periods may activate an autophagy process [29,30]. Autophagy is a non-selective degradation of organelles or intracellular macromolecules, a recycling process that allows the amino acid supply and survival. *S. cerevisiae* can do mitophagy (removal of damaged mitochondria), therefore, releasing mitochondrial ROS into the cytoplasm [29,30]. On the other hand, when the cells have damage in the membrane, this is no longer impermeable to PI, dying cells in red [23].

Thus, as a positive control of both processes, we carried out an induction to the oxidative stress and membrane damage by zymolyase and H_2O_2 treatment (Figure 1D–F). As observed in this figure the treatment produced a 63.39 ± 6.92% permeabilization to cell surface membrane, allowing the penetration of PI into the cell (compared with control sample 1B). Besides, a 63.61 ± 8.17% of cells show a rise of intracellular ROS, which allowed the observation of green fluorescence derived from C400 oxidation (Figure 1F). Additionally, when yeasts were exposed to *C. intermedia* supernatant at 12h, it was observed a rise in the number of cells which oxide C400 (Figure 1I) which is sustained at 24 h of incubation (Figure 1L), while it is observed a little rise of permeable cells of PI to 24 h of incubation (Figure 1H,K). When the *C. intermedia* supernatant is treated with protease, a decrease decrease in the number of cells that oxidize C400 and the permeable cells of PI (Figure 1N,O) was observed, confirming that antifungal compounds have protein nature [16].

By comparing the percentage of fluorescent yeast in different conditions (Figure 2), it can be observed that the incubation of *B. bruxellensis* with the antifungal supernatant produce a sustaining little rise in the number of permeable cells to PI at incubation time, is not statistically different from the negative control (Figure 2A).

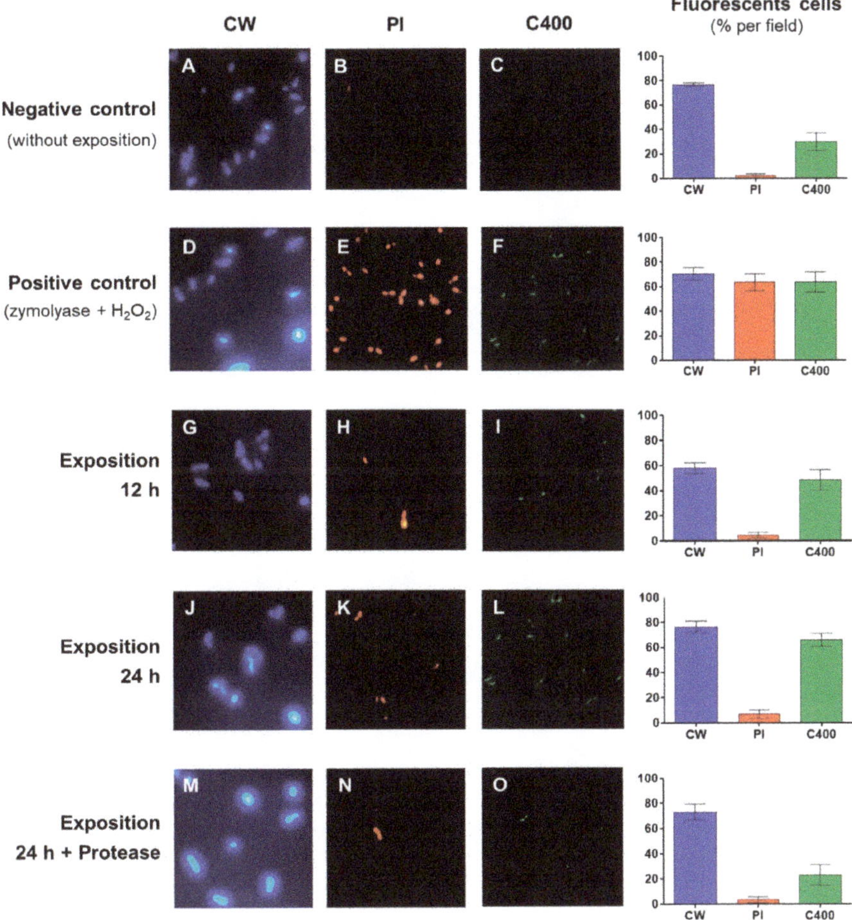

Figure 1. Evaluation of permeability and reactive oxygen species (ROS) accumulation in *B. bruxellensis* LAMAP2480 cells exposed to *C. intermedia* LAMAP1790 antifungal supernatant at different times, using epifluorescence microscopy. Graphics at the right side of each line represents a percentage of fluorescent cells per field counted in each treatment. (**A–C**): untreated yeasts (Negative control). (**D–F**): yeasts exposed to H_2O_2 30% *v/v* for 30 min, after treatment with zymolyase (600 μg/mL) for 2 h at 37 °C (Positive control). (**G–I**): yeasts exposed to *C. intermedia* antifungal supernatant for 12 h. (**J–L**): yeasts exposed to *C. intermedia* antifungal supernatant for 24 h. (**M–O**): yeasts exposed to *C. intermedia* antifungal supernatant for 24 h after a proteolytic treatment to the supernatant for 4 h at 37 °C with 2 mg/mL of *Streptomyces griseus* protease (Sigma®). CW: calcofluor white staining, PI: propidium iodide staining, C400: 6-carboxy-2′,7′-dichlorodihydrofluorescein diacetate staining.

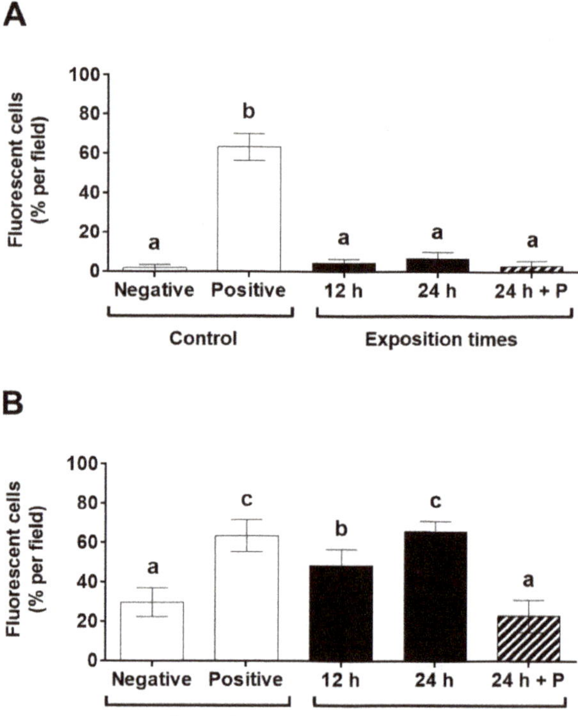

Figure 2. Evaluation of damage produced in *B. bruxellensis* cells after 12 h and 24 h of exposition to antifungal supernatant from *C. intermedia* LAMAP1790. (**A**): Membrane damage observed by cell permeability to propidium iodide staining, (**B**): Oxidative damage derived of ROS accumulation measured with 6-carboxy-2′′,7′-dichlorodihydrofluorescein diacetate staining. The evaluation was performed comparing the percentage of stained fluorescent cells of *B. bruxellensis* after each treatment. White bars correspond to control treatments, black bars correspond to yeasts exposed to antifungal supernatant at labeled times and bars in striped lines (labeled 24 h + P) correspond to 24 h of yeasts exposition to antifungal supernatant treated previously with 2 mg/mL of *Streptomyces griseus* protease (Sigma®). Different letter above each column represents a significative difference ($p < 0.05$).

Nevertheless, by evaluating the percentage of cells which oxidize C400, they increase significantly as incubation time increases, even achieving similar values to the obtained due to the introduction to oxidative stress with H_2O_2 (Figure 2B). These results allow to demonstrate that the sustaining increase of ROS in *B. bruxellensis* is related to the presence of peptides of mass under 10 kDa in the antifungal supernatant of *C. intermedia* LAMAP1790 and propose that its antifungal action mechanism would be related to the oxidative damage that the exposed cell suffers to the supernatant. This should be proved by the non-significant rise of the observed permeability to PI in *B. bruxellensis* between 12 h and 24 h of exposure to antifungal supernatant, because it has been demonstrated that the induction of ROS in yeasts such as *H. guilliermondii* produces the permeabilization of its cellular membrane [11]. Similar effects have been reported to synthetic peptide PAF26 and other similar peptides in which have been demonstrated that they can penetrate the cytoplasm of *S. cerevisiae*, without affecting firstly the integrity of the cellular membrane [13,14,31]. Thus, it determined that the synthetic peptide PAF26 would have a multistep mechanism of action, where it first interacts with the wall or cellular membrane, then it would be endocytosed and accumulated in the vacuoles, and finally, it would be transported to the cytoplasm and perform its antifungal action [14]. This mechanism would be like the observations made for *B. bruxellensis* by means of fluorescence microscopy.

It was carried out a qualitative antifungal test with *S. cerevisiae*, *B. bruxellensis*, and two strains of *P. guilliermondii* in solid MBA agar plates to determine the formation of inhibition halos produced by *C. intermedia* LAMAP1790. The two strains of *P. guilliermondii* selected was previously studied by Sangorrín et al. (2013) [20]. In that work, from a pool of 15 strains, it was possible to conclude that strains LAMAP3202 and LAMAP3203 (labeled by Sangorrín as P7 and P8) have the highest transformation efficiencies of *p*-coumaric acid in 4-vinylphenol (more aggressive wine-spoilage phenomena). For these reasons, we considered these strains as the best model to our study. As shown in Table 1, *S. cerevisiae* EC1118 strain does not show growth inhibition, while the *B. bruxellensis* LAMAP2480 strain shows a clear inhibition halo surrounding culture of *C. intermedia*, whose diameter reached 19.00 ± 0.62 mm, as it was described by Peña et al. [16,17]. By analyzing the behavior of *P. guilliermondii*, LAMAP3202 and LAMAP3203 strains can be observed that an inhibition halo appears, whose diameters reached 15.33 ± 0.82 mm and 16.17 ± 0.75 mm, respectively (Table 1). Then, *B. bruxellensis* shows a greater sensitivity to the presence of *C. intermedia* than *P. guilliermondii*. Similar studies carried out by Lopes and Sangorrín (2010) [32] have demonstrated that *P. guilliermondii* sensitivity depends on the yeast strains to which it is exposed. On the other hand, Villalba et al. (2016) [23] demonstrated that the production of antifungal compounds of protein nature produced by *Torulaspora delbrueckii*, which has a molecular mass above 30 kDa, shows glucanase and chitinase activity. Therefore, the authors conclude that this would be a killer toxin rather than an antimicrobial peptide (AMP). Thus, this work would constitute the first qualitative evidence which shows the sensitivity of *P. guilliermondii* strains to antimicrobial peptides produced by non-*Saccharomyces* yeasts.

Table 1. Inhibition halos obtained after the exposure of *C. intermedia* LAMAP1790 against strains of *S. cerevisiae*, *B. bruxellensis* and *Pichia guilliermondii* in MBA medium.

		C. intermedia **LAMAP1790** Inhibition Halo (mm)
S. cerevisiae	EC1118	[†] ND [a]
B. bruxellensis	LAMAP2480	19.00 ± 0.62 [c]
P. guilliermondii	LAMAP3202	15.33 ± 0.82 [b]
	LAMAP3203	16.17 ± 0.75 [b]

Values with the same superscript letter are not significantly different ($p < 0.05$). [†] ND: Non-Detected.

With the purpose of determining whether the antifungal effect of *C. intermedia* LAMAP1790 is similar in winemaking conditions, it was carried out assays on synthetic must [12]. We decided to use this media to avoid the antimicrobial influence on yeast described to the hydroxycinnamic acids present in the natural grape must (mainly *p*-coumaric and ferulic acid) [33–35]. Thus, the viability of the spoilage strains *B. bruxellensis* LAMAP2480 and *P. guilliermondii* LAMAP3202 were assessed in mixed culture with *S. cerevisiae* EC1118 for 21 days (Figure 3).

The synthetic must was supplemented with 1 µg of low-mass protein fraction obtained from *C. intermedia* supernatant, and then was inoculated using the spoilage yeasts. Posteriorly, it was inoculated with *S. cerevisiae* EC1118 (fermentation starter). As can be seen in Figure 3 (A, B), the growth of *S. cerevisiae* is not affected, demonstrating the harmlessness of the antifungal peptides against this yeast. In the case of the effect on *B. bruxellensis*, it was observed that its growth decreases in one magnitude order of difference compared to the control (3A), while in the case of *P. guilliermondii* L3202, minimal changes between the treatment and control were observed (Figure 3B). Despite having growth inhibition of *P. guilliermondii* in solid medium (Table 1), this effect was not seen in synthetic must, which can be related to a greater concentration of an antifungal compound, possibly requiring a greater concentration for this specie compared to *B. bruxellensis*. To date, there are no previous studies that assess the antifungal capacity of a compound of protein nature (AMP or killer toxin) on the growth of

P. guilliermondii in mixed cultures with *S. cerevisiae* in synthetic wine must. Thus, it would be necessary to further study the action of *C. intermedia* peptides in winemaking conditions.

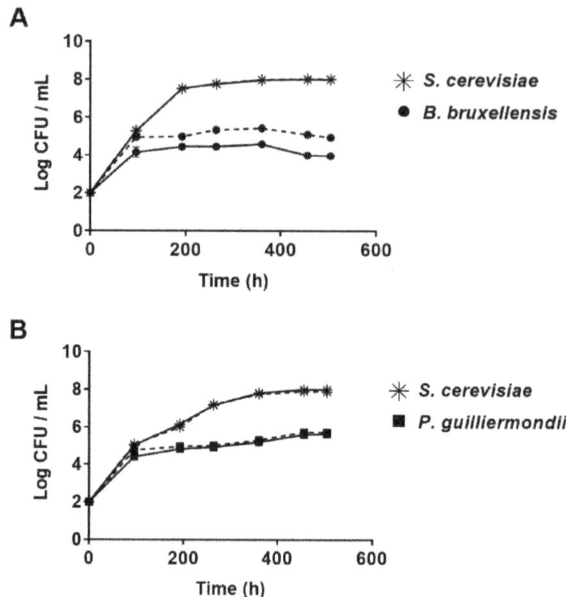

Figure 3. Antifungal activity of 1 µg of low mass protein fraction (under 10 kDa) concentrated from *C. intermedia* L1790 supernatant (solid lines) used synthetic wine must. (**A**) *S. cerevisiae* with *B. bruxellensis* L2480 (**B**) *S. cerevisiae* with *P. guilliermondii* 3202. The controls (stripped lines) corresponds to the concentrated supernatant of *S. cerevisiae* BY4741 (not antifungal activity). All assays were performed in triplicate.

4. Conclusions

The antifungal supernatant obtained from de culture media of *C. intermedia* LAMAP1790 produces a continuous rise of oxygen reactive species (ROS) in *B. bruxellensis*, without trigger a significant effect on its membrane damage. This effect was totally avoided when the supernatant was treated with a proteolytic enzyme, proving that low mass peptides contained in this fraction are responsible for this effect. Herewith, *C. intermedia* L1790 showed antimicrobial effect on *B. bruxellensis* LAMAP2480, *Pichia guilliermondii* LAMAP3202, and LAMAP3203 when laboratory medium was used; however, similar effect was not observed when synthetic must was used. Therefore, it is necessary to identify the peptides with antifungal activity produced by *C. intermedia* LAMAP1790 and study how some enological factors (pH, ethanol, sugars, etc.) may affect their antifungal capacity. This will allow us to determine its possible technological application in the control of yeast contaminants in the wine industry.

Author Contributions: Conceptualization: R.P. and M.A.G.; Methodology: R.P. and M.A.G.; Formal analysis: R.P. and J.V.; Data analysis: R.P., J.V., and C.G.-P.; Supervision: M.A.G.; Writing-Original Draft: R.P., C.G.-P. and M.A.G.; Project Administration: M.A.G. All authors have read and agreed to the published version of the manuscript.

Funding: Rubén Peña was funded by the Comisión Nacional de Investigación Científica y Tecnológica CONICYT-PCHA/DoctoradoNacional/2013-21130439 Doctoral Fellowship. Camila G-Poblete was funded by Facultad Tecnológica Fellowship from Universidad de Santiago de Chile.

Conflicts of Interest: The authors hereby declare they do not have any conflict of interest associated to this work.

References

1. Liu, Y.; Rousseaux, S.; Tourdot-Maréchal, R.; Sadoudi, M.; Gougeon, R.; Schmitt-Kopplin, P.; Alexandre, H. Wine microbiome: A dynamic world of microbial interactions. *Crit. Rev. Food Sci. Nutr.* **2017**, *57*, 856–873. [CrossRef] [PubMed]

2. Loureiro, V.; Malfeito-Ferreira, M. Spoilage yeasts in the wine industry. *Int. J. Food Microbiol.* **2003**, *86*, 23–50. [CrossRef]

3. Oelofse, A.; Pretorius, I.S.; du Toit, M. Significance of Brettanomyces and Dekkera during Winemaking: A Synoptic Review. *S. Afr. J. Enol. Vitic.* **2008**, *29*, 128–144. [CrossRef]

4. Chatonnet, P.; Dubourdie, D.; Boidron, J.; Pons, M. The origin of ethylphenols in wines. *J. Sci. Food Agric.* **1992**, *60*, 165–178. [CrossRef]

5. Godoy, L.; García, V.; Peña, R.; Martínez, C.; Ganga, M.A. Identification of the Dekkera bruxellensis phenolic acid decarboxylase (PAD) gene responsible for wine spoilage. *Food Control* **2014**, *45*, 81–86. [CrossRef]

6. Dias, L.; Dias, S.; Sancho, T.; Stender, H.; Querol, A.; Malfeito-Ferreira, M.; Loureiro, V. Identification of yeasts isolated from wine-related environments and capable of producing 4-ethylphenol. *Food Microbiol.* **2003**, *20*, 567–574. [CrossRef]

7. Devalia, J.L.; Rusznak, C.; Herdman, M.J.; Trigg, C.J.; Davies, R.J.; Tarraf, H. Effect of nitrogen dioxide and sulphur dioxide on airway response of mild asthmatic patients to allergen inhalation. *Lancet* **1994**, *344*, 1668–1671. [CrossRef]

8. Enrique, M.; Marcos, J.F.; Yuste, M.; Martínez, M.; Vallés, S.; Manzanares, P. Antimicrobial action of synthetic peptides towards wine spoilage yeasts. *Int. J. Food Microbiol.* **2007**, *118*, 318–325. [CrossRef]

9. Enrique, M.; Marcos, J.F.; Yuste, M.; Martínez, M.; Vallés, S.; Manzanares, P. Inhibition of the wine spoilage yeast Dekkera bruxellensis by bovine lactoferrin-derived peptides. *Int. J. Food Microbiol.* **2008**, *127*, 229–234. [CrossRef]

10. Albergaria, H.; Francisco, D.; Gori, K.; Arneborg, N.; Gírio, F. Saccharomyces cerevisiae CCMI 885 secretes peptides that inhibit the growth of some non-Saccharomyces wine-related strains. *Appl. Microbiol. Biotechnol.* **2010**, *86*, 965–972. [CrossRef]

11. Branco, P.; Viana, T.; Albergaria, H.; Arneborg, N. Antimicrobial peptides (AMPs) produced by Saccharomyces cerevisiae induce alterations in the intracellular pH, membrane permeability and culturability of Hanseniaspora guilliermondii cells. *Int. J. Food Microbiol.* **2015**, *205*, 112–118. [CrossRef]

12. Branco, P.; Francisco, D.; Monteiro, M.; Almeida, M.G.; Caldeira, J.; Arneborg, N.; Prista, C.; Albergaria, H. Antimicrobial properties and death-inducing mechanisms of saccharomycin, a biocide secreted by Saccharomyces cerevisiae. *Appl. Microbiol. Biotechnol.* **2017**, *101*, 159–171. [CrossRef]

13. Muñoz, A.; Harries, E.; Contreras-Valenzuela, A.; Carmona, L.; Read, N.D.; Marcos, J.F. Two Functional Motifs Define the Interaction, Internalization and Toxicity of the Cell-Penetrating Antifungal Peptide PAF26 on Fungal Cells. *PLoS ONE* **2013**, *8*, e54813. [CrossRef]

14. Muñoz, A.; Gandía, M.; Harries, E.; Carmona, L.; Read, N.D.; Marcos, J.F. Understanding the mechanism of action of cell-penetrating antifungal peptides using the rationally designed hexapeptide PAF26 as a model. *Fungal Biol. Rev.* **2013**, *26*, 146–155. [CrossRef]

15. Zhang, L.; Gallo, R.L. Antimicrobial peptides. *Curr. Biol.* **2016**, *26*, R14–R19. [CrossRef]

16. Peña, R.; Ganga, M.A. Novel antimicrobial peptides produced by Candida intermedia LAMAP1790 active against the wine-spoilage yeast Brettanomyces bruxellensis. *Antonie van Leeuwenhoek* **2018**, *9*. [CrossRef]

17. Peña, R.; Chávez, R.; Rodríguez, A.; Ganga, M.A. A Control Alternative for the Hidden Enemy in the Wine Cellar. *Fermentation* **2019**, *5*, 25. [CrossRef]

18. Ganga, M.A.; Martínez, C. Effect of wine yeast monoculture practice on the biodiversity of non-Saccharomyces yeasts. *J. Appl. Microbiol.* **2004**, *96*, 76–83. [CrossRef]

19. Valdes, J.; Tapia, P.; Cepeda, V.; Varela, J.; Godoy, L.; Cubillos, F.A.; Silva, E.; Martinez, C.; Ganga, M.A. Draft genome sequence and transcriptome analysis of the wine spoilage yeast Dekkera bruxellensis LAMAP2480 provides insights into genetic diversity, metabolism and survival. *FEMS Microbiol. Lett.* **2014**, *361*, 104–106. [CrossRef]

20. Sangorrín, M.P.; García, V.; Lopes, C.A.; Sáez, J.S.; Martínez, C.; Ganga, M.A. Molecular and physiological comparison of spoilage wine yeasts. *J. Appl. Microbiol.* **2013**, *114*, 1066–1074. [CrossRef]

21. Roostita, L.B.; Fleet, G.H.; Wendry, S.P.; Apon, Z.M.; Gemilang, L.U. Determination of yeasts antimicrobial activity in milk and meat products. *Adv. J. Food Sci. Technol.* **2011**, *3*, 442–445.

22. Kaiserer, L.; Oberparleiter, C.; Weiler-Görz, R.; Burgstaller, W.; Leiter, E.; Marx, F. Characterization of the Penicillium chrysogenum antifungal protein PAF. *Arch. Microbiol.* **2003**, *180*, 204–210. [CrossRef] [PubMed]

23. Villalba, M.L.; Susana Sáez, J.; del Monaco, S.; Lopes, C.A.; Sangorrín, M.P. TdKT, a new killer toxin produced by Torulaspora delbrueckii effective against wine spoilage yeasts. *Int. J. Food Microbiol.* **2016**, *217*, 94–100. [CrossRef] [PubMed]

24. Bradford, M.M. A rapid and sensitive method for the quantitation of microgram quantities of protein utilizing the principle of protein-dye binding. *Anal. Biochem.* **1976**, *72*, 248–254. [CrossRef]

25. Gutiérrez, A.; Chiva, R.; Sancho, M.; Beltran, G.; Arroyo-López, F.N.; Guillamon, J.M. Nitrogen requirements of commercial wine yeast strains during fermentation of a synthetic grape must. *Food Microbiol.* **2012**, *31*, 25–32. [CrossRef] [PubMed]

26. Coronado, P.; Aguilera, S.; Carmona, L.; Godoy, L.; Martínez, C.; Ganga, M.A. Comparison of the behaviour of Brettanomyces bruxellensis strain LAMAP L2480 growing in authentic and synthetic wines. *Antonie van Leeuwenhoek* **2015**, *107*, 1217–1223. [CrossRef] [PubMed]

27. Sáez, J.S.; Lopes, C.A.; Kirs, V.C.; Sangorrín, M.P. Enhanced volatile phenols in wine fermented with Saccharomyces cerevisiae and spoiled with Pichia guilliermondii and Dekkera bruxellensis. *Lett. Appl. Microbiol.* **2010**, *51*, 170–175. [CrossRef] [PubMed]

28. Harrington, B.J.; Hageage, G.J. your lab focus: Calcofluor White: A Review of its Uses and Applications in Clinical Mycology and Parasitology. *Lab. Med.* **2003**, *34*, 361–367. [CrossRef]

29. Suzuki, S.W.; Onodera, J.; Ohsumi, Y. Starvation induced cell death in autophagy-defective yeast mutants is caused by mitochondria dysfunction. *PLoS ONE* **2011**, *6*, e17412. [CrossRef]

30. Reggiori, F.; Klionsky, D.J. Autophagic processes in yeast: Mechanism, machinery and regulation. *Genetics* **2013**, *194*, 341–361. [CrossRef]

31. López-García, B.; Gandía, M.; Muñoz, A.; Carmona, L.; Marcos, J.F. A genomic approach highlights common and diverse effects and determinants of susceptibility on the yeast Saccharomyces cerevisiae exposed to distinct antimicrobial peptides. *BMC Microbiol.* **2010**, *10*. [CrossRef]

32. Lopes, C.A.; Sangorrín, M.P. Optimization of killer assays for yeast selection protocols. *Rev. Argent. Microbiol.* **2010**, *42*, 298–306. [CrossRef]

33. Stead, D. The effect of hydroxycinnamic acids and potassium sorbate on the growth of 11 strains of spoilage yeasts. *J. Appl. Bacteriol.* **1995**, *78*, 82–87. [CrossRef]

34. Harris, V.; Jiranek, V.; Ford, C.M.; Grbin, P.R. Inhibitory effect of hydroxycinnamic acids on *Dekkera* spp. *Appl. Microbiol. Biotechnol.* **2010**, *86*, 721–729. [CrossRef]

35. Lentz, M.; Harris, C. Analysis of Growth Inhibition and Metabolism of Hydroxycinnamic Acids by Brewing and Spoilage Strains of Brettanomyces Yeast. *Foods* **2015**, *4*, 581–593. [CrossRef]

 *fermentation*

Review

Non-*Saccharomyces* Yeasts and Organic Wines Fermentation: Implications on Human Health

Alice Vilela

CQ-VR, Chemistry Research Centre, School of Life Sciences and Environment, Dep. of Biology and Environment, University of Trás-os-Montes and Alto Douro (UTAD), 5000-801 Vila Real, Portugal; avimoura@utad.pt

Received: 12 May 2020; Accepted: 21 May 2020; Published: 25 May 2020

Abstract: A relevant trend in winemaking is to reduce the use of chemical compounds in both the vineyard and winery. In organic productions, synthetic chemical fertilizers, pesticides, and genetically modified organisms must be avoided, aiming to achieve the production of a "safer wine". Safety represents a big threat all over the world, being one of the most important goals to be achieved in both Western society and developing countries. An occurrence in wine safety results in the recovery of a broad variety of harmful compounds for human health such as amines, carbamate, and mycotoxins. The perceived increase in sensory complexity and superiority of successful uninoculated wine fermentations, as well as a thrust from consumers looking for a more "natural" or "organic" wine, produced with fewer additives, and perceived health attributes has led to more investigations into the use of non-*Saccharomyces* yeasts in winemaking, namely in organic wines. However, the use of copper and sulfur-based molecules as an alternative to chemical pesticides, in organic vineyards, seems to affect the composition of grape microbiota; high copper residues can be present in grape must and wine. This review aims to provide an overview of organic wine safety, when using indigenous and/or non-*Saccharomyces* yeasts to perform fermentation, with a special focus on some metabolites of microbial origin, namely, ochratoxin A (OTA) and other mycotoxins, biogenic amines (BAs), and ethyl carbamate (EC). These health hazards present an increased awareness of the effects on health and well-being by wine consumers, who also enjoy wines where terroir is perceived and is a characteristic of a given geographical area. In this regard, vineyard yeast biota, namely non-*Saccharomyces* wine-yeasts, can strongly contribute to the uniqueness of the wines derived from each specific region.

Keywords: ochratoxin A (OTA); mycotoxins; biogenic amines (BAs); ethyl carbamate (EC); organic wines; non-*Saccharomyces*

---

1. General Introduction

As defined at the European level by the European Council Regulations on organic production, organic grapes come from vineyards grown under organic farming methods. Indeed, the International Federation of Organic Agriculture Movement (IFOAM) defines organic viticulture and winemaking as a "holistic production management system which promotes and enhances agro-ecosystem health, including biodiversity, biological cycles, and soil biological activity. It emphasizes the use of management practices in preference to the use of off-farm inputs, considering that regional conditions require locally adapted systems" [1,2]. As of August 2012, organic wines can be labeled "organic" with the EU organic logo. This means the wine can now be properly recognized as an organic product [3]. However, and because the laws regulating organic wine production vary worldwide, the definition "organic wine" does not have the same meaning in all places. Usually, the most important purpose is to avoid synthetic chemical fertilizers and pesticides and genetically modified organisms. In many countries such as the USA, Canada, and Australia, this category of wines has been regulated from

2000, while in Europe, organic wines have been regulated by law since 2012 (EC Regulation No. 203/2012 [4]). Nowadays, many countries, despite having some different regulations, allow organic wine agronomists and winemakers to use the term "organic wine" along with the organic logo on their label after certification (Table 1). In Europe, the organic wines are certified by private structures authorized by a public authority. This regulation allows consumers to distinguish organic wines from conventional wines [5,6].

Table 1. Difference about regulation in terms of maximum use of SO_2 (mg/L) allowed during vinification, the percentage of the organic vineyard in the country (data from 2015 to 2016), and the organic wine label used on their local market [7].

Country	Maxim SO_2 during Vinification	% of Organic Vineyard	Label: Organic or Sustainable
Chile	Red: 75 White: 100	3	Sustainable [b]
Argentina	Red: 70 White: 80 [a]	2	
USA	Use is forbidden	4.1	
Europe	Red: 100 White: 150	8.5	
Australia	Red: 100 White: 100	No data available	
New Zealand	Red: 100 White: 150	7	
South Africa	Red: 90 White: 100	2	Sustainable [b]

[a] 100 mg/L in aged wines; [b] no specific label for organic wines.

As Europe vineyards constituted over 80% of the world's total organic grape growing area in 2014 [6], the European Union (EU) regulations on "organic wine" were an important measure for the global organic wine market. Thus, since the organic certification and standards defined in the EU regulation, it is possible to define exclusive standards with additional detailed production rules. Private standards are appreciated by many winemakers and many consumers as indications of quality wine that authentically express terroir, and that aim to strengthen the subsequent aspects of viticulture and enology: (i) biodiversity in grape production; (ii) attention to soil fertility and soil life; (iii) alternative approaches to pests and diseases; (iv) sustainability of grape production and wine processing and storage; (v) quality and source of wine ingredients, including further limitations on enrichment and requirements for ingredients to be fully organic; (vi) quality of yeasts, including wild yeasts and spontaneous fermentation; (vii) further limitations on additives and further reduction or total ban of sulfites; (viii) further limitations on processing techniques; and (ix) requirements or limitations on tools and equipment [3]. Undeniably, the yeasts present on grape berries from organic vineyards have an inimitable composition and may deliver distinct regional characteristics to this kind of wine [8].

Moreover, according to European regulations, organic wine must be made of organic ingredients. Thus, additional rules for oenological practices, processes, treatments, and substances such as additives and processing aids must be considered. Many practices and substances used in conventional production are unsuitable for organic wine production (Table 2), and there are specific restrictions and limitations, requiring that organic products and substances be used if they are available.

Table 2. List of substances forbidden in organic wines production [3].

Substance	Application
Sorbic acid and sorbates Lysozyme Chitosan	Microbiological stabilization
L-malic acid, D, L-malic acid	Acidification
Ammonium bisulfite	Protection of harvesting
Ammonium sulfate	Management of alcoholic fermentation
Chitin-glucan Chitosan Calcium alginate Co-polymer of vinylimidazole and vinylpyrrolidone (PVI-PVP)	Wine finning
Carboxymethylcellulose (CMC) Yeast mannoproteins	Tartrate/Color stabilization
Polyvinylpolypyrrolidone (PVPP)	Correction of color
Beta-glucanase enzymes	Glucan elimination
Chitin-glucan Chitosan Calcium phytate Potassium ferrocyanide	Clarification elimination of heavy metals (iron, copper)
Urease	Treatment, elimination of ochratoxin A and urea
Caramel	Various

However, despite all these wine private standards, wine safety, for winemakers and consumers, relies upon a complex equilibrium from good winemaking practices, quality of grapes, fermentation, and post-fermentation events. An occurrence in wine safety results in the recuperation of a broad variety of harmful compounds for human health such as amines, carbamate, methanol, mycotoxins, and other dangerous compounds [9].

The perceived increase in sensory complexity and superiority of successful uninoculated wine fermentations, as well as a thrust from consumers looking for a more "natural" or "organic" wine, produced with fewer additives and perceived health attributes, has led to more investigations into the uses of non-*Saccharomyces* yeasts in wine [10–14]. Research in this field aims to understand how to use only the positive contributions of non-*Saccharomyces* yeasts while avoiding negative contributions.

This review aims to provide an overview of organic wine safety, when using indigenous and/or inoculated non-*Saccharomyces* yeasts to perform fermentation, with a special focus on some metabolites of microbial origin, namely, ochratoxin A (OTA) and other mycotoxins, biogenic amines (BAs), and ethyl carbamate (EC). These health hazards present an increased awareness of the effects on health and well-being by wine consumers.

2. Wine Contamination by Ochratoxin A (OTA) and Other Mycotoxins

Human health issues and scientific attention are focused mainly on carcinogenic/toxic mycotoxins [15,16]. More than 300 mycotoxins have been identified, and they are produced by filamentous fungi, mainly *Aspergillus* spp, *Fusarium* spp., and *Penicillium* spp. [17].

OTA is produced from fungi, namely *Aspergillus* spp. and *Penicillium* spp., and derives from 3,4-dihydrocumarin linked to an amide bond with an amino group of L-β-phenylalanine [18,19], and it can appear in grapes (pre-harvest) and/or during pre-fermentation [20,21]. Its presence in wines is mainly found in red wine, followed by rosé and white wines [22,23].

Ochratoxin A is classified by the International Agency for Research on Cancer (IARC) [24] in group 2B (possible human carcinogen), so it is a great threat for humans. It accumulates in several tissues in the body, with the kidneys being its main target, causing Balkan endemic nephropathy (BEN), chronic interstitial nephritis, and karyomegalic interstitial nephritis [9]. The presence of OTA in blood from healthy humans confirms continuous and widespread exposure, thus the Scientific Panel on Contaminants in the Food Chain from the European Food Safety Authority [25] set OTA tolerable weekly intake (TWI) to 120 ng/kg body weight [9]. OTA levels in wines depend on various factors such as weather and vineyard location, the period of harvest, pesticide treatments, wine fermentation, and duration of grape maceration [9]. The European Union allows a maximum limit for OTA in the wine of 2 ng/g [23].

The wines/musts decontamination of OTA has been revised by Quintela et al. [26] and, for conventional wines, physical, chemical, or biological methods can be applied. For organic wines, owing to the restrictions imposed by IFAOM, most of the chemical treatments cannot be used once the chemical products recommended for conventional wines (chitin and chitosan, urease, polyvinylpolypyrrolidone (PVPP) [23]) are forbidden for this kind of wine.

A possible way for wine decontamination could be the bioremediation [21,26] through toxin degradation and adsorption. Several enzymes may be involved in the microbiological degradation of OTA, but not much information is available and only a few have been characterized, including the pancreatic enzyme carboxypeptidase A (CPA) (EC 3.4.17.1) from bovine, the first protease reported to be able to hydrolyze OTA [27]. Toxin degradation can be performed by the bacteria *Pediococcus parvulus* [28], the bacteria *Acinetobacter calcoaceticus* [29], and the soil bacteria *Cupriavidus basilensis* [19]. These microorganisms hydrolyze the OTA amide bond and produce ochratoxin α (OT α), a non-toxic compound (Figure 1). This pathway is promising; however, the production of OTα could also be a threat, because the implication of the accumulation of this compound in the body is yet unclear.

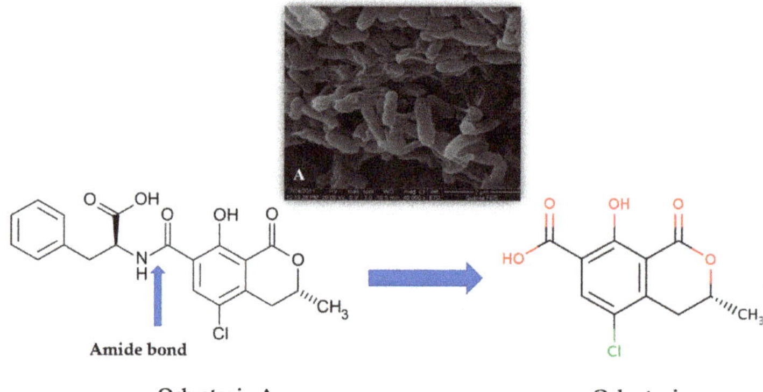

Figure 1. Proposed cleavage of ochratoxin A by *Cupriavidus basilensis* ŐR16. The amide bond hydrolysis forming ochratoxin α as the major degradation product [19]. A— Scanning electron microscope (SEM) micrograph, magnification of 40,000 of a *Cupriavidus basilensis* biofilm [30].

A second way for wine decontamination is OTA adsorption on the yeast cell wall during fermentation. For oenological strains, the parietal adsorption activity is a new selection feature that is attractive because it can enhance wine safety and quality [31]. Several authors proposed the yeasts as adsorbing tools under both in vitro and in vivo conditions [21,32,33]. Several studies have also reported the interaction of yeast cells with a diversity of wine compounds, from coloring pigments such as anthocyanins [34] to sulfur compounds [35] or detrimental components such as octanoic and decanoic acids [36], pesticides [37], geosmin [38], and 4-ethylphenol [39].

The yeast parietal adsorption activity is different from yeast to yeast, depending on the structural characteristics and chemical composition of the outermost layer of the cell wall. This layer is made up of mannoproteins, which represent 25–50% of the entire cell wall [40]. Parietal mannoproteins relate to an inner matrix of amorphous β-1,3 glucan and are partly released in wine (Figure 2).

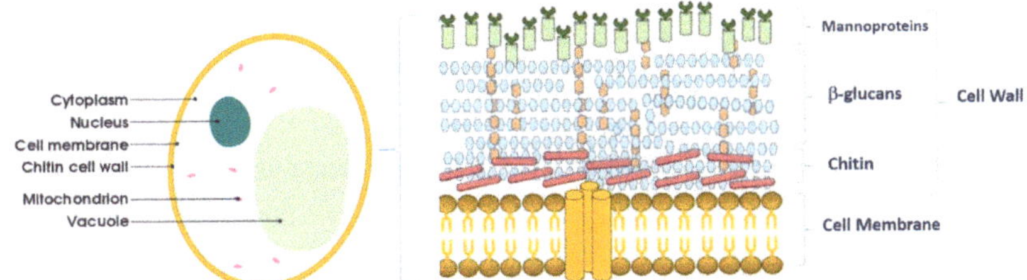

Figure 2. Constitution of yeast and yeast cell wall. Yeast β-glucans form long chains of a β-(1→3) linked glucan backbone with β-(1→6) linked glucose side branches Adapted from Anwar et al. [40].

Total charge, charge distribution, and accessible surface area of mannoproteins are the most important features determining adsorption, and these features differ among strains [41]. Mannoproteins from *Schizosaccharomyces pombe* generally contain β-1,3-linked pyruvylated galactose residues [42], whereas mannoproteins from *Pichia pastoris* mainly contain mannose phosphate diesters [43]. The percentage of acidic oligosaccharides, containing mannosyl phosphate, varies from strain to strain in *Saccharomyces*, *Kloeckera brevis*, and *Candida albicans*, whereas the oligosaccharides of *S. pombe* and *Kluyveromyces lactis* do not contain mannosyl phosphate [44]. The presence of acidic oligosaccharides and negative charges modifies the electrostatic and ionic interactions between the yeast's cell wall and wine components. Moreover, wine yeasts may exhibit a different ratio of neutral/acidic oligosaccharides in mannoproteins. This oenological effect can help us to choose a specific selection of wine yeasts, even though the extent of mannosyl phosphorylation also depends on culture conditions such as the media and cultivation period [44].

Parietal mannoproteins, which perform various oenological functions [31], regarding their adsorption activity, adsorb ochratoxin A from grape must and wine. Numerous investigators studied the removal of OTA by yeasts during fermentation [21,33,45,46]. Truly, ochratoxin A removal depends on yeast macromolecules, such as mannoproteins, and corresponds to a spontaneous adsorption mechanism [47,48] where mannoproteins act like a sponge, removing ochratoxin A [49]. Moreover, this phenomenon can be strongly affected by some factors that also affect the fermentation process, like pH, temperature, sugar, and nitrogen supplementation [50]. Furthermore, this phenomenon is somewhat reversible, as the toxin can be released back into the wine [50]. Yeast immobilization into alginate beads is an interesting technique, aiming to promote a better absorptivity. According to Farbo et al. [51], immobilized yeasts were able to remove 80% of OTA in 48 h and toxin release by beads could be better controlled than in free cells, and, additionally, the entrapped cells could be re-usable.

Yeast mycotoxin adsorption, considered to be performed by the surface of cell walls, varies with the yeast species as they are diverse in cell wall composition, varying in adsorption capacity [52]. Many yeast species can absorb mycotoxins, including *S. cerevisiae*, *Candida tropicalis*, *Pichia pastoris*, and *Phaffia rhodozyma* [45,53,54].

Yeast cells having integrated cell walls are generally more effective in their mycotoxin adsorption capacity than other yeast cells, indicating that toxin adsorption requires the structural integrity of the yeast cell wall [55]. The interaction mechanism studies of mycotoxins and yeast cells focused on the toxin adsorption capability associated with cell wall physical structure and the morphology, chemical components, and complicated interactions between structure and these components have been studied. Armando et al. [56] suggested that the cells with the greatest cell wall content seem to present the highest mycotoxin removal percentage, in contrast with those with less content. Luo et al. [57] investigated patulin adsorption capabilities of four yeast strains, among them were two non-*Saccharomyces*, *Candida tropicalis* N-10 and *Pichia anomala* B-2p, with different cell wall thicknesses and cell morphologies. The mycotoxin patulin adsorption capability decreased or disappeared when the cell wall three-dimensional network was damaged or removed.

3. Organic Wines Contamination with Biogenic Amines

Biogenic amines (BAs) are low-molecular-weight organic molecules originating in fermented foods from the microbial catabolism of the corresponding amino acids. Wine BA includes putrescine (from arginine and ornithine), cadaverine (from lysine), tyramine (from tyrosine), histamine (from histidine), and tryptamine (from tryptophane) [9]; Figure 3. The production of BAs is a strategy to obtain metabolic advantages to face certain stress conditions [58].

BAs are present as salts, but, at the mouth pH, they are partly in free form, becoming reactive with other compounds responsible for the aroma of the wine, thus they can be responsible for sensory changes like loss of varietal character and the appearance of musty smell and flavor [59,60].

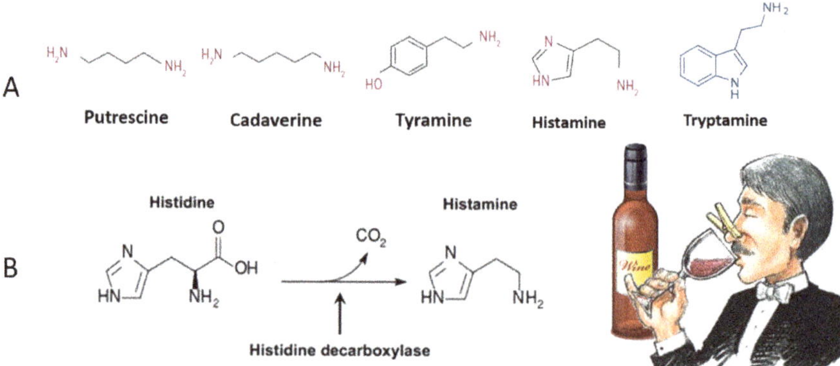

Figure 3. Biogenic amines (BAs) in wine that, besides being a healthy treat, negatively affect the aromatic quality of wines owing to their unpleasant smells (**A**). The decarboxylase enzyme transforms amino acid into a biogenic amine by removing its carboxyl group. The example presented is the formation of histamine (**B**).

The intake of high amounts of dietary BA can lead to several disorders, from minor symptoms resembling allergic reactions to death in severe cases of histaminosis or tyraminosis [20]. Moreover, the synergistic effect of inhibitors of the amino oxidases, such as some drugs, putrescine, and alcohol, lead them to act as histamine enhancers [9]. Humans' high sensitivity toward biogenic amines ingested with the diet depends on insufficient amino oxidase activity caused by drugs, genetic predisposition (histamine intolerance), gastrointestinal disease, inhibition by alcohol, acetaldehyde, and other amines like putrescine and cadaverine [61,62].

Arginine and histidine are the most abundant amino acids in grapes. Consequently, histamine production in wines is a huge concern, as its toxicity is amplified by the alcohol and high levels of putrescine [63]. Besides, high levels of putrescine and cadaverine negatively affect the aromatic quality of wines owing to their unpleasant smells [63].

Some factors of agronomic practice as well as of the winemaking process can cause discrete levels of biogenic amines in the wine; that is, the fertilization of the soil (nitrogen level), the poor state of health of the grapes and presence of molds, non-regular lowering of the pH of the must and development of some non-*Saccharomyces* yeasts, and the activity of lactic acid bacteria responsible for malolactic fermentation (MLF) [64]. As MLF especially occurs in red wines, higher BA amounts are usually found in red wine than in rosé, white, or sparkling wines [65]. *O. oeni* is the main lactic acid bacteria (LAB) species carrying out the MLF, and its capability to produce histamine has been reported [62].

Yildirim et al. [66] compared organic and conventional Turkish wines from several grape varieties (Cabernet Sauvignon, Carignan, Colombard, Merlot, and Semillon) for their BA content. The highest average values were found in putrescine (5.55 mg/L), ethylamine (0.825 mg/L), and histamine (0.628 mg/L) in organic wines, and putrescine (3.68 mg/L), histamine (1.14 mg/L), and agmatine (0.662 mg/L) in non-organic wines. No β-phenylethylamine was detected. Putrescine was more predominant in organic wines than in non-organic wines ($p = 0.008$). Changes of BAs were previously studied by Garcia-Marino et al. [67] during the winemaking process of red wine, including an organic wine. Even though organic foods were popular in consumers, organic wines produced higher levels of BAs than conventional wines. This may be related to the fact that, in organic wines, MLF normally occurs spontaneously; moreover, in this kind of wine, low levels of SO_2 are added owing to the legal restrictions [62].

The effect of organic or conventional agriculture on the BA content of wines was also evaluated by Tassoni et al. [68]. The authors analyzed the BA content in Lambrusco (red) and Albana (white) wines, and they compared conventional, organic, and biodynamic agricultural and oenological practices. In all the samples, putrescine was the most abundant polyamine, but its content was lower in biodynamic wines than in conventional wines. Samples from Albana organic wines and Lambrusco biodynamic wines contained the highest BA amounts, with histamine and tryptamine being the most abundant amines in both wines. Moreover, in Lambrusco, spermidine was present in organic and biodynamic samples, but it was absent in conventional samples; in Albana, this amine was present in the same amount in all of the samples.

Although biogenic amines formation during the alcoholic fermentation (AF) is considered irrelevant, during AF, yeasts consume amino acids as a nitrogen source, which plays a key role in creating aroma compounds. Caruso et al. [69] studied the BA production capability of fifty yeast strains isolated from grapes and wines. Among them, five species, *Brettanomyces bruxellensis*, *Candida krusei*, *Metschnikowia pulcherrima*, *Kloeckera apiculate*, and *S. cerevisiae*, were BAs producers. They also found that *B. bruxellensis* produced the highest levels of total BA (15 mg/L), followed by *S. cerevisiae* (12.14 mg/L). Moreover, these yeast species were able to significantly produce putrescine, phenylethylamine, and ethanolamine. These results suggest that correct yeast management during winemaking is important [61].

Benito and co-workers [70] in a study aiming to evaluate the influence of *Lachancea thermotolerans* on low-acidity Airén grape must from the south of Spain, proving that *L. thermotolerans* does not produce higher levels of biogenic amines than *S. cerevisiae*. Moreover, the lower concentration of histidine (precursor of histamine) found during *L. thermotolerans* and *S. cerevisiae* fermentation can contribute to reducing the potential risk of histamine formation by bacterial metabolism. Other authors have also reported reductions of histamine of up to 2.2 mg/L during alcoholic fermentation with the non-*Saccharomyces* species *Hanseniaspora vineae* [71].

4. Wines and Ethyl Carbamate Contamination

Wine, including organic wines, possess distinct nutrients, in which a variety of microorganisms, namely yeasts and bacteria, exist. The fermentation processes may unavoidably produce toxic products because of metabolism and side reactions, including biogenic amines (BAs) and ethyl carbamate (EC). Curiously, these compounds are generated owing to the incomplete metabolism of nitrogen-containing compounds during the fermentation process [72]. EC is mainly produced by lactic acid bacteria and through the chemical combination of urea with ethanol during wine aging. EC has been upgraded by the IARC to a "probable human carcinogen", Group 2A [73,74].

The carcinogenicity of EC has been verified in several animal species from rats, hamsters, and monkeys [73,75]. In rodents, EC has been found to cause a dose-dependent increase in carcinomas of the liver, lungs, heart, mammary gland, ovaries, skin, and forestomach, among which hepatocellular tumors appear to linearly increase with EC concentration [74,76].

During fermentation, five metabolic pathways were identified for the formation of EC. The major precursors of the formation of EC contain a carbamyl group, and these include urea, citrulline, and carbamoyl phosphate. Furthermore, it has been shown that cyanic acid and diethylpyrocarbonate are involved in EC formation [74].

The reaction between urea and ethanol is the most common metabolic pathway of EC formation found in wine. The abundance of urea in grapes makes it the most common precursor. Moreover, during ethanol fermentation, the accumulation of urea originated from the catabolism of arginine contributes to the reaction between urea and ethanol. Additionally, urea mainly results from the metabolism of arginine by *S. cerevisiae* [74,77].

EC can also be formed by the reaction between citrulline and ethanol. Grape juice already contains a certain amount of citrulline, but much of this compound has its origin in the catabolism of arginine. Moreover, the generation of citrulline is assigned to the metabolism of arginine by lactic acid bacteria (LAB) via malolactic fermentation [78].

EC, in some alcoholic beverages, may also appear as a result of the reaction between cyanic acid and ethanol, and/or the reaction between carbamyl phosphate and ethanol. However, these are rare phenomena in wine [74].

The reaction between diethylpyrocarbonate and ammonia nowadays occurs less frequently, mainly in organic wines. The appearance of diethylpyrocarbonate stems from artificial additives. This compound was known to reduce contamination and spoilage by microorganisms (yeasts or bacteria). However, the use of diethylpyrocarbonate was abandoned owing to its toxicity and the undesirable side effect of EC formation [79].

To better understand the metabolic formation of EC in *S. cerevisiae*, transport and metabolic regulation of urea in *S. cerevisiae* must be studied. Intracellular urea mainly results from the degradation of arginine through catalysis by arginase (Figure 4). As a toxic and poor nitrogen source for *S. cerevisiae*, the generated urea is usually accumulated and exported to the nearby medium via a facilitated transport diffusion system (Figure 4, [74]). *S. cerevisiae* metabolizes urea in two steps. First, urea is carboxylated to form allophanate by urea carboxylase. Then, allophanate is degraded to CO_2 and NH_4^+ by allophanate hydrolase. The activities of urea carboxylase and allophanate hydrolase are performed by a bifunctional enzyme, urea amidolyase, encoded by the DUR1,2 genes, and silenced by nitrogen catabolic repression (NCR) [80]. The DUR3 gene encodes urea permease. Under fermentation conditions, degradation is obstructed by arginine, which is abundant in fermented sources and acts as a superior nitrogen supply compared with urea [74].

Several methods have been proposed for decreasing EC in wines: (i) the modification of raw materials (established recommendations on vineyard fertilization, cultivars, and nutrient status/additions, including avoiding excessive fertilization with urea, ammonia, and other N-fertilizers) and the optimization of fermentation processing parameters (such as temperature, light irradiation, pH, oxygen, and storage time); (ii) the addition of acid urease (commercial grade acid ureases are currently acquired mainly from *L. fermentum*) [81]; and (iii) the modification of the fermentation bacterium. All these approaches aim to reduce EC precursors [74]. However, the most common type of management in the wine industry is based on the use of a commercial urease enzyme, able to remove all of the urea that can evolve into ethyl carbamate [82]. Researchers have also focused on the immobilization of acid urease, possessing the advantages of facilitating enzyme recycling, reducing cost, and improving stability and resistance to inhibitory compounds [83].

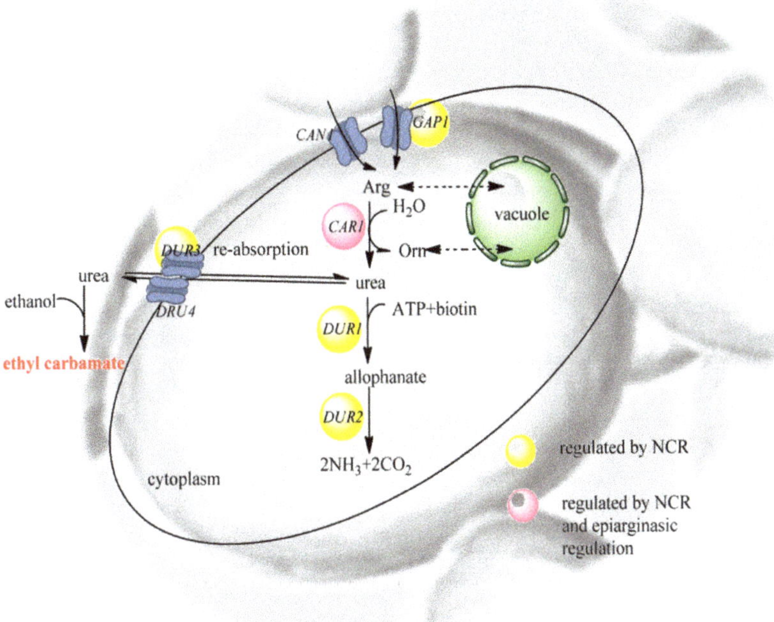

Figure 4. Schematic metabolism of urea and arginine by *S. cerevisiae*. Intracellular urea mainly results from the degradation of arginine through catalysis by arginase (CAR1). As a toxic and poor nitrogen source for *S. cerevisiae*, the generated urea is usually accumulated and exported to the surrounding medium via a facilitated diffusion system. NCR—nitrogen catabolic repression; ATP—adenosine triphosphate; CAN1—arginine transporter; GAP1—general amino acid permease [74].

However, owing to organic wine fermentation restrictions, urease cannot be used to treat this kind of wine [3]. Thus, the use of non-*Saccharomyces* species with urease activity allows the removal of the main ethyl carbamate precursor from wine, making it virtually impossible for ethyl carbamate to appear during wine aging [84].

Past studies of *Schizosaccharomyces* focused on malic acid degradation [85]. Later, the genus *Schizosaccharomyces* also showed the ability to reduce levels of hazardous compounds for human health, such as ochratoxin A, biogenic amines, and ethyl carbamate [86,87]. Thus, this seems to be a promising strain for EC control in organic wines.

5. Other Benefits of the Use of Non-*Saccharomyces* in Organic Winemaking

The yeast population of vineyard and grape berries surface is significant as these yeasts may contribute to the fermentative process of organic wines. It is well known that the microflora composition of grape berries surface is influenced by factors such as climate, UV radiation, nutritive limitations, and agrochemical treatments [88]. One of the objectives of organic viticulture is the use of copper- and sulfur-based molecules to protect vines as an alternative to synthetic chemical pesticides. However, these "more friendly" compounds may influence the occurrence and abundance of yeasts on the surface of grape berries. Several studies have been made, and contradictory findings have been reported. Cordero-Bueso et al. [89] found that conventional phytosanitary treatments reduced both the number and diversity of yeasts, whereas Grangeteau et al. [90] reported a lower yeast concentration and biodiversity, on grape berry surface, in organic grape samples in comparison with conventional ones.

It is well known that indigenous yeasts can play a role in the development of distinctive terroir-related characteristics, thus creating specific traits specifically in organic wine. Moreover, after numerous studies, during the last recent years, it is believed that non-*Saccharomyces* are important tools for the wine fermentation process. As these yeasts are usually unable to complete alcoholic fermentation, they are mostly used in co- or sequential inoculation with *S. cerevisiae* [14,91] in conventional or organic wines production.

Non-*Saccharomyces* are important to improve wine complexity and pleasantness [14,92], either alone or in co/sequential inoculation with *S. cerevisiae* strains. For example, *Hanseniaspora vineae* enhanced benzenoid compounds and phenyl ethyl acetate with an agreeable rose-like aroma [93]; *Torulaspora delbrueckii* increased the concentration of 3-Sulfanylhexan-1-ol (tropical fruit nuances) in the mixed fermentation with an *S. cerevisiae* strain [94]. Mixed fermentations with *Starmerella bacillaris*, *Zygotorulaspora florentina*, and *Hanseniaspora uvarum* enhanced the flavor [95], floral notes and lower perception of astringency [96], and wine organoleptic quality reducing the volatile acidity [97], respectively.

Non-*Saccharomyces* can also be used for controlling wild undesired microflora once many species can produce active extracellular molecules that neutralize the development of wild spoilage microorganisms. One example is the investigations focused on the biological control of the wine spoilage *Brettanomyces/Dekkera* yeasts [98]. In the former work, Oro and collaborators [98] demonstrated the practical application of zymocins produced by *Kluyveromyces wickerhamii* and *Wickerhanomyces anomalus* in wines contaminated by *Brett*, thus avoiding the use of huge amounts of sulfur dioxide. Similar studies were performed by Mehlomakulu et al. [99], finding two novel killer toxins, CpKT1 and CpKT2, produced by *Candida pyralidae*, active and stable under winemaking conditions; furthermore, new zymocins from *T. delbrueckii* able to control spoilage by *B. bruxellensis* were identified and characterized [100].

One interesting phenomenon, also recently studied, is cell-to-cell contact and quorum sensing. Quorum sensing is a mechanism in which the production of a small molecule known as an autoinducer accumulates in the extracellular environment and, on reaching a critical concentration, activates the transcription of target genes [101]. Quorum sensing was recently analyzed in *S. cerevisiae*, *H Hanseniaspora uvarum*, *Torulaspora pretoriensis*, *Zygosaccharomyces bailii*, *Candida zemplinina*, and *Dekkera bruxellensis*, and 2-phenylethanol, tryptophol, and tyrosol, produced by these non-*Saccharomyces* yeasts, were found to be the main molecules involved in the quorum-sensing mechanism [102].

The use of non-*Saccharomyces* can be a biological way to reduce the ethanol content in wines. Different mechanisms of some of these yeasts, balanced between respiration and fermentation, compared with *S. cerevisiae*, could be explored to reduce ethanol production through partial and controlled aeration of the grape juice. In this way, sugar is consumed via respiration rather than fermentation [103,104].

Non-*Saccharomyces* yeasts can also be a valuable tool for production low-sulfur wines. Owing to the restrictions imposed by the EU on additives used in organic wines, the main concern in the organic winemaking process is the risks of oxidation combined with microbial contamination and H_2S production. Although new starters 'low H_2S-SO_2-acetaldehyde producers', obtained by selective breeding, are already available in the market, indigenous tailor-made yeasts are needed to imprint the wines with the specific terroir, and simultaneously, to avoid the production of compounds undesirable in organic wines [88,105]. Moreover, non-*Saccharomyces* are also able to reduce copper, which can appear in organic wines owing to the use of copper fungicide in organic agricultural practices. These metals in concentrations higher than 64 mg/L cause sluggish or stuck fermentation and a reduction in alcohol production [88].

6. Final Remarks

The growing social interest in organic wines stimulated several studies aiming at improving organic grape production and organic wine quality during the entire chain production. A change in consumer preference towards organic wines, perceived to be more natural, made using less or no chemical additions as compared with conventional wine, has encouraged bioprospecting for naturally occurring microorganisms, yeast, and bacteria, which can be applied in winemaking as an alternative to such additions.

Furthermore, the increasing request of organic wines determines changes in yeast microbiota and fermentation requirements, demanding starter strains, particularly non-*Saccharomyces*, with peculiar features, such as low production of H_2S, SO_2, and acetaldehyde; the reduction of ethanol content; the ability to reduce copper content; the bio-control of undesirable spoilage yeasts; and more importantly, the possible control of mycotoxins (like OTA), biogenic amines (BAs), and ethyl carbamate (EC), harmful compounds for human health.

In conclusion, the use of selected cultures of non-*Saccharomyces* wine yeasts in organic wines production, further than the enhancement of wine complexity and typicity, offers other advantages, related to wine safety and consumer's health.

Funding: FCT and COMPETE, grants numbers UIDB/00616/2020 and UIDP/00616/2020.

Acknowledgments: The author appreciates the financial support provided to the Research Unit in Vila Real (CQ-VR).

Conflicts of Interest: The author declares no conflict of interest.

References

1. IFOAM. Basic Standards for Organic Production and Processing. Bonn-Germany. 2014. Available online: http://www.ifoam.orghttps://www.ifoam.bio/sites/default/files/ifoam_norms_july_2014_t.pdf (accessed on 15 April 2020).
2. Trioli, G.; Hofmann, U. ORWINE: Code of good organic viticulture and winemaking. In *ECOVIN-Federal Association of Organic Wine-Producer*; Ecovin: Oppenheim, Germany, 2009.
3. IFAOM. EU Rules for Organic Wine Production: Background, Evaluation, and Further Sector Development. 2013. Available online: https://orgprints.org/29867/1/ifoameu_reg_wine_dossier_201307.pdf (accessed on 17 April 2020).
4. Commission Implementing Regulation (EU) No 203/2012 of 8 March 2012 Amending Regulation (EC) No 889/2008 Laying Down Detailed Rules for the Implementation of Council Regulation (EC) No 834/2007, as Regards Detailed Rules on Organic Wine. OJ L 71. 9 March 2012, pp. 42–47. Available online: https://eur-lex.europa.eu/legal-content/EN/TXT/?uri=celex%3A32012R0203 (accessed on 10 May 2020).
5. Schäufele, I.; Hamm, U. Consumers' perceptions, preferences, and willingness-to-pay for wine with sustainability characteristics: A review. *J. Clean. Prod.* **2017**, *147*, 379–394. [CrossRef]
6. Cravero, M.C. Organic, and biodynamic wines quality and characteristics: A review. *Food Chem.* **2019**, *295*, 334–340. [CrossRef] [PubMed]
7. DIVA. Overview of the Organic Wine Market. 2007. Available online: https://divawine.com/overview-organic-market/ (accessed on 7 May 2020).
8. Tofalo, R.; Schirone, M.; Telera, G.C.; Manetta, A.C.; Corsetti, A.; Suzzi, G. Influence of organic viticulture on non-*Saccharomyces* wine yeast populations. *Ann. Microbiol.* **2011**, *61*, 57–66. [CrossRef]
9. Russo, P.; Capozzi, V.; Spano, G.; Corbo, M.R.; Sinigaglia, M.; Antonio, B. Metabolites of Microbial Origin with an Impact on Health: Ochratoxin A and Biogenic Amines. *Front. Microbiol.* **2016**, *7*, 482. [CrossRef]
10. Ciani, M.; Comitini, F. Non-Saccharomyces wine yeasts have a promising role in biotechnological approaches to winemaking. *Ann. Microbiol.* **2011**, *61*, 25–32. [CrossRef]
11. Jolly, N.P.; Varela, C.; Pretorius, I.S. Not your ordinary yeast: Non-*Saccharomyces* yeasts in wine production uncovered. *FEMS Yeast Res.* **2014**, *14*, 215–237. [CrossRef]
12. Maurizio, C.; Francesca, C.; Ilaria, M.; Paola, D. Controlled mixed culture fermentation: A new perspective on the use of non-*Saccharomyces* yeasts in winemaking. *FEMS Yeast Res.* **2010**, *10*, 123–133. [CrossRef]
13. Vilela, A. The Importance of Yeasts on Fermentation Quality and Human Health-Promoting Compounds. *Fermentation* **2019**, *5*, 46. [CrossRef]

14. Vilela, A. Modulating Wine Pleasantness Throughout Wine-Yeast Co-Inoculation or Sequential Inoculation. *Fermentation* **2020**, *6*, 22. [CrossRef]

15. Walker, R. Risk assessment of ochratoxin: Current views of the European Scientific Committee on Food, the JECFA and the Codex Committee on Food Additives and Contaminants. *Adv. Exp. Med. Biol.* **2002**, *504*, 249–255. [CrossRef]

16. Ji, C.; Fan, Y.; Zhao, L. Review on biological degradation of mycotoxins. *Anim. Nutr.* **2016**, *2*, 127–133. [CrossRef]

17. Barreira, M.J.; Alvito, P.C.; Almeida, C.M. Occurrence of patulin in apple-based foods in Portugal. *Food Chem.* **2010**, *121*, 653–658. [CrossRef]

18. Peraica, M.; Radic, B.; Lucić, P.; Pavlovic, M. Toxic effects of mycotoxins in humans. *Int. J. Public Health* **1999**, *77*, 754–766.

19. Ferenczi, S.; Cserháti, M.; Krifaton, C.; Szoboszlay, S.; Kukolya, J.; Szőke, Z.; Kovács, K.J. A New Ochratoxin A Biodegradation Strategy Using *Cupriavidus basilensis* Őr16 Strain. *PLoS ONE* **2014**, *9*, e109817. [CrossRef] [PubMed]

20. Spano, G.; Russo, P.; Lonvaud-Funel, A.; Lucas, P.; Alexandre, H.; Grandvalet, C.; Rattray, F. Biogenic amines in fermented foods. *Eur. J. Clin. Nutr.* **2010**, *64*, 95–100. [CrossRef]

21. Petruzzi, L.; Sinigaglia, M.; Corbo, M.R.; Campaniello, D.; Speranza, B.; Bevilacqua, A. Decontamination of Ochratoxin A by yeasts: Possible approaches and factor leading to toxin removal in wine. *Appl. Microbiol. Biotechnol.* **2014**, *98*, 6555–6567. [CrossRef]

22. Battilani, P.; Giorni, P.; Bertuzzi, T.; Formenti, S.; Pietri, A. Black *Aspergilli* and Ochratoxin A in grapes in Italy. *Int. J. Food Microbiol.* **2006**, *111*, S53–S60. [CrossRef]

23. Bellver Soto, J.; Fernández-Franzón, M.; Ruiz, M.J.; García, A.J. Presence of Ochratoxin A (OTA) mycotoxin in alcoholic drinks from southern European countries: Wine and beer. *J. Agric. Food Chem.* **2014**, *62*, 7643–7651. [CrossRef]

24. IARC. Mycotoxins and Human Health (Chapter 6). 2020; pp. 87–104. Available online: file:///C:/Users/avimo/Downloads/IARC_SP158_Chapter%206.pdf (accessed on 17 April 2020).

25. EFSA (European Food Safety Authority). Opinion of the scientific panel on contaminants in the food chain on a request. Commission related to Ochratoxin A in food. *EFSA J.* **2006**, *365*, 1–56.

26. Quintela, S.; Villarán, M.C.; Armentia, I.L.; Elejalde, E. Ochratoxin A removal in wine: A review. *Food Control* **2013**, *30*, 439–445. [CrossRef]

27. Pitout, M.J. The hydrolysis of ochratoxin A by some proteolytic enzymes. *Biochem. Pharmacol.* **1969**, *18*, 485–491. [CrossRef]

28. Abrunhosa, L.; Inês, A.; Rodrigues, A.I.; Guimarães, A.; Pereira, V.L.; Paropt, P.; Venâncio, A. Biodegradation of Ochratoxin A by *Pediococcus parvulus* isolated from Douro wines. *Int. J. Food Microbiol.* **2014**, *188*, 45–52. [CrossRef] [PubMed]

29. De Bellis, P.; Tristezza, M.; Haidukowski, M.; Fanelli, F.; Sisto, A.; Mulè, G.; Grieco, F. Biodegradation of Ochratoxin A by Bacterial Strains Isolated from Vineyard Soils. *Toxins (Basel)* **2015**, *7*, 5079–5093. [CrossRef] [PubMed]

30. Friman, H.; Schechter, A.; Ioffe, Y.; Nitzan, Y.; Cahan, R. Electricity formation in a microbial fuel cell. *Microb. Biotechnol.* **2013**, *6*, 425–434. [CrossRef]

31. Caridi, A.; Cufari, A.; Lovino, R.; Palumbo, R.; Tedesco, I. Influence of yeast on polyphenol composition of wine. *Food Technol. Biotechnol.* **2004**, *42*, 37–40.

32. Petruzzi, L.; Baiano, A.; De Gianni, A.; Sinigaglia, M.; Corbo, M.R.; Bevilacqua, A. Differential Adsorption of Ochratoxin A and Anthocyanins by Inactivated Yeasts and Yeast Cell Walls during Simulation of Wine Aging. *Toxins (Basel)* **2015**, *7*, 4350–4365. [CrossRef]

33. Bevilacqua, A.; Petruzzi, L.; Corbo, M.R.; Baiano, A.; Garofalo, C.; Sinigaglia, M. Ochratoxin A released back into the medium by *Saccharomyces cerevisiae* as a function of the strain, washing medium, and fermentative conditions. *J. Sci. Food Agric.* **2014**, *94*, 3291–3295. [CrossRef]

34. Mazauric, J.P.; Salmon, J.M. Interactions between yeast lees and wine polyphenols during simulation of wine aging: II. Analysis of desorbed polyphenol compounds from yeast lees. *J. Agric. Food Chem.* **2006**, *54*, 3876–3881. [CrossRef]

35. Palacios, S.; Vasserot, Y.; Maujean, A. Evidence for sulfur volatile products adsorption by yeast lees. *Am. J. Enol. Vitic.* **1997**, *48*, 525–526.

36. Alexandre, H.; Lubbers, S.; Charpentier, C. Interactions between toxic fatty acids for yeasts and colloids, cellulose and yeast ghost using the equilibrium dialysis method in a model wine system. *Food Biotechnol.* **1997**, *11*, 89–99. [CrossRef]

37. Navarro, S.; Barba, A.; Oliva, J.; Navarro, G.; Pardo, F. Evolution of residual levels of six pesticides during elaboration of red wines. Effect of wine-making procedures in their disappearance. *J. Agric. Food Chem.* **1999**, *47*, 264–270. [CrossRef]

38. Pradelles, R.; Chassagne, D.; Vichi, S.; Gougeon, R.; Alexandre, H. (−) Geosmin sorption by enological yeasts in model wine and FTIR spectroscopy characterization of the sorbent. *Food Chem.* **2010**, *120*, 531–538. [CrossRef]

39. Palomero, F.; Ntanos, K.; Morata, A.; Benito, S.; Suárez-Lepe, J.A. Reduction of wine 4-ethylphenol concentration using lyophilized yeast as a bioadsorbent: Influence on anthocyanin content and chromatic variables. *Eur. Food Res. Technol.* **2011**, *232*, 971–977. [CrossRef]

40. Anwar, M.I.; Muhammad, F.; Awais, M.M.; Akhtar, M. A review of β-glucans as a growth promoter and antibiotic alternative against enteric pathogens in poultry. *World's Poult. Sci. J.* **2017**, *73*, 651–661. [CrossRef]

41. Huwig, A.; Freimund, S.; Käppeli, O.; Dutler, H. Mycotoxin detoxication of animal feed by different adsorbents. *Toxicol. Lett.* **2001**, *122*, 179–188. [CrossRef]

42. Domizio, P.; Liu, Y.; Bisson, L.F.; Barile, D. Cell wall polysaccharides released during the alcoholic fermentation by *Schizosaccharomyces pombe* and *S. japonicus*: Quantification and characterization. *Food Microbiol.* **2017**, *61*, 136–149. [CrossRef]

43. Marx, H.; Sauer, M.; Resina, D.; Vai, M.; Porro, D.; Valero, F.; Ferrer, P.; Mattanovich, D. Cloning, disruption and protein secretory phenotype of the GAS1 homologue of *Pichia pastoris*. *FEMS Microbiol. Lett.* **2006**, *264*, 40–47. [CrossRef]

44. Jigami, Y.; Odani, T. Mannosylphosphate transfer to yeast mannan. Molecular and Cell Biology of Lipids. *Biochim. Biophys. Acta* **1999**, *1426*, 335–345. [CrossRef]

45. Esti, M.; Benucci, I.; Liburdi, K.; Acciaro, G. Monitoring of ochratoxin A fate during alcoholic fermentation of wine must. *Food Control* **2012**, *27*, 53–56. [CrossRef]

46. Luo, Y.; Liu, X.; Yuan, L.; Li, J. Complicated interactions between bio-adsorbents and mycotoxins during mycotoxin adsorption: Current research and future prospects. *Trends Food Sci. Technol.* **2020**, *96*, 127–134. [CrossRef]

47. Bejaoui, H.; Mathieu, F.; Taillandier, P.; Lebrihi, A. Ochratoxin A removal in synthetic and natural grape juices by selected oenological *Saccharomyces* strains. *J. Appl. Microbiol.* **2004**, *97*, 1038–1044. [CrossRef] [PubMed]

48. Ringot, D.; Lerzy, B.; Bonhoure, J.P.; Auclair, E.; Oriol, E.; Larondelle, Y. Effect of temperature on in vitro ochratoxin A biosorption onto yeast cell wall derivatives. *Proc. Biochem.* **2005**, *40*, 3008–3016. [CrossRef]

49. Caridi, A. New perspectives in safety and quality enhancement of wine through selection of yeasts based on the parietal adsorption activity. *Int. J. Food Microbiol.* **2007**, *120*, 167–172. [CrossRef]

50. Piotrowska, M.; Masek, A. *Saccharomyces Cerevisiae* Cell Wall Components as Tools for Ochratoxin A Decontamination. *Toxins* **2015**, *7*, 1151–1162. [CrossRef] [PubMed]

51. Farbo, M.G.; Urgeghe, P.P.; Fiori, S.; Marceddu, S.; Jaoua, S.; Migheli, Q. Adsorption of Ochratoxin A from grape juice by yeast cells immobilised in calcium alginate beads. *Int. J. Food Microbiol.* **2016**, *217*, 29–34. [CrossRef] [PubMed]

52. Pfliegler, W.P.; Pusztahelyi, T.; Pócsi, I. Mycotoxins-prevention and decontamination by yeasts. *J. Basic Microbiol.* **2015**, *55*, 805–818. [CrossRef]

53. Cecchini, F.; Morassut, M.; Saiz, J.C.; Moruno, E.G. Anthocyanins enhance yeast's adsorption of Ochratoxin A during the alcoholic fermentation. *Eur. Food Res. Technol.* **2019**, *245*, 309–314. [CrossRef]

54. Luo, Y.; Wang, Z.L.; Yuan, Y.H.; Zhou, Z.K.; Yue, T.L. Patulin adsorption of a superior microorganism strain with low flavor-affection of kiwi fruit juice. *World Mycotoxin J.* **2016**, *9*, 195–203. [CrossRef]

55. Nunez, Y.P.; Pueyo, E.; Carrascosa, A.V.; Martínez-Rodríguez, A.J. Effects of Aging and Heat Treatment on Whole Yeast Cells and Yeast Cell Walls and on Adsorption of Ochratoxin A in a Wine Model System. *J. Food Prot.* **2008**, *71*, 1496–1499. [CrossRef]

56. Armando, M.R.; Pizzolitto, R.P.; Dogi, C.A.; Cristofolini, A.; Merkis, C.; Poloni, V.; Cavaglieri, L.R. Adsorption of ochratoxin A and zearalenone by potential probiotic *Saccharomyces cerevisiae* strains and its relation with cell wall thickness. *J. Appl. Microbiol.* **2012**, *113*, 256–264. [CrossRef]

57. Luo, Y.; Wang, J.G.; Liu, B.; Wang, Z.L.; Yuan, Y.H.; Yue, T.L. Effect of yeast cell morphology, cell wall physical structure and chemical composition on patulin adsorption. *PLoS ONE* **2015**, *21*, e0136045. [CrossRef] [PubMed]

58. Wolken, W.A.M.; Lucas, P.M.; Lonvaud-Funel, A.; Lolkema, J.S. The mechanism of the tyrosine transporter TyrP supports a proton motive tyrosine decarboxylation pathway in Lactobacillus brevis. *J. Bacteriol.* **2006**, *188*, 2198–2206. [CrossRef] [PubMed]

59. Smit, A.Y.; du Toit, W.J.; du Toit, M. Biogenic amines in wine: Understanding the headache. *South Afr. J. Enol. Vitic.* **2008**, *29*, 109–127. [CrossRef]

60. Cappello, M.S.; Zapparoli, G.; Logrieco, A.; Bartowsky, E.J. Linking wine lactic acid bacteria diversity with wine aroma and flavour. Review article. *Int. J. Food Microbiol.* **2017**, *243*, 16–27. [CrossRef] [PubMed]

61. Guo, Y.-Y.; Yang, Y.-P.; Peng, Q.; Han, Y. Biogenic amines in wine: A review. *Int. J. Food Sci. Technol.* **2015**, *50*, 1523–1532. [CrossRef]

62. Martuscelli, M.; Mastrocola, D. *Biogenic Amines: A Claim for Wines, Biogenic Amines, Charalampos Proestos*; IntechOpen: London, UK, 2018. [CrossRef]

63. Beneduce, L.; Romano, A.; Capozzi, V.; Lucas, P.; Barnavon, L.; Bach, B.; Spano, G. Biogenic amines in regional wines. *Ann. Microbiol.* **2010**, *60*, 573–578. [CrossRef]

64. Ancín-Azpilicueta, C.; González-Marco, A.; Jiménez-Moreno, N. Current knowledge about the presence of amines in wine. *Crit. Rev. Food Sci. Nutr.* **2008**, *48*, 257–275. [CrossRef]

65. Tassoni, A.; Tango, N.; Ferri, M. Comparison of biogenic amine and polyphenol profiles of grape berries and wines obtained following conventional, organic and biodynamic agricultural and oenological practices. *Food Chem.* **2013**, *139*, 405–413. [CrossRef]

66. Yildirim, H.K.; Üren, A.; Yücel, U. Evaluation of biogenic amines in organic and non-organic wines by HPLC OPA derivatization. *Food Technol. Biotechnol.* **2007**, *45*, 62–68.

67. Garcia-Marino, M.; Trigueros, Á.; Escribano-Bailón, T. Influence of oenological practices on the formation of biogenic amines in quality red wines. *J. Food Comp. Anal.* **2010**, *23*, 455–462. [CrossRef]

68. Tassoni, A.; Tango, N.; Ferri, M. Polyphenol and Biogenic Amine Profiles of Albana and Lambrusco Grape Berries and Wines Obtained Following Different Agricultural and Oenological Practices. *Food Nutr. Sci.* **2014**, *5*, 8–16. [CrossRef]

69. Caruso, M.; Fiore, C.; Contursi, M.; Salzano, G.; Paparella, A.; Romano, P. Formation of biogenic amines as criteria for the selection of wine yeasts. *World J. Microbiol. Biotechnol.* **2002**, *18*, 159–163. [CrossRef]

70. Benito, Á.; Calderón, F.; Palomero, F.; Benito, S. Quality and Composition of Airén Wines Fermented by Sequential Inoculation of *Lachancea thermotolerans* and *Saccharomyces cerevisiae*. *Food Technol. Biotechnol.* **2016**, *54*, 135–144. [CrossRef] [PubMed]

71. Medina, K.; Boido, E.; Fariña, L.; Gioia, O.; Gomez, M.E.; Barquet, M.; Gaggero, C.; Dellacassa, E.; Carrau, F. Increased flavour diversity of Chardonnay wines by spontaneous fermentation and co-fermentation with *Hanseniaspora vineae*. *Food Chem.* **2013**, *141*, 2513–2521. [CrossRef]

72. Thibon, C.; Marullo, P.; Claisse, O.; Cullin, C.; Dubourdieu, D.; Tominaga, T. Nitrogen catabolic repression controls the release of volatile thiols by *Saccharomyces cerevisiae* during wine fermentation. *Fems Yeast Res.* **2008**, *8*, 1076–1086. [CrossRef]

73. Thorgeirsson, U.P.; Dalgard, D.W.; Reeves, J.; Adamson, R.H. Tumor incidence in a chemical carcinogenesis study of nonhuman primates. *Regul. Toxicol. Pharm.* **1994**, *19*, 130–151. [CrossRef]

74. Jiao, Z.; Dong, Y.; Chen, Q. Ethyl Carbamate in Fermented Beverages: Presence, Analytical Chemistry, Formation Mechanism, and Mitigation Proposals. *Compr. Rev. Food Sci. Food Saf.* **2014**, *13*, 611–626. [CrossRef]

75. Salmon, A.G.; Zeise, L. *Risks of Carcinogenesis from Urethane Exposure*; CRC Press: Boca Raton, FL, USA, 1991; p. 115.

76. Beland, F.A.; Benson, R.W.; Mellick, P.W.; Kovatch, R.M.; Roberts, D.W.; Fang, J.-L.; Doerge, D.R. Effect of ethanol on the tumorigenicity of urethane (ethyl carbamate) in B6C3F1 mice. *Food Chem. Toxicol.* **2005**, *43*, 1–19. [CrossRef]

77. Dahabieh, M.; Husnik, J.; Van Vuuren, H. Functional enhancement of sake yeast strains to minimize the production of ethyl carbamate in sake wine. *J. Appl. Microbiol.* **2010**, *109*, 963–973. [CrossRef]

78. Arena, M.; Saguir, F.; Manca de Nadra, M. Arginine, citrulline and ornithine metabolism by lactic acid bacteria from wine. *Int. J. Food Microbiol.* **1999**, *52*, 155–161. [CrossRef]

79. Polychroniadou, E.; Kanellaki, M.; Iconomopoulou, M.; Koutinas, A.; Marchant, R.; Banat, I. Grape and apple wines volatile fermentation products and possible relation to spoilage. *Bioresour. Technol.* **2003**, *87*, 337–339. [CrossRef]

80. Zhao, X.; Zou, H.; Fu, J.; Chen, J.; Zhou, J.; Du, G. Nitrogen regulation involved in the accumulation of urea in *Saccharomyces cerevisiae*. *Yeast* **2013**, *30*, 437–447. [CrossRef] [PubMed]

81. Fidaleo, M.; Esti, M.; Moresi, M. Assessment of urea degradation rate in model wine solutions by acid urease from *Lactobacillus fermentum*. *J. Agric. Food Chem.* **2006**, *54*, 6226–6235. [CrossRef] [PubMed]

82. Benito, S. The Management of Compounds that Influence Human Health in Modern Winemaking from an HACCP Point of View. *Fermentation* **2019**, *5*, 33. [CrossRef]

83. Andrich, L.; Esti, M.; Moresi, M. Urea removal in model wine solutions by immobilized acid urease in a stirred bioreactor. *Chem. Eng. Trans.* **2009**, *17*, 915–920. [CrossRef]

84. Pflaum, T.; Hausler, T.; Baumung, C.; Ackermann, S.; Kuballa, T.; Rehm, J.; Lachenmeier, D.W. Carcinogenic compounds in alcoholic beverages: An update. *Arch. Toxicol.* **2016**, *90*, 2349–2367. [CrossRef] [PubMed]

85. Silva, S.; Ramón-Portugal, F.; Andrade, P.; Abreu, S.; de Fatima, T.M.; Strehaiano, P. Malic acid consumption by dry immobilized cells of *Schizosaccharomyces pombe*. *Am. J. Enol. Vitic.* **2003**, *54*, 50–55.

86. Cecchini, F.; Morassut, M.; Moruno, E.G.; Di Stefano, R. Influence of yeast strain on ochratoxin A content during fermentation of white and red must. *Food Microbiol.* **2006**, *23*, 411–417. [CrossRef]

87. Benito, Á.; Calderón, F.; Benito, S. Combined use of *S. pombe* and *L. thermotolerans* in winemaking. Beneficial effects determined through the study of wines' analytical characteristics. *Molecules* **2016**, *21*, 1744. [CrossRef]

88. Comitini, F.; Capece, A.; Ciani, M.; Romano, P. New insights on the use of wine yeasts. *Curr. Opin. Food Sci.* **2017**, *13*, 44–49. [CrossRef]

89. Cordero-Bueso, G.; Arroyo, T.; Serrano, A.; Tello, J.; Aporta, I.; Vélez, M.D.; Valero, E. Influence of the farming system and vine variety on yeast communities associated with grape berries. *Int. J. Food Microbiol.* **2011**, *145*, 132–139. [CrossRef] [PubMed]

90. Grangeteau, C.; Gerhards, D.; von Wallbrunn, C.; Alexandre, N.; Rousseaux, S.; Guilloux-Benatier, M. Persistence of two non-*Saccharomyces* yeasts (*Hanseniaspora* and *Starmerella*) in the cellar. *Front. Microbiol.* **2016**, *7*, 11. [CrossRef]

91. Comitini, F.; Gobbi, M.; Domizio, P.; Romani, C.; Lencioni, L.; Mannazzu, I.; Ciani, M. Selected non-Saccharomyces wine yeasts in controlled multistarter fermentations with *Saccharomyces cerevisiae*. *Food Microbiol.* **2011**, *28*, 873–882. [CrossRef] [PubMed]

92. Morata, A.; Benito, S.; Loira, I.; Palomero, F.; González, M.C.; Suárez-Lepe, J.A. Formation of pyranoanthocyanins by *Schizosaccharomyces* pombe during the fermentation of red must. *Int. J. Food Microbiol.* **2012**, *159*, 47–53. [CrossRef]

93. Lleixà, J.; Martín, V.; Portillo, M.C.; Carrau, F.; Beltran, G.; Mas, A. Comparison of fermentation and wines produced by inoculation of *Hanseniaspora vineae* and *Saccharomyces cerevisiae*. *Front. Microbiol.* **2016**, *7*, 338. [CrossRef] [PubMed]

94. Renault, P.; Coulon, J.; Moine, V.; Thibon, C.; Bely, M. Enhanced 3-sulfanylhexan-1-ol production in sequential mixed fermentation with *Torulaspora delbrueckii/Saccharomyces cerevisiae reveals* a situation of synergistic interaction between two industrial strains. *Front. Microbiol.* **2016**, *7*, 22. [CrossRef]

95. Englezos, V.; Torchio, F.; Cravero, F.; Marengo, F.; Giacosa, S.; Gerbi, V.; Rantsiou, K.; Rolle, L.; Cocolin, L. Aroma profile and composition of Barbera wines obtained by mixed fermentations of *Starmerella bacillaris* (synonym *Candida zemplinina*) and *Saccharomyces cerevisiae*. *LWT—Food Sci. Technol.* **2016**, *73*, 567–575. [CrossRef]

96. Lencioni, L.; Romani, C.; Gobbi, M.; Comitini, F.; Ciani, M.; Domizio, P. Controlled mixed fermentation at winery scale using *Zygotorulaspora florentina* and *Saccharomyces cerevisiae*. *Int. J. Food Microbiol.* **2016**, *234*, 36–44. [CrossRef]

97. Tristezza, M.; Tufariello, M.; Capozzi, V.; Spano, G.; Mita, G.; Grieco, F. The oenological potential of *Hanseniaspora uvarum* in simultaneous and sequential co-fermentation with *Saccharomyces cerevisiae* for industrial wine production. *Front. Microbiol.* **2016**, *7*, 670. [CrossRef]

98. Oro, L.; Ciani, M.; Bizzaro, D.; Comitini, F. Evaluation of damage induced by Kwkt and Pikt zymocins against *Brettanomyces/Dekkera* spoilage yeast, as compared to sulphur dioxide. *J. Appl. Microbiol.* **2016**, *121*, 207–214. [CrossRef]

99. Mehlomakulu, N.N.; Setati, M.E.; Divol, B. Characterization of novel killer toxins secreted by wine- related non-*Saccharomyces* yeasts and their action on *Brettanomyces* spp. *Int. J. Food Microbiol.* **2014**, *8*, 83–91. [CrossRef] [PubMed]

100. Villalba, M.L.; Susana- Sáez, J.; Del Monaco, S.; Lopes, C.A.; Sangorrín, M.P. TdKT, a new killer toxin produced by *Torulaspora delbrueckii* effective against wine spoilage yeasts. *Int. J. Food Microbiol.* **2016**, *217*, 94–100. [CrossRef] [PubMed]

101. Zupan, J.; Avbelj, M.; Butinar, B.; Kosel, J.; Šergan, M.; Raspor, P. Monitoring of quorum-sensing molecules during minifermentation studies in wine yeast. *J. Agric. Food Chem.* **2013**, *61*, 2496–2505. [CrossRef] [PubMed]

102. Avbelj, M.; Zupan, J.; Raspor, P. Quorum-sensing in yeast and its potential in wine making. *Appl. Microbiol. Biotechnol.* **2016**, *100*, 7841–7852. [CrossRef]

103. Contreras, A.; Hidalgo, C.; Schmidt, S.; Henschke, P.A.; Curtin, C.; Varela, C. The application of non-*Saccharomyces* yeast in fermentations with limited aeration as a strategy for the production of wine with reduced alcohol content. *Int. J. Food Microbiol.* **2015**, *205*, 7–15. [CrossRef]

104. Ciani, M.; Morales, P.; Comitini, F.; Tronchoni, J.; Canonico, L.; Curiel, J.A.; Oro, L.; Rodrigues, A.J.; Gonzalez, R. Non-conventional yeast species for lowering ethanol content of wines. *Front. Microbiol.* **2016**, *7*, 642. [CrossRef]

105. Balboa-Lagunero, T.; Arroyo, T.; Cabellos, J.M.; Aznar, M. Yeast selection as a tool for reducing key oxidation notes in organic wines. *Food Res. Int.* **2013**, *53*, 252–259. [CrossRef]

 *fermentation*

Review

Non-*Saccharomyces* in Winemaking: Source of Mannoproteins, Nitrogen, Enzymes, and Antimicrobial Compounds

Ricardo Vejarano

Dirección de Investigación y Desarrollo, Universidad Privada del Norte (UPN), Trujillo 13001, Peru; ricardo.vejarano@upn.edu.pe

Received: 21 June 2020; Accepted: 21 July 2020; Published: 29 July 2020

Abstract: Traditionally, non-*Saccharomyces* yeasts have been considered contaminants because of their high production of metabolites with negative connotations in wine. This aspect has been changing in recent years due to an increased interest in the use of these yeasts in the winemaking process. The majority of these yeasts have a low fermentation power, being used in mixed fermentations with *Saccharomyces cerevisiae* due to their ability to produce metabolites of enological interest, such as glycerol, fatty acids, organic acids, esters, higher alcohols, stable pigments, among others. Additionally, existing literature reports various compounds derived from the cellular structure of non-*Saccharomyces* yeasts with benefits in the winemaking process, such as polysaccharides, proteins, enzymes, peptides, amino acids, or antimicrobial compounds, some of which, besides contributing to improving the quality of the wine, can be used as a source of nitrogen for the fermentation yeasts. These compounds can be produced exogenously, and later incorporated into the winemaking process, or be uptake directly by *S. cerevisiae* from the fermentation medium after their release via lysis of non-*Saccharomyces* yeasts in sequential fermentations.

Keywords: non-*Saccharomyces*; winemaking; aging-on-lees; yeast assimilable nitrogen

---

1. Introduction

The genus *Saccharomyces* has been the most industrially used in the production of wine. This aspect has been changing in recent years due to an increased interest in the use of non-*Saccharomyces* yeasts [1]. The high potential of these yeasts makes them a useful tool for improving oenological parameters such as nutrients content, stability, aromatic profile, or bioactive profile, in spite of their low fermentative power, and, in some cases, their high production of certain metabolites with negative connotations in the wine [2,3].

Currently, species such as *Torulaspora delbrueckii*, *Lachancea thermotolerans*, *Metschnikowia pulcherrima*, *Schizosaccharomyces pombe*, *Pichia kluyveri*, among others, are sold for wine production [4]. These yeasts are often isolated from the grape, grape-must, and wine, as well as from the soil, winery surfaces, harvesting machinery, and other objects associated with the wine production. In the case of grape-must/wine, the majority are isolated during the first stages of fermentative process given their low fermentation ability and low tolerance to ethanol, with respect to *S. cerevisiae*, so that its use would mainly be associated with the contribution of the specific enological metabolites such as enzymes, aromatic compounds, glycerol, organic acids, fatty acids, proteins, amino acids, polysaccharides, among others [3].

Figure 1 summarizes some of the potential applications of non-*Saccharomyces* yeasts, taking as reference the available commercial preparations for wine production. These preparations are sold in the form of (a) dry yeast, inactivated through thermal treatment and then dried; (b) cell autolysates, which include soluble and insoluble yeast components, partially degraded by the endogenous enzymes;

(c) soluble cell extracts, derived from the cytoplasm; and (d) insoluble cell hulls, mainly containing cell walls [5]. These preparations may be the source of compounds of interest such as polysaccharides, proteins, enzymes, or other metabolites with the potential for use in wine production and can be produced exogenously [6].

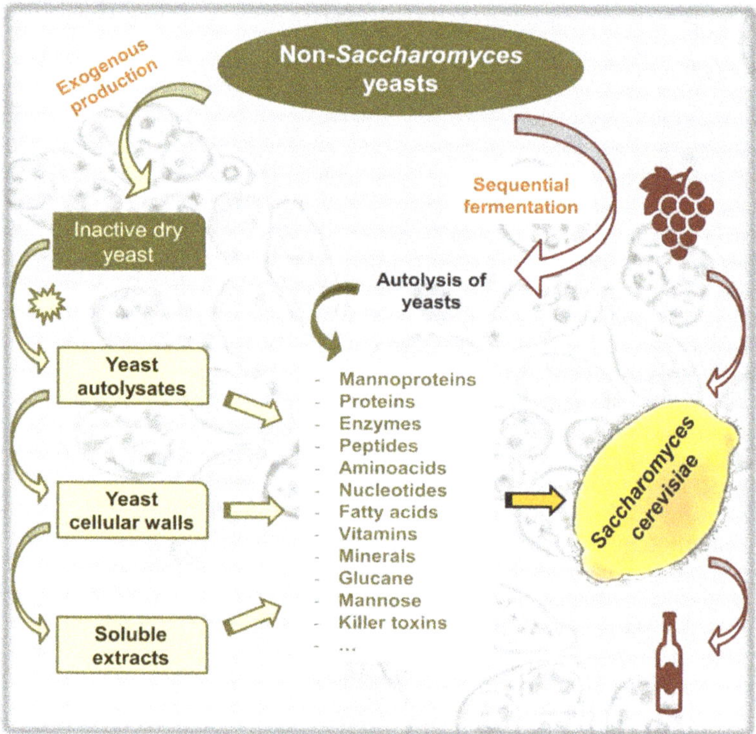

Figure 1. Main cell components of non-*Saccharomyces* yeasts with the potential for use in wine production.

In addition to these preparations, compounds of interest can be obtained directly from the fermentation process as a result of their release from the yeast during cell lysis, for example, in mixed cultures.

2. Non-*Saccharomyces* Yeasts as a Source of Polysaccharides

During aging-on-lees (AOL), a series of changes take place with a direct impact on the wine's properties, due to the interaction between the wine's components, the enzymatic activity, and the polysaccharides released by the lees [7]. The use of non-*Saccharomyces* yeasts for the exogenous production of polysaccharides [6], which can be added to wine, is an interesting alternative. The addition of yeast polysaccharides is a practice authorized by the International Code of Oenological Practice of the International Organization of Vine and Wine (OIV).

These polysaccharides, particularly in red wines, can improve the mouthfullness and body [8], sweetness and roundness [9], aromatic persistence [10], protein and tartaric stability [11,12], and interact with tannins and reduce astringency [13], as well as have an antioxidant effect (due to glutathione) that protects the aromatic compounds and anthocyanins [14], enabling its antioxidant and anti-inflammatory potential to be maintained [15]. They also interact with the tertiary aromatic compounds, reducing the perception of woody aromas in long-aged wines [16], encourage malolactic fermentation [17], and adsorb undesirable and dangerous compounds such as ochratoxin A (OTA) [18]. In sparkling wines, they improve the quality of the foam [19]. However, during AOL, a reduction of aromatic compounds such as terpenes, esters, and aldehydes [20], and, in some cases, anthocyanins in red wines [16,21], has also been reported.

The most important polysaccharides released during AOL are the mannoproteins (Table 1), which are fixed on a three-dimensional network of glucan and chitin in the cell wall [21,22]. The mannoproteins represent between 80% and 100% of the fraction of polysaccharides founded in the wine (molecular mass 100–2000 kDa: 90% mannose and 10% protein), while glucomannoproteins (molecular mass 20–90 kDa: 25% glucose, 25% mannose, and 50% protein) represent between 10% and 20% of the total [23]. Furthermore, α-galactomannose rather than mannose has been found as part of the structure of polysaccharides in *S. pombe* [22].

The enzymes involved in the cell autolysis and subsequent releasing of polysaccharides into the wine are endo $\beta(1\rightarrow3)$-glucanases, endo β-$(1\rightarrow6)$-glucanases, exo-$(1\rightarrow6)$-α-D-mannose, exo-$(1\rightarrow2)$-α-mannose, and α-D-mannosidase [23].

2.1. Non-Saccharomyces Species as a Source of Polysaccharides

Even though *S. cerevisiae* is the most commonly used yeast as a source of polysaccharides, the literature reports different non-*Saccharomyces* species with the potential to produce and release polysaccharides during AOL: *S. pombe*, *Saccharomycodes ludwigii*, *Wickerhamomyces anomalus*, *Hanseniaspora vineae*, *L. thermotolerans*, *M. pulcherrima*, *Starmerella bacillaris*, *T. delbrueckii*, *Zygosaccharomyces rouxii*, and *Zygotorulaspora florentina* (formerly *Zygosaccharomyces florentinus*), among others.

The genus *Schizosaccharomyces* has shown high rates of release of polysaccharides during alcoholic fermentation, of up to seven times higher than *S. cerevisiae* [24]. During AOL, *S. pombe* together with *S'codes ludwigii* are found among the species with high potentials for releasing polysaccharides [21]. *S. pombe* primarily releases galactomannoproteins (Table 1), at concentrations up to 10 times higher than some strains of *Saccharomyces* and *Pichia* [21]. This potential could mean an advantage with regard to accelerating aging processes.

S'codes ludwigii has also shown a high potential for releasing polysaccharides during fermentation and AOL [21,25,26]. Palomero et al. [21] obtained higher rates of polysaccharide release from *S'codes. ludwigii* (110.51 gm/L) and *S. pombe* (103.61 mg/L) with respect to *S. cerevisiae* (36.65 mg/L). Furthermore, the polysaccharides from these non-*Saccharomyces* yeasts show a greater molecular size and may potentially impact the wine's palatability.

However, one aspect to take into account during AOL in red wines is color loss, due to the interaction between the lees and anthocyanins, mainly through adsorption of pigments by the lees [16] and through degradation due to anthocyanin-β-glucosidase activity [27]. The lees from *M. pulcherrima*, *S'codes ludwigii*, or *S. pombe* have shown a low adsorption of anthocyanins with respect to the lees of *S. cerevisiae* and other non-*Saccharomyces* yeasts such as *T. delbrueckii* or *L. thermotolerans* [21,28]. Furthermore, there may be less color loss in wines in the presence of higher pyranoanthocyanins content, which are more stable due to its chemical structure [29], with *S. pombe* being one of the yeasts with the greatest synthesis of these pigments [30].

Additionally, *S. pombe* has also shown the ability to decrease the biogenic amine content in sparkling wines, with better results than *S. cerevisiae* [30], which would be related to the adsorption of the amines on the lees during second fermentation and subsequent aging in the bottle [31]. However, considering the short duration of AOL studies, it is necessary to evaluate the evolution of amines over longer aging periods, such as those carried out in traditional sparkling wine production, given that other studies have reported an increase in the biogenic amines content in wines in contact with lees [32].

In terms of other non-*Saccharomyces* yeasts, *Hanseniaspora vineae* has shown a higher rate of cell lysis with respect to commercial *S. cerevisiae* strains [33]. Other yeasts that have shown a notable polysaccharide contribution are *L. thermotolerans* [34], *Z. rouxii* [26], and *Z. florentina*, particularly in mixed fermentations with *S. cerevisiae* [35].

Table 1. Some non-*Saccharomyces* yeasts with the potential to release mannoproteins.

Yeast	Protein and Monosaccharide Content in Mannoproteins [25]				Nitrogen Requirement
	Protein (%) [a]	Mannose (%) [b]	Glucose (%) [b]	Galactose (%) [b]	
Saccharomyces cerevisiae	24	88	12	-	
Metschnikowia pulcherrima	25	86	14	-	Slow ammonium uptake [4]. Weak or no growth in nitrate agar, and unable to develop in YPD agar at 37 °C [28].
Wickerhamomyces anomalus [c]	9	74	26	-	Capable to uptake nitrate [36].
Saccharomycodes ludwigii [d]	12	93	7	-	Unable to uptake nitrate [37]. Capable to uptake cadaverine and ethylamine [37].
Schizosaccharomyces pombe [e]	11	55	22	23	
Starmerella bombicola [f]	14	73	27	-	
Pichia fermentans	15	87	13	-	
Hanseniaspora uvarum [g]	23	81	19	-	
Hanseniaspora valbyensis	20	75	25	-	
Lachancea thermotolerans	16	82	18	-	
Torulaspora delbrueckii	18	85	15	-	
Zygosaccharomyces bailii	29	79	21	-	
Brettanomyces bruxellensis	16	88	12	-	

[a] Percentage of dry matter. [b] Sugars (%) in the polysaccharide fraction. [c] *Wickerhamomyces anomalus* (formerly *Pichia anomala*). [d] *Saccharomycodes ludwigii*: high autolytic activity: polysaccharide-releasing [21,25]. [e] *Schizosaccharomyces pombe*: high autolytic activity: polysaccharide-releasing [21,38]. [f] *Starmerella bombicola* (formerly *Candida stellata* DBVPG 3827). [g] *Hanseniaspora uvarum* (formerly *Kloeckera apiculata*).

For its part, the information regarding the use of other yeasts such as *W. anomalus* as a source of mannoproteins is scarce [26]. This yeast has a high potential for the production of polysaccharides and other metabolites of interest, given its ability to metabolize a large variety of nitrogen sources, including nitrate [36], which would allow production costs to be reduced.

2.2. Accelerated Release of Polysaccharides

One disadvantage of AOL is that it is a very slow process, which requires up to 9 months to obtain the desired effects in the treated wine. Among the strategies to improve AOL, existing literature reports (a) the selection of yeast species and strains with rapid autolysis [21,39], (b) acceleration of the cell autolysis through mixed cultures involving sensitive and killer yeasts [40], mixed cultures among different yeast species, which enable the regulation of cell death [41], the addition of β-glucanase [42], and the application of ultrasound [39,43].

Several studies have reported a higher rate of polysaccharide release with ultrasound (US) treatments [39,43,44]. Lees from *S'codes ludwigii*, *S. pombe*, *M. pulcherrima*, and *S. cerevisiae*, among others, were evaluated over a 7-week aging period in a hydroalcoholic medium, applying US for 10 min a day, with *S'codes ludwigii* lees showing the highest rate of polysaccharide release after the third week (around 460 mg/L) [39]. In the same study [39], a decrease in anthocyanins content was observed in the treated red wine, without affecting the pyranoanthocyanins content [21,29], particularly with *S'codes ludwigii* lees, which also allowed to reduce proanthocyanidins content, and consequently, astringency and bitterness.

More recently, Del Fresno et al. [43] evaluated the effect of AOL (lees from *S. cerevisiae*) in the presence of oak chips in red wines. The evaluation period was 135 days at 14°C, under dark conditions, agitating the samples once a week to simulate "bâtonnage". The samples were treated with US twice a week for 5 min for the first 5 weeks. From this moment, the process was accelerated increasing to two US treatments a week (15 min per treatment). In a parallel experiment, a hydroalcoholic medium was used to adequately quantify the polysaccharides released from the lees. In general, the polysaccharide release increased after using US for 135 days, releasing around 11.8 mg/L, more than double that of the untreated samples (approx. 5.3 mg/L). Additionally, an increase in the protein content of the US-treated samples was observed after 120 days of AOL.

However, the same authors [43] reported the increase in dissolved oxygen in the treated red wines as a disadvantage, whose effect was evident in the lower anthocyanins content in the US-treated wines in the absence of lees. This finding reveals the protective effect of polysaccharides against oxidation.

3. Non-*Saccharomyces* Yeasts as a Source of Nitrogen for *Saccharomyces cerevisiae*

Yeast assimilable nitrogen (YAN) is nitrogen that yeasts can assimilate and metabolize, preferably from sources such as ammonium (NH_4^+), amino acids, and peptides of up to five amino acids [45]. Its content in the grape-must varies between 50 and 500 mg/L [46].

Among YAN's functions are reproduction and cell growth, the protein for sugar transport synthesis, and enzyme synthesis, as well as the functions accomplished by the amino acids as the precursors of aromatic compounds, mainly higher alcohols, produced by deamination and decarboxylation [47].The most predominant amino acids as NH_4^+ transporters in grape-must are α-alanine, serine, arginine, proline, glutamic acid, and glutamine [48]. During fermentation, their use varies, with higher uptake of glutamic acid, glutamine, and arginine [4], whereas proline, a proteinogenic amino acid not metabolized by *S. cerevisiae*, is among the least used [4]. Additionally, the demand for arginine is increased by lactic acid bacteria during malolactic fermentation [48].

S. cerevisiae has shown a preference for NH_4^+ and amino acids, thanks to its nitrogen catabolite repression (NCR) mechanism [49], through which the genes involved in the transport and metabolism of NH_4^+ and glutamine are activated and, once they are depleted, the genes involved in the transport and metabolism of other sources such as arginine, glutamate, and alanine, among others, become activated.

Concentrations of YAN below 150 mg/L in the grape-must carry the risk of sluggish or stuck fermentation [4], as well as a low synthesis of some aromatic compounds such as of esters, volatile fatty acids, and higher alcohols [50]. This N deficiency is remedied in the winery through the addition of extra N sources such as diammonium phosphate (DAP). However, the excessive use of DAP can lower phenylpropanoid production (affecting the complexity of the wine) [51,52], increase the wine acidification, produce high levels of residual phosphate, stimulate the production of esters such as ethyl acetate, and increase the levels of hydrogen sulfide (H_2S), especially when there is a deficit of other essential nutrients such as vitamins, minerals, and lipids; further, excessive levels of DAP can increase turbidity, promote microbial instability, and facilitate the production of unpleasant aromas and harmful compounds such as ethyl carbamate and biogenic amines [53–56].

3.1. Non-Saccharomyces Species as a Source of Nitrogen

As an alternative N source, yeast cell structures can be used in the form of hulls, hydrolyzed, or extracts, which can be produced exogenously [6], and then added to the fermentative medium. Additionally, this source of N can be obtained from the rest of the cells after the death and lysis of non-*Saccharomyces* yeasts used at the beginning of the sequential fermentations with *S. cerevisiae*.

Aureobasidium pullulans (an yeast-like fungus) is a potential source of essential amino acids [57], and it can growth in low-cost carbon sources like agricultural and food waste, due to its amylase [58], cellulase [59], lipase [60], xylanase [61], laccase [62], mannanase [63], and protease [64] activities. However, the use of *A. pullulans* for the production of protein on a large scale has still not been explored [65].

Similarly, the use of species such as *S'codes ludwigii*, *S. pombe*, *Candida stellata*, *M. pulcherrima*, *W. anomalus*, *H. vineae*, *Z. rouxii*, *Zygosaccharomyces bailii*, *L. thermotolerans*, and *Z. florentina*, among others, as potential N sources is limited, considering their high capacity for production and release of mannoproteins [21,25,26,33–35], which is up to 29% of protein (Table 1). From the little background information available, *W. anomalus* has been used as a source of single cell protein at industrial level, specifically to produce protein for fish farming [66]. This demonstrates the potential of *W. anomalus* as source of N, taking into account the ability of this yeast to use a wide range of N sources, including nitrate [36].

One aspect to consider is the depletion of nutrients during fermentation with different yeasts. Difficulties have been reported in the introduction of *S. cerevisiae* after *Hanseniaspora/Kloeckera* in sequential fermentations [67], an effect that is associated with the depletion of thiamine and calcium pantothenate [68,69], reducing the availability of these nutrients for *S. cerevisiae* (second inoculum).

Similarly, sequential fermentations can cause a depletion of YAN by first phase yeasts, especially those with high nutrient demands and a low ability to release nitrogenous compounds. This situation can be overcome with the use of yeasts such as *M. pulcherrima*, with high proteolytic activity and amino acid release as a source of N for *S. cerevisiae* (second inoculum) [70,71].

In a recent study, Prior et al. [4] carried out sequential fermentations of *L. thermotolerans/S. cerevisiae* and *T. delbrueckii/S. cerevisiae* to evaluate whether the rest of these non-*Saccharomyces* yeasts (first phase) can be used as a source of N for *S. cerevisiae*. After 48 h of fermentation with non-*Saccharomyces*, the medium was filtered and then inoculated with *S. cerevisiae*. A reduction in sugar consumption by *S. cerevisiae* was observed in the filtered medium, in other words, fermentative yield decreased.

The results obtained in the aforementioned studies show that in sequential fermentations, non-*Saccharomyces* yeast rests (first phase) after death and lysis can release various cell components into the fermentative medium to be used as N sources for *S. cerevisiae* [4]. This strategy can be used more efficiently with non-*Saccharomyces* species with high ability to release mannoproteins (Table 1) and with high β-glucanase and protease activities.

3.2. Nitrogen Requirements for Sparkling Wines Production

The selection of yeasts with high rates of nitrogenous compounds release can ensure an adequate supply of N. Two of the non-*Saccharomyces* species which have shown the greatest ability to release amino acids during the second fermentation are *S'codes ludwigii* and *S. pombe* [30], increasing the amino acids content with respect to the base wine. The advantage of these yeasts over *S. cerevisiae* is related to their amino acid release mechanisms, with their different consumption rates and their cell composition [21,47].

One of the most important properties in sparkling wine is the ability to form foam or foamability. Mannoproteins improve foam formation and stability thanks to their hydrophobicity, high glycosylation, and high molecular mass, which enable them to surround and stabilize the gas bubbles in the foam [72]. The proteins also participate in this process which, together with the peptides and amino acids, are released mainly during aging in the bottle as a result of the enzymatic degradation of the cell walls and other cell structures [7,73], contributing to the complexity of sparkling wines.

The higher production of mannoproteins in sequential fermentations by *S. bacillaris/S. cerevisiae* [74], *L. thermotolerans/S. cerevisiae*, and *T. delbrueckii/S. cerevisiae* can also be utilized, which also contributes to reducing volatile acidity and increasing 2-phenylethanol [75,76], and improving foamability and foam stability (sequential *T. delbrueckii/S. cerevisiae* and *M. pulcherrima/S. cerevisiae* fermentations) [77,78].

Finally, during AOL (in the bottle), it has been observed that the content of free amino acids and peptides depends on the yeast species and strains [73], which have an influence on the flocculation ability to facilitate lees movement during riddling [79]. A potential field for future studies is the optimization of aging and disgorging operation in sparkling wines, for example, by inserting magnetic nanoparticles to accelerate sedimentation and lees removal [80].

4. Non-*Saccharomyces* Yeasts as a Source of Exogenous Enzymes

Another potential application of non-*Saccharomyces* yeasts is enzyme production. In general, enzymes of microbial origin are considered to have greater activity and stability than those of plant and animal origin [81]. Considering that the genus *Saccharomyces* is not characterized as being a good producer of exogenous enzymes [82], there is increasing interest in finding sources of enzymes of enological interest among non-*Saccharomyces* species, some of which are summarized in Table 2.

Table 2. Summary of the enzymes produced by non-*Saccharomyces* yeasts for wine production.

Enzyme	Yeast	Application
β-glucosidase	*Lachancea thermotolerans*	Release of terpenes and thiols from their precursors: improvement of the aromatic profile [83–85].
	Torulaspora delbrueckii	Release of thiols from their cysteinylated precursors: improvement of the aromatic profile [84].
	Wickerhamomyces anomalus	- High stability of β-glucosidase excreted by *W. anomalus* MDD24 [86]. - Release of terpenes from their glycosylated precursors: improvement of the aromatic profile [85,86].
	Metschnikowia pulcherrima	- Release of terpenes from their glycosylated precursors: improvement of the aromatic profile [71,83]. - Release of thiols from their cysteinylated precursors: improvement of the aromatic profile [84]. - Foaming and release of aromatic compounds in sparkling wines [87].
	Candida stellata	Release of terpenes (β-myrcene, limonene, linalool, α-terpineol, and farnesol) from their glycosylated precursors: improvement of the aromatic profile [88].
	Hanseniaspora uvarum	- Release of terpenes and C13-norisoprenoids from their precursors: improvement of the aromatic profile [89,90]. - Activity up to 6.6-fold higher than some *S. cerevisiae* strains [90].
	Saccharomycodes ludwigii	Activity up to 46% higher than *S. cerevisiae* at 30 °C [91].
	Aureobasidium pullulans	Release of terpenes from their glycosylated precursors: improvement of the aromatic profile [92,93].
Protease	*Wickerhamomyces anomalus*	Aspartic protease WaAPr1 excreted by *W. anomalus* 227 [94].
	Metschnikowia pulcherrima	- Aspartic protease MpAPr1 excreted by *M. pulcherrima* IWBT Y1123 [95]. - Degradation of proteins: improvement of clarification and stabilization [95]. - Degradation of proteins and peptides: source of N for fermentative yeast [70,71,83]. - Improvement of grape-must extraction and clarification, wine filtration [71], and stabilization [83]. - Improvement of foam stability in sparkling wines [71,87,96]. - Increase in the amino acids content: production of aromatic compounds [83,95].
	Candida stellata	- Degradation of proteins: improvement of clarification and stabilization [97]. - Degradation of proteins and peptides: source of N for fermentative yeast [97].

Table 2. *Cont.*

Enzyme	Yeast	Application
	Hanseniaspora uvarum	Degradation of proteins: improvement of clarification and stabilization [89,97].
	Lachancea thermotolerans, Torulaspora delbrueckii, Zygosaccharomyces bailii, Pichia kluyveri	- Increase in the amino acids content: source of N for fermentative yeast [83]. - Increase in the amino acids content: production of aromatic compounds [83]. - Degradation of proteins: improvement of clarification and stabilization [83].
Glucanase	*Wickerhamomyces anomalus*	- β-1,3-glucanase excreted by *W. anomalus* AS1: viscosity reduction in grape-musts [98,99]. - β-1,3-glucanase excreted by *W. anomalus* AS1: improvement of bioactive profile by release of *trans*-resveratrol from its glycosylated precursor polydatin [98]. - Antimicrobial control against *Dekkera/Brettanomyces*: attack at the cell wall level [100].
Hydroxycinnamate decarboxylase (HCDC)	*Schizosaccharomyces pombe* *Saccharomycodes ludwigii* *Lachancea thermotolerans, Metschnikowia pulcherrima, Debaryomyces hansenii* *Metschnikowia pulcherrima, Hanseniaspora guilliermondii, Hanseniaspora opuntiae, Hanseniaspora vineae, Hanseniaspora clermontiae, Pichia guillermondii*	High mannoproteins-releasing during AOL [21,38,99]. High mannoproteins-releasing during AOL [21,25]. Release of mannoproteins [25,83]. Involved in the synthesis of vinylphenolic pyranoanthocyanins: improvement of color stability in red wines [83,101–103].
Urease	*Schizosaccharomyces pombe*	- High urease activity [104]. - Urea hydrolysis: reduced synthesis of ethyl carbamate (carcinogenic [105]).
Carboxypeptidase	*Aureobasidium pullulans*	- Degradation of ochratoxin A (OTA) [106]. - Drawback: possible action against fermentative yeasts [107].
Pectinase	*Wickerhamomyces anomalus*	Degradation of pectins: improvement of clarification and turbidity reduction [85].
	Metschnikowia pulcherrima	- Degradation of pectins: improvement of clarification and turbidity reduction [87,97]. - Improvement of extraction of anthocyanins and polyphenols from the skins of the grape berries [108].
	Candida stellata	Degradation of pectins: improvement of clarification and turbidity reduction [85,97].
	Hanseniaspora uvarum	Degradation of pectins: improvement of clarification and turbidity reduction [89,97].
	Aureobasidium pullulans	- Degradation of pectins: improvement of clarification and turbidity reduction [92,93]. - Production of cold-active and acid-tolerant pectinases, suitable for low-temperature winemaking [92].

Table 2. *Cont.*

Enzyme	Yeast	Application
Cellulase	*Lachancea thermotolerans, Metschnikowia pulcherrima, Candida stellata, Hanseniaspora uvarum, Aureobasidium pullulans, Debaryomyces hansenii*	- Degradation of cellulose released from the grape cell walls: improvement of extraction, filtration and clarification [83,92,97]. - Improvement of extraction of pigments and aromatic compounds from the skins of the grape berries [83,92,97].
Xylanase	*Lachancea thermotolerans, Candida stellata, Hanseniaspora uvarum, Aureobasidium pullulans Torulaspora delbrueckii,*	Degradation of hemicellulose: improvement of wine aroma by increasing of monoterpenyl diglycoside precursors in the grape-must [83,92,97].
β-lyase	*Kluyveromyces marxianus, Meyerozyma guilliermondii* (formerly *Pichia guilliermondii*)	Release of thiols from their cysteinylated precursors: improvement of the aromatic profile [109,110].
Lipase	*Lachancea thermotolerans*	Increase on free fatty acids concentration [83].
	Aureobasidium pullulans	Improvement of wine aroma: synthesis of ethyl esters and ethyl acetates from lipid cleavage [111].

There is little information regarding the application of purified β-glucosidase, β-lyase, xylanase, cellulase, among others enzymes, produced by non-*Saccharomyces* yeasts for winemaking processes, leaving open the possibility of future research which focuses on improving the release of terpenes, thiols, norisoprenoids, and/or their precursors, with positive impacts on the aromatic profile, especially in white wines, considering that approximately 90% of these compounds are conjugated in the grape skin [99]. This is in addition to the selection of suitable strains, because high levels of β-glucosidase can increase the synthesis of undesirable volatile phenols [101], as well as hydrolyze the anthocyanins in red wines, exposing them to oxidation and/or transformation into colorless forms [27,112].

In contrast, one problem in white wines is the protein haze (wine turbidity), which is usually corrected by removing the proteins from the grape-must with bentonite, with the disadvantage of removing other compounds of enological interest, mainly aromatic compounds. The protease activity of non-*Saccharomyces* yeasts can be used (Table 2) and thus reduce the protein content in the grape-must and therefore prevent the wine haze. The proteases hydrolyze the proteins from the grape-must into smaller molecules like peptides and amino acids, with the consequent clarification and subsequent stability of the wine obtained, and with the additional advantage of providing YAN for the fermentative yeasts [87,89].

5. Non-*Saccharomyces* as Biocontrol Agents Against Contaminating Yeasts

5.1. Antimicrobial Peptides

Some peptides produced by yeasts have shown antimicrobial effects against several grape-must/wine contaminating yeasts. In general, these peptides show lengths of up to 100 amino acids, sorted into variable sequences [113]. Their mechanism of action would be related to changes in the integrity of the cell wall of the target yeasts [114]. Peptides with molecular mass below 10 kDa have shown greater antimicrobial effects [115], such as those produced by *Candida intermedia*, especially the LAMAP1790 strain, with an effect on several strains of *Brettanomyces bruxellensis* and without affecting the growth of *S. cerevisiae* [115].

S. cerevisiae CCMI885 is another yeast that produces antimicrobial peptides with molecular mass lower than 10 kDa and with an effect on *B. bruxellensis*, *Hanseniaspora uvarum*, *Hanseniaspora guilliermondii*, *C. stellata*, *L. thermotolerans*, *Kluyveromyces marxianus*, and *T. delbrueckii* [116]. However, as these peptides have not shown total inhibition over *B. bruxellensis*, their application may require the use of other usual winemaking treatments, such as the addition of SO_2.

According to Peña et al. [117], to achieve the application of these peptides at an industrial level, it is necessary to understand their behavior in mediums with different pH values and sugar levels, as well as high alcohol levels and in the presence of other winemaking yeasts and/or bacteria. Additionally, the implementation of procedures that enable high production and, therefore, satisfy a potential industrial level demand is needed.

5.2. Killer Toxins

Some yeasts produce molecules called "killer toxins", which are glycosylated proteins with effect against sensitive yeast strains [118]. For instance, the action of killer toxins CpKT1 and CpKT2 produced by *Candida pyralidae* (YWBT Y1140 strain) on the cell wall of *B. bruxellensis* have been reported [119]. These toxins have a molecular mass of over 50 kDa, are stable at acidic pH values (3.5–4.5), temperatures between 15 and 25 °C, at high alcoholic content, and at different sugar concentrations. In other words, they are stable under normal winemaking conditions.

In the same way, a killer toxin produced by *T. delbrueckii* NPCC 1033 (TdKT) has shown potential to control yeasts such as *Brettanomyces bruxellensis*, *Pichia guilliermondii*, *Pichia mandshurica*, and *Pichia membranifaciens* [120], being the mechanism of action, the attack at the cell wall level, related to their glucanase and chitinase enzymatic activities. The toxin has a molecular mass of over 30 kDa, its killer activity is stable at pH values of 4.2–4.8, and is inactivated at temperature above 40 °C, confirming their potential use as a biocontrol tool at oenological conditions.

Similarly, Kwkt toxins produced by *Kluyveromyces wickerhamii* [121] and PMKT2 produced by *P. membranifaciens* have shown effects on *B. bruxellensis* [122], although in the case of PMKT2, effects on *S. cerevisiae* have also been observed.

W. anomalus also produces killer toxins with effects on *Dekkera/Brettanomyces*, especially the Pikt toxin [121,123] produced by the D2 and DBVPG 3003 strains, whose fungicidal effect on wine can be sustained for 10 days [121]. The mechanism of action would be the attack at the cell wall level, specifically of β-1,6 glucans. [124], in a similar manner to the *W. anomalus* Cf20 toxin (KTCf20) that binds to the β-1,3 and β-1,6 glucans. β-1,3-glucanase activity has been previously reported in toxins secreted by *W. anomalus* [100]. Additionally, several toxins secreted by this yeast has shown stability and high activity at pH values of 3.0-5.0 and temperatures up to 30 °C, i.e., these toxins are compatible with the winemaking conditions [100,125].

Other studies have reported the antimicrobial effect of *W. anomalus* on species such as *Pichia guilliermondii* or *P. membranifaciens* [125,126] during the first stages of fermentation, and even on *S. cerevisiae* [125,127]. This indicates the need for more studies to evaluate the compatibility of the *W. anomalus* strains that produce these toxins with other yeasts to avoid technological problems such as sluggish and stuck fermentations, as well as take advantage of the potential of the different non-*Saccharomyces* species involved in the fermentation process, to obtain wines with greater complexity and stability.

Although it is not a non-*Saccharomyces* yeast, *Saccharomyces eubayanus* has also shown the ability to produce the killer toxin SeKT, with effect on spoilage yeasts such as *B. bruxellensis*, *P. membranifaciens*, *Meyerozyma guilliermondii* and *P. manshurica* [128]. The mechanism of action comprises cell wall disruption through β-glucanase and chitinase activities. The toxin have a molecular mass of around 70 kDa, and has shown stability at high glucose and ethanol concentrations (300 g/L and 16% *v/v*, respectively), at SO_2 concentrations of up to 100 mg/L, and at temperatures and pH values less than 26 °C and 5.0, respectively.

The results obtained in the aforementioned studies show the potential use of the killer toxins as a biocontrol tool at oenological conditions, especially against the spoilage yeast *B. bruxellensis*, for example, during wine aging and storing.

5.3. Other Molecules as Biocontrol Agents

One feature of *M. pulcherrima* is its potential effect against different contaminating yeast species. In the context of wine production, this activity would be mainly related to the production of the pigment pulcherrimin, from the chelation of Fe in the fermentative medium [129], reducing the availability of this mineral, with harmful effects on *Brettanomyces/Dekkera*, *Pichia*, *Hanseniaspora*, *S'codes ludwigii*, and *Candida* [130].

However, the most significant advantage of this antimicrobial mechanism is the absence of harmful effects on *S. cerevisiae* [130]. In other words, it may be compatible with the main yeast used for wine production, for example, in mixed fermentations, with the consequent reduction of the SO_2 dose, usually used as an antimicrobial agent [2].

6. Future Perspectives

One of the greatest challenges related to the industrial use of non-*Saccharomyces* yeasts for AOL is achieving acceptance from producers, especially in regions with a deep-rooted winemaking tradition and taking into account the unfavorable background of these yeasts, as well as the economic impact of prolonged storage of the wines in AOL and the possible risks of contamination. The optimization of aging conditions is one aspect that requires special attention, especially those conditions which enable the acceleration of the release of polysaccharides from the lees.

Another aspect that requires more research is the use of exogenously produced lees added to the wine [6], especially for the aging of red wines, which involves the identification of species, and especially strains with high rates of polysaccharide release which, in turn, present low anthocyanin adsorption [28] and low expression of anthocyanase activity (anthocyanin-β-glucosidase) that causes hydrolysis of anthocyanins [27]. Additionally, this requires strategies, which can be simultaneously implemented with the selection of yeasts with high capacity to produce pyranoanthocyanins, more stable against degradation [27], thereby minimizing color loss. One of the yeasts that has shown these characteristics is *S. pombe*.

There is also a need for additional studies on fermentation with natural grape-musts to verify the properties of mannoproteins released by non-*Saccharomyces* yeasts in synthetic mediums, given that in these last mediums, there is no interaction between the mannoproteins and the components which are naturally present in the grape-must, and to take advantage of the potential of these yeasts for the production of new products with industrial applicability [74].

Similarly, the effect of the addition of inactive yeasts, hulls, or lees as sources of N on parameters related to protein haze and turbidity formation is still not clear. One possible alternative is the addition of protease-hydrolyzed hulls/lees, which are also produced by non-*Saccharomyces* yeasts. *M. pulcherrima* has shown high protease activity and, therefore, the release of amino acids as a source of N for *S. cerevisiae*, especially in mixed cultures. This activity has also been shown by *W. anomalus*, with the advantage of using a wide range of N sources, including nitrate [36], and by *A. pullulans*, with the ability to use low-cost carbon sources such as agricultural and food waste, and whose protein production on a large scale has still not been studied [65].

It has been reported that the species and strain of the yeast influence the free amino acids and peptide content during AOL [57,58], thus reducing the content of amino acids in aging periods of over 9 months [73]. Therefore, it is necessary to conduct studies, which simulate normal aging conditions (even up to 10 years in sparkling wines) by evaluating the impact of the proteins, mannose, glucose, and galactose present in the lees on the quality of treated wine, which, until now, has only been studied in model mediums [131]. The effect of proteins from non-*Saccharomyces* yeasts on the quality of sparkling wines, their effect on lees movement during riddling, as well as the effect on the sensorial profile must also be studied.

Finally, most of the positive contributions of non-*Saccharomyces* yeasts with regard to *S. cerevisiae* are related to a higher presence of active enzymes, which depends, in part, on the carbon and nitrogen sources present in the fermentative medium. Small changes in the composition of the medium can affect the nature, quantity, and diversity of the secreted enzymes [132]. Therefore, proper maintenance of N levels in the fermentative medium is of vital importance, which can be achieved by identifying yeast species and strains with high release rates of nitrogenous compounds, mainly amino acids.

Funding: Project UPN-20201003 "Non-*Saccharomyces* yeasts: potential applications at industrial level" (Universidad Privada del Norte, Peru).

Conflicts of Interest: The author declares no conflict of interest.

References

1. Morata, A. Enological repercussions of non-*Saccharomyces* species in wine biotechnology. *Fermentation* **2019**, *5*, 72. [CrossRef]
2. Vejarano, R. *Saccharomycodes ludwigii*, control and potential uses in winemaking processes. *Fermentation* **2018**, *4*, 71. [CrossRef]
3. Ivit, N.N.; Kemp, B. The impact of non-*Saccharomyces* yeast on traditional method sparkling wine. *Fermentation* **2018**, *4*, 73. [CrossRef]
4. Prior, K.J.; Bauer, F.F.; Divol, B. The utilisation of nitrogenous compounds by commercial non-*Saccharomyces* yeasts associated with wine. *Food Microbiol.* **2019**, *79*, 75–84. [CrossRef]
5. Del Barrio-Galán, R.; Ortega-Heras, M.; Sánchez-Iglesias, M.; Pérez-Magariño, S. Interactions of phenolic and volatile compounds with yeast lees, commercial yeast derivatives and non toasted chips in model solutions and young red wines. *Eur. Food Res. Technol.* **2012**, *234*, 231–244. [CrossRef]
6. Suárez-Lepe, J.A.; Morata, A. Nuevo Método de Crianza Sobre Lías. Oficina Española de Patentes y Marcas ES2311372B2, 7 October 2009.
7. Alexandre, H.; Guilloux-Benatier, M. Yeast autolysis in sparkling wine—A review. *Aust. J. Grape Wine Res.* **2006**, *12*, 119–127. [CrossRef]
8. Vidal, S.; Francis, L.; Williams, P.; Kwiatkowski, M.; Gawel, R.; Cheynier, V.; Waters, E. The mouth-feel properties of polysaccharides and anthocyanins in a wine like medium. *Food Chem.* **2004**, *85*, 519–525. [CrossRef]
9. Guadalupe, Z.; Palacios, A.; Ayestarán, B. Maceration enzymes and mannoproteins: A possible strategy to increase colloidal stability and color extraction in red wines. *J. Agric. Food Chem.* **2007**, *55*, 4854–4862. [CrossRef]
10. Chalier, P.; Angot, B.; Delteil, D.; Doco, T.; Gunata, Z. Interactions between aroma compounds and whole mannoprotein isolated from *Saccharomyces cerevisiae* strains. *Food Chem.* **2007**, *100*, 22–30. [CrossRef]
11. Gonzalez-Ramos, D.; Cebollero, E.; Gonzalez, R. A recombinant *Saccharomyces cerevisiae* strain overproducing mannoproteins stabilizes wine against protein haze. *Appl. Environ. Microbiol.* **2008**, *74*, 5533–5540. [CrossRef]
12. Lubbers, S.; Leger, B.; Charpentier, C.; Feuillat, M. Effet colloide-protecteur d'extraits de parois de levures sur la stabilité tartrique d'une solution hydro-alcoolique modèle. *OENO One* **1993**, *27*, 13. [CrossRef]
13. Rodrigues, A.; Ricardo-Da-Silva, J.M.; Lucas, C.; Laureano, O. Effect of commercial mannoproteins on wine colour and tannins stability. *Food Chem.* **2012**, *131*, 907–914. [CrossRef]
14. Morata, A.; Loira, I.; Suárez-Lepe, J.A. Influence of yeasts in wine colour. In *Grape and Wine Biotechnology*; Morata, A., Loira, I., Eds.; IntechOpen: London, UK, 2016; pp. 285–305.
15. Vejarano, R.; Gil-Calderón, A.; Díaz-Silva, V.; León-Vargas, J. Improvement of the bioactive profile in wines and its incidence on human health: Technological strategies. In *Advances in Grape and Wine Biotechnology*; Morata, A., Loira, I., Eds.; IntechOpen: London, UK, 2019.
16. Loira, I.; Vejarano, R.; Morata, A.; Ricardo-Da-Silva, J.M.; Laureano, O.; González, M.C.; Suárez-Lepe, J.A. Effect of *Saccharomyces* strains on the quality of red wines aged on lees. *Food Chem.* **2013**, *139*, 1044–1051. [CrossRef]
17. Rosi, I.; Gheri, A.; Domizio, P.; Fia, G. Production de macromolecules parietals de *Saccharomyces cerevisiae* au cours de la fermentation et leur influence sur la fermentation malolactique. *Rev. des Œnologues* **2000**, *94*, 18–20.
18. Moruno, E.G.; Sanlorenzo, C.; Boccaccino, B.; Di Stefano, R. Treatment with yeast to reduce the concentration of ochratoxin A in red wine. *Am. J. Enol. Vitic.* **2005**, *56*, 73–76.

19. Moreno-Arribas, V.; Pueyo, E.; Nieto, F.J.; Martín-Álvarez, P.J.; Polo, M.C. Influence of the polysaccharides and the nitrogen compounds on foaming properties of sparkling wines. *Food Chem.* **2000**, *70*, 309–317. [CrossRef]

20. Gallardo-Chacón, J.J.; Vichi, S.; López-Tamames, E.; Buxaderas, S. Changes in the sorption of diverse volatiles by *Saccharomyces cerevisiae* lees during sparkling wine aging. *J. Agric. Food Chem.* **2010**, *58*, 12426–12430. [CrossRef]

21. Palomero, F.; Morata, A.; Benito, S.; Calderón, F.; Suárez-Lepe, J.A. New genera of yeasts for over-lees aging of red wine. *Food Chem.* **2009**, *112*, 432–441. [CrossRef]

22. Kopecká, M.; Fleet, G.H.; Phaff, H.J. Ultrastructure of the cell wall of *Schizosaccharomyces pombe* following treatment with various glucanases. *J. Struct. Biol.* **1995**, *114*, 140–152. [CrossRef]

23. Hidalgo, J. *Tratado de Enología, Tomo I*, 2nd ed.; Mundiprensa: Madrid, Spain, 2011; ISBN 84-8476-119-3.

24. Domizio, P.; Liu, Y.; Bisson, L.F.; Barile, D. Cell wall polysaccharides released during the alcoholic fermentation by *Schizosaccharomyces pombe* and *S. japonicus*: Quantification and characterization. *Food Microbiol.* **2017**, *61*, 136–149. [CrossRef]

25. Giovani, G.; Rosi, I.; Bertuccioli, M. Quantification and characterization of cell wall polysaccharides released by non-*Saccharomyces* yeast strains during alcoholic fermentation. *Int. J. Food Microbiol.* **2012**, *160*, 113–118. [CrossRef] [PubMed]

26. Domizio, P.; Romani, C.; Lencioni, L.; Comitini, F.; Gobbi, M.; Mannazzu, I.; Ciani, M. Outlining a future for non-*Saccharomyces* yeasts: Selection of putative spoilage wine strains to be used in association with *Saccharomyces cerevisiae* for grape juice fermentation. *Int. J. Food Microbiol.* **2011**, *147*, 170–180. [CrossRef] [PubMed]

27. Wightman, J.D.; Wrolstad, R.E. β-glucosidase activity in juice-processing enzymes based on anthocyanin analysis. *J. Food Sci.* **1996**, *61*, 544–548. [CrossRef]

28. Morata, A.; Loira, I.; Escott, C.; del Fresno, J.M.; Bañuelos, M.A.; Suárez-Lepe, J.A. Applications of *Metschnikowia pulcherrima* in wine biotechnology. *Fermentation* **2019**, *5*, 63. [CrossRef]

29. Morata, A.; Gómez-Cordovés, M.C.; Calderón, F.; Suárez, J.A. Effects of pH, temperature and SO_2 on the formation of pyranoanthocyanins during red wine fermentation with two species of *Saccharomyces*. *Int. J. Food Microbiol.* **2006**, *106*, 123–129. [CrossRef]

30. Ivit, N.N.; Loira, I.; Morata, A.; Benito, S.; Palomero, F.; Suárez-Lepe, J.A. Making natural sparkling wines with non-*Saccharomyces* yeasts. *Eur. Food Res. Technol.* **2018**, *244*, 925–935. [CrossRef]

31. Granchi, L.; Romano, P.; Mangani, S.; Guerrini, S.; Vincenzini, M. Production of biogenic amines by wine microorganisms. *Bull. OIV* **2005**, *78*, 595–609.

32. Marcobal, A.; Martín-Álvarez, P.J.; Polo, M.C.; Muñoz, R.; Moreno-Arribas, M.V. Formation of biogenic amines throughout the industrial manufacture of red wine. *J. Food Prot.* **2006**, *69*, 397–404. [CrossRef]

33. Martin, V. *Hanseniaspora Vineae: Caracterización y su Uso en la Vinificación*; Universidad de la Republica: Montevideo, Uruguay, 2016.

34. Domizio, P.; Liu, Y.; Bisson, L.F.; Barile, D. Use of non-*Saccharomyces* wine yeasts as novel sources of mannoproteins in wine. *Food Microbiol.* **2014**, *43*, 5–15. [CrossRef]

35. Lencioni, L.; Romani, C.; Gobbi, M.; Comitini, F.; Ciani, M.; Domizio, P. Controlled mixed fermentation at winery scale using *Zygotorulaspora florentina* and *Saccharomyces cerevisiae*. *Int. J. Food Microbiol.* **2016**, *234*, 36–44. [CrossRef]

36. Zha, Y.; Hossain, A.H.; Tobola, F.; Sedee, N.; Havekes, M.; Punt, P.J. *Pichia anomala* 29X: A resistant strain for lignocellulosic biomass hydrolysate fermentation. *FEMS Yeast Res.* **2013**, *13*, 609–617. [CrossRef] [PubMed]

37. Fugelsang, K.C.; Edwards, C.G. *Wine Microbiology: Practical Applications and Procedures*; Springer Science & Business Media: Berlin, Germany, 2007; ISBN 038733341X.

38. Loira, I.; Morata, A.; Palomero, F.; González, C.; Suárez-Lepe, J.A. *Schizosaccharomyces pombe*: A promising biotechnology for modulating wine composition. *Fermentation* **2018**, *4*, 70. [CrossRef]

39. Kulkarni, P.; Loira, I.; Morata, A.; Tesfaye, W.; González, M.C.; Suárez-Lepe, J.A. Use of non-*Saccharomyces* yeast strains coupled with ultrasound treatment as a novel technique to accelerate ageing on lees of red wines and its repercussion in sensorial parameters. *LWT* **2015**, *64*, 1262. [CrossRef]

40. Todd, B.E.N.; Fleet, G.H.; Henschke, P.A. Promotion of autolysis through the interaction of killer and sensitive yeasts: Potential application in sparkling wine production. *Am. J. Enol. Vitic.* **2000**, *51*, 65–72.

41. Englezos, V.; Rantsiou, K.; Giacosa, S.; Río Segade, S.; Rolle, L.; Cocolin, L. Cell-to-cell contact mechanism modulates *Starmerella bacillaris* death in mixed culture fermentations with *Saccharomyces cerevisiae*. *Int. J. Food Microbiol.* **2019**, *289*, 106–114. [CrossRef]
42. Palomero, F.; Morata, A.; Benito, S.; González, M.C.; Suárez-Lepe, J.A. Conventional and enzyme-assisted autolysis during ageing over lees in red wines: Influence on the release of polysaccharides from yeast cell walls and on wine monomeric anthocyanin content. *Food Chem.* **2007**, *105*, 838–846. [CrossRef]
43. Del Fresno, J.M.; Loira, I.; Morata, A.; González, C.; Suárez-Lepe, J.A.; Cuerda, R. Application of ultrasound to improve lees ageing processes in red wines. *Food Chem.* **2018**, *261*, 157–163. [CrossRef]
44. Del Fresno, J.M.; Morata, A.; Escott, C.; Loira, I.; Cuerda, R.; Suárez-Lepe, J.A. Sonication of yeast biomasses to improve the ageing on lees technique in red wines. *Molecules* **2019**, *24*, 635. [CrossRef]
45. Jackson, R.S. Fermentation. In *Wine Science: Principles and Applications*; Jackson, R.S., Ed.; Academic Press/Elsevier: Hoboken, NJ, USA, 2014; ISBN 9780123814692.
46. Bely, M.; Rinaldi, A.; Dubourdieu, D. Influence of assimilable nitrogen on volatile acidity production by *Saccharomyces cerevisiae* during high sugar fermentation. *J. Biosci. Bioeng.* **2003**, *96*, 507–512. [CrossRef]
47. Gobert, A.; Tourdot-Maréchal, R.; Morge, C.; Sparrow, C.; Liu, Y.; Quintanilla-Casas, B.; Vichi, S.; Alexandre, H. Non-*Saccharomyces* yeasts nitrogen source preferences: Impact on sequential fermentation and wine volatile compounds profile. *Front. Microbiol.* **2017**, *8*, 2175. [CrossRef]
48. Ribéreau-Gayon, P.; Glories, Y.; Maujean, A.; Dubourdieu, D. Nitrogen compounds. In *Handbook of Enology: The Chemistry of Wine Stabilization and Treatments*; John Wiley and Sons Ltd.: Hoboken, NJ, USA, 2006; ISBN 978-0-470-01037-2.
49. Cooper, T.G. Transmitting the signal of excess nitrogen in *Saccharomyces cerevisiae* from the Tor proteins to the GATA factors: Connecting the dots. *FEMS Microbiol. Rev.* **2002**, *26*, 223–238. [CrossRef] [PubMed]
50. De Koker, S. *Nitrogen Utilisation of Selected Non-Saccharomyces Yeasts and the Impact on Volatile Compound Production*; Stellenbosch University: Stellenbosch, South Africa, 2015.
51. Martin, V.; Giorello, F.; Fariña, L.; Minteguiaga, M.; Salzman, V.; Boido, E.; Aguilar, P.S.; Gaggero, C.; Dellacassa, E.; Mas, A.; et al. De novo synthesis of benzenoid compounds by the yeast *Hanseniaspora vineae* increases the flavor diversity of wines. *J. Agric. Food Chem.* **2016**, *64*, 4574–4583. [CrossRef]
52. Martin, V.; Boido, E.; Giorello, F.; Mas, A.; Dellacassa, E.; Carrau, F. Effect of yeast assimilable nitrogen on the synthesis of phenolic aroma compounds by *Hanseniaspora vineae* strains. *Yeast* **2016**, *33*, 323–328. [CrossRef] [PubMed]
53. Burin, V.M.; Gomes, T.M.; Caliari, V.; Rosier, J.P.; Bordignon Luiz, M.T. Establishment of influence the nitrogen content in musts and volatile profile of white wines associated to chemometric tools. *Microchem. J.* **2015**, *122*, 20–28. [CrossRef]
54. Mendes-Ferreira, A.; Barbosa, C.; Lage, P.; Mendes-Faia, A. The impact of nitrogen on yeast fermentation and wine quality. *Cienc. Tec. Vitivinic.* **2011**, *26*, 17–32.
55. Sturgeon, J.Q.; Bohlscheid, J.C.; Edwards, C.G. The effect of nitrogen source on yeast metabolism and H_2S formation. *J. Wine Res.* **2013**, *24*, 182–194. [CrossRef]
56. Ribereau-Gayon, P.; Dubourdieu, D.; Doneche, B.; Lonvaud, A. *Handbook of Enology: The Microbiology of Wine and Vinifications*, 2nd ed.; John Wiley & Sons, Ltd.: Chichester, UK, 2006; Volume 1, ISBN 978-0-470-01034-1.
57. Chi, Z.; Yan, K.; Gao, L.; Li, J.; Wang, X.; Wang, L. Diversity of marine yeasts with high protein content and evaluation of their nutritive compositions. *J. Mar. Biol. Assoc. United Kingdom* **2008**, *88*, 1347–1352. [CrossRef]
58. Manitchotpisit, P.; Skory, C.D.; Leathers, T.D.; Lotrakul, P.; Eveleigh, D.E.; Prasongsuk, S.; Punnapayak, H. α-Amylase activity during pullulan production and α-amylase gene analyses of *Aureobasidium pullulans*. *J. Ind. Microbiol. Biotechnol.* **2011**, *38*, 1211–1218. [CrossRef]
59. Leite, R.S.R.; Bocchini, D.A.; Martins, E.S.; Silva, D.; Gomes, E.; Da Silva, R. Production of cellulolytic and hemicellulolytic enzymes from *Aureobasidium pullulans* on solid state fermentation. *Appl. Biochem. Biotechnol.* **2007**, *137–140*, 281–288.
60. Leathers, T.D.; Rich, J.O.; Anderson, A.M.; Manitchotpisit, P. Lipase production by diverse phylogenetic clades of *Aureobasidium pullulans*. *Biotechnol. Lett.* **2013**, *35*, 1701–1706. [CrossRef]

61. Manitchotpisit, P.; Leathers, T.D.; Peterson, S.W.; Kurtzman, C.; Li, X.L.; Eveleigh, D.E.; Lotrakul, P.; Prasongsuk, S.; Dunlap, C.A.; Vermillion, K.E.; et al. Multilocus phylogenetic analyses, pullulan production and xylanase activity of tropical isolates of *Aureobasidium pullulans*. *Mycol. Res.* **2009**, *113*, 1107–1120. [CrossRef]

62. Rich, J.O.; Leathers, T.D.; Anderson, A.M.; Bischoff, K.M.; Manitchotpisit, P. Laccases from *Aureobasidium pullulans*. *Enzyme Microb. Technol.* **2013**, *53*, 33–37. [CrossRef] [PubMed]

63. Chi, Z.; Wang, F.; Chi, Z.; Yue, L.; Liu, G.; Zhang, T. Bioproducts from *Aureobasidium pullulans*, a biotechnologically important yeast. *Appl. Microbiol. Biotechnol.* **2009**, *82*, 793–804. [CrossRef] [PubMed]

64. Ni, X.; Chi, Z.; Ma, C.; Madzak, C. Cloning, characterization, and expression of the gene encoding alkaline protease in the marine yeast *Aureobasidium pullulans* 10. *Mar. Biotechnol.* **2008**, *10*, 319–327. [CrossRef] [PubMed]

65. Bozoudi, D.; Tsaltas, D. The multiple and versatile roles of *Aureobasidium pullulans* in the vitivinicultural sector. *Fermentation* **2018**, *4*, 85. [CrossRef]

66. Huyben, D.; Nyman, A.; Vidaković, A.; Passoth, V.; Moccia, R.; Kiessling, A.; Dicksved, J.; Lundh, T. Effects of dietary inclusion of the yeasts *Saccharomyces cerevisiae* and *Wickerhamomyces anomalus* on gut microbiota of rainbow trout. *Aquaculture* **2017**, *473*, 528–537. [CrossRef]

67. Medina, K.; Boido, E.; Dellacassa, E.; Carrau, F. Growth of non-*Saccharomyces* yeasts affects nutrient availability for *Saccharomyces cerevisiae* during wine fermentation. *Int. J. Food Microbiol.* **2012**, *157*, 245–250. [CrossRef]

68. Bataillon, M.; Rico, A.; Sablayrolles, J.M.; Salmon, J.M.; Barre, P. Early thiamin assimilation by yeasts under enological conditions: Impact on alcoholic fermentation kinetics. *J. Ferment. Bioeng.* **1996**, *82*, 145–150. [CrossRef]

69. Wang, X.D.; Bohlscheid, J.C.; Edwards, C.G. Fermentative activity and production of volatile compounds by *Saccharomyces* grown in synthetic grape juice media deficient in assimilable nitrogen and/or pantothenic acid. *J. Appl. Microbiol.* **2003**, *94*, 349–359. [CrossRef]

70. Romano, P.; Capece, A.; Jespersen, L. Taxonomic and ecological diversity of food and beverage yeasts. In *Yeasts in Food and Beverages*; Springer: Berlin/Heidelberg, Germany, 2006; pp. 13–53.

71. Barbosa, C.; Lage, P.; Esteves, M.; Chambel, L.; Mendes-Faia, A.; Mendes-Ferreira, A. Molecular and phenotypic characterization of *Metschnikowia pulcherrima* strains from Douro Wine Region. *Fermentation* **2018**, *4*, 8. [CrossRef]

72. Vincenzi, S.; Crapisi, A.; Curioni, A. Foamability of Prosecco wine: Cooperative effects of high molecular weight glycocompounds and wine PR-proteins. *Food Hydrocoll.* **2014**, *34*, 202–207. [CrossRef]

73. Martínez-Rodríguez, A.J.; Carrascosa, A.V.; Martín-Álvarez, P.J.; Moreno-Arribas, V.; Polo, M.C. Influence of the yeast strain on the changes of the amino acids, peptides and proteins during sparkling wine production by the traditional method. *J. Ind. Microbiol. Biotechnol.* **2002**, *29*, 314–322. [CrossRef]

74. Lemos Junior, W.J.F.; Nadai, C.; Rolle, L.; da Silva Gulao, E.; Miguez da Rocha Leão, M.H.; Giacomini, A.; Corich, V.; Vincenzi, S. Influence of the mannoproteins of different strains of *Starmenella bacillaris* used in single and sequential fermentations on foamability, tartaric and protein stabilities of wines. *OENO One* **2020**, *54*, 231–243. [CrossRef]

75. Del Fresno, J.M.; Morata, A.; Loira, I.; Bañuelos, M.A.; Escott, C.; Benito, S.; González Chamorro, C.; Suárez-Lepe, J.A. Use of non-*Saccharomyces* in single-culture, mixed and sequential fermentation to improve red wine quality. *Eur. Food Res. Technol.* **2017**, *243*, 2175–2185. [CrossRef]

76. Comitini, F.; Gobbi, M.; Domizio, P.; Romani, C.; Lencioni, L.; Mannazzu, I.; Ciani, M. Selected non-*Saccharomyces* wine yeasts in controlled multistarter fermentations with *Saccharomyces cerevisiae*. *Food Microbiol.* **2011**, *28*, 873–882. [CrossRef]

77. González-Royo, E.; Pascual, O.; Kontoudakis, N.; Esteruelas, M.; Esteve-Zarzoso, B.; Mas, A.; Canals, J.M.; Zamora, F. Oenological consequences of sequential inoculation with non-*Saccharomyces* yeasts (*Torulaspora delbrueckii* or *Metschnikowia pulcherrima*) and *Saccharomyces cerevisiae* in base wine for sparkling wine production. *Eur. Food Res. Technol.* **2015**, *240*, 999–1012. [CrossRef]

78. Medina-Trujillo, L.; González-Royo, E.; Sieczkowski, N.; Heras, J.; Canals, J.M.; Zamora, F. Effect of sequential inoculation (*Torulaspora delbrueckii/Saccharomyces cerevisiae*) in the first fermentation on the foaming properties of sparkling wine. *Eur. Food Res. Technol.* **2017**, *243*, 681–688. [CrossRef]

79. Perpetuini, G.; Di Gianvito, P.; Arfelli, G.; Schirone, M.; Corsetti, A.; Tofalo, R.; Suzzi, G. Biodiversity of autolytic ability in flocculent *Saccharomyces cerevisiae* strains suitable for traditional sparkling wine fermentation. *Yeast* **2016**, *33*, 303–312. [CrossRef]

80. Loira, I.; Morata, A.; Escott, C.; Del Fresno, J.M.; Tesfaye, W.; Palomero, F.; Suárez-Lepe, J.A. Applications of nanotechnology in the winemaking process. *Eur. Food Res. Technol.* **2020**, *246*, 1533–1541. [CrossRef]

81. Anbu, P.; Gopinath, S.C.B.; Chaulagain, B.P.; Lakshmipriya, T. Microbial enzymes and their applications in industries and medicine 2016. *Biomed Res. Int.* **2017**, *2017*, 2195808. [CrossRef]

82. García, M.; Esteve-Zarzoso, B.; Cabellos, J.M.; Arroyo, T. Advances in the study of *Candida stellata*. *Fermentation* **2018**, *4*, 74. [CrossRef]

83. Escribano, R.; González-Arenzana, L.; Garijo, P.; Berlanas, C.; López-Alfaro, I.; López, R.; Gutiérrez, A.R.; Santamaría, P. Screening of enzymatic activities within different enological non-*Saccharomyces* yeasts. *J. Food Sci. Technol.* **2017**, *54*, 1555–1564. [CrossRef] [PubMed]

84. Zott, K.; Thibon, C.; Bely, M.; Lonvaud-Funel, A.; Dubourdieu, D.; Masneuf-Pomarede, I. The grape must non-*Saccharomyces* microbial community: Impact on volatile thiol release. *Int. J. Food Microbiol.* **2011**, *151*, 210–215. [CrossRef] [PubMed]

85. Cordero-Bueso, G.; Esteve-Zarzoso, B.; Cabellos, J.M.; Gil-Díaz, M.; Arroyo, T. Biotechnological potential of non-*Saccharomyces* yeasts isolated during spontaneous fermentations of Malvar (*Vitis vinifera* cv. L.). *Eur. Food Res. Technol.* **2013**, *236*, 193–207. [CrossRef]

86. Swangkeaw, J.; Vichitphan, S.; Butzke, C.E.; Vichitphan, K. Characterization of β-glucosidases from *Hanseniaspora* sp. and *Pichia anomala* with potentially aroma-enhancing capabilities in juice and wine. *World J. Microbiol. Biotechnol.* **2011**, *27*, 423–430. [CrossRef]

87. Fernández, M.; Úbeda, J.F.; Briones, A.I. Typing of non-*Saccharomyces* yeasts with enzymatic activities of interest in wine-making. *Int. J. Food Microbiol.* **2000**, *59*, 29–36. [CrossRef]

88. Hock, R.; Benda, I.; Schreier, P. Formation of terpenes by yeasts during alcoholic fermentation. *Z. Lebensm. Unters. Forsch.* **1984**, *179*, 450–452. [CrossRef]

89. Andorrà, I.; Berradre, M.; Rozès, N.; Mas, A.; Guillamón, J.M.; Esteve-Zarzoso, B. Effect of pure and mixed cultures of the main wine yeast species on grape must fermentations. *Eur. Food Res. Technol.* **2010**, *231*, 215–224. [CrossRef]

90. Hu, K.; Jin, G.J.; Xu, Y.H.; Tao, Y.S. Wine aroma response to different participation of selected *Hanseniaspora uvarum* in mixed fermentation with *Saccharomyces cerevisiae*. *Food Res. Int.* **2018**, *108*, 119–127. [CrossRef] [PubMed]

91. Bovo, B.; Carlot, M.; Lombardi, A.; Lomolino, G.; Lante, A.; Giacomini, A.; Corich, V. Exploring the use of *Saccharomyces cerevisiae* commercial strain and *Saccharomycodes ludwigii* natural isolate for grape marc fermentation to improve sensory properties of spirits. *Food Microbiol.* **2014**, *41*, 33–41. [CrossRef] [PubMed]

92. Merín, M.G.; Morata de Ambrosini, V.I. Highly cold-active pectinases under wine-like conditions from non-*Saccharomyces* yeasts for enzymatic production during winemaking. *Lett. Appl. Microbiol.* **2015**, *60*, 467–474. [CrossRef] [PubMed]

93. Onetto, C.A.; Borneman, A.R.; Schmidt, S.A. Investigating the effects of *Aureobasidium pullulans* on grape juice composition and fermentation. *Food Microbiol.* **2020**, *90*, 103451. [CrossRef]

94. Schlander, M.; Distler, U.; Tenzer, S.; Thines, E.; Claus, H. Purification and properties of yeast proteases secreted by *Wickerhamomyces anomalus* 227 and *Metschnikovia pulcherrima* 446 during growth in a white grape juice. *Fermentation* **2017**, *3*, 2. [CrossRef]

95. Theron, L.W.; Bely, M.; Divol, B. Monitoring the impact of an aspartic protease (MpAPr1) on grape proteins and wine properties. *Appl. Microbiol. Biotechnol.* **2018**, *102*, 5173–5183. [CrossRef] [PubMed]

96. Reid, V.J.; Theron, L.W.; Toit, M.D.; Divol, B. Identification and partial characterization of extracellular aspartic protease genes from *Metschnikowia pulcherrima* IWBT Y1123 and *Candida apicola* IWBT Y1384. *Appl. Environ. Microbiol.* **2012**, *78*, 6838–6849. [CrossRef]

97. Strauss, M.L.A.; Jolly, N.P.; Lambrechts, M.G.; Van Rensburg, P. Screening for the production of extracellular hydrolytic enzymes by non-*Saccharomyces* wine yeasts. *J. Appl. Microbiol.* **2001**, *91*, 182–190. [CrossRef]

98. Schwentke, J.; Sabel, A.; Petri, A.; König, H.; Claus, H. The yeast *Wickerhamomyces anomalus* AS1 secretes a multifunctional exo-β-1,3-glucanase with implications for winemaking. *Yeast* **2014**, *31*, 349–359. [CrossRef]

99. Claus, H.; Mojsov, K. Enzymes for wine fermentation: Current and perspective applications. *Fermentation* **2018**, *4*, 52. [CrossRef]

100. Izgü, F.; Altinbay, D.; Acun, T. Killer toxin of *Pichia anomala* NCYC 432; purification, characterization and its exo-β-1,3-glucanase activity. *Enzyme Microb. Technol.* **2006**, *39*, 669–676. [CrossRef]

101. Morata, A.; Vejarano, R.; Ridolfi, G.; Benito, S.; Palomero, F.; Uthurry, C.; Tesfaye, W.; González, C.; Suárez-Lepe, J.A. Reduction of 4-ethylphenol production in red wines using HCDC+ yeasts and cinnamyl esterases. *Enzyme Microb. Technol.* **2013**, *52*, 99–104. [CrossRef]

102. Medina, K.; Boido, E.; Dellacassa, E.; Carrau, F. Effects of non-*Saccharomyces* yeasts on color, anthocyanin, and anthocyanin-derived pigments of Tannat grapes during fermentation. *Am. J. Enol. Vitic.* **2018**, *69*, 148–156. [CrossRef]

103. Benito, S.; Morata, A.; Palomero, F.; González, M.C.; Suárez-Lepe, J.A. Formation of vinylphenolic pyranoanthocyanins by *Saccharomyces cerevisiae* and *Pichia guillermondii* in red wines produced following different fermentation strategies. *Food Chem.* **2011**, *124*, 15–23. [CrossRef]

104. Lubbers, M.W.; Rodriguez, S.B.; Honey, N.K.; Thornton, R.J. Purification and characterization of urease from *Schizosaccharomyces pombe*. *Can. J. Microbiol.* **1996**, *42*, 132–140. [CrossRef] [PubMed]

105. Uthurry, C.A.; Suárez Lepe, J.A.; Lombardero, J.; García Del Hierro, J.R. Ethyl carbamate production by selected yeasts and lactic acid bacteria in red wine. *Food Chem.* **2006**, *94*, 262–270. [CrossRef]

106. De Felice, D.V.; Solfrizzo, M.; De Curtis, F.; Lima, G.; Visconti, A.; Castoria, R. Strains of *Aureobasidium pullulans* can lower ochratoxin a contamination in wine grapes. *Phytopathology* **2008**, *98*, 1261–1270. [CrossRef]

107. Zhang, H.; Apaliya, M.T.; Mahunu, G.K.; Chen, L.; Li, W. Control of ochratoxin A-producing fungi in grape berry by microbial antagonists: A review. *Trends Food Sci. Technol.* **2016**, *51*, 88–97. [CrossRef]

108. Belda, I.; Conchillo, L.B.; Ruiz, J.; Navascués, E.; Marquina, D.; Santos, A. Selection and use of pectinolytic yeasts for improving clarification and phenolic extraction in winemaking. *Int. J. Food Microbiol.* **2016**, *223*, 1–8. [CrossRef]

109. Belda, I.; Ruiz, J.; Navascués, E.; Marquina, D.; Santos, A. Improvement of aromatic thiol release through the selection of yeasts with increased β-lyase activity. *Int. J. Food Microbiol.* **2016**, *225*, 1–8. [CrossRef]

110. Belda, I.; Ruiz, J.; Alastruey-Izquierdo, A.; Navascués, E.; Marquina, D.; Santos, A. Unraveling the enzymatic basis of wine "Flavorome": A phylo-functional study of wine related yeast species. *Front. Microbiol.* **2016**, *7*, 12. [CrossRef]

111. Lin, M.M.H.; Boss, P.K.; Walker, M.E.; Sumby, K.M.; Grbin, P.R.; Jiranek, V. Evaluation of indigenous non-*Saccharomyces* yeasts isolated from a South Australian vineyard for their potential as wine starter cultures. *Int. J. Food Microbiol.* **2020**, *312*, 108373. [CrossRef]

112. Manzanares, P.; Rojas, V.; Genovés, S.; Vallés, S. A preliminary search for anthocyanin-β-D-glucosidase activity in non-*Saccharomyces* wine yeasts. *Int. J. Food Sci. Technol.* **2000**, *35*, 95–103. [CrossRef]

113. Zhang, L.J.; Gallo, R.L. Antimicrobial peptides. *Curr. Biol.* **2016**, *26*, R14–R19. [CrossRef] [PubMed]

114. Branco, P.; Viana, T.; Albergaria, H.; Arneborg, N. Antimicrobial peptides (AMPs) produced by *Saccharomyces cerevisiae* induce alterations in the intracellular pH, membrane permeability and culturability of *Hanseniaspora guilliermondii* cells. *Int. J. Food Microbiol.* **2015**, *205*, 112–118. [CrossRef] [PubMed]

115. Peña, R.; Ganga, M.A. Novel antimicrobial peptides produced by *Candida intermedia* LAMAP1790 active against the wine-spoilage yeast *Brettanomyces bruxellensis*. *Antonie van Leeuwenhoek* **2019**, *112*, 297–304.

116. Albergaria, H.; Francisco, D.; Gori, K.; Arneborg, N.; Gírio, F. *Saccharomyces cerevisiae* CCMI 885 secretes peptides that inhibit the growth of some non-*Saccharomyces* wine-related strains. *Appl. Microbiol. Biotechnol.* **2010**, *86*, 965–972. [CrossRef]

117. Peña, R.; Chávez, R.; Rodríguez, A.; Ganga, M.A. A control alternative for the hidden enemy in the wine cellar. *Fermentation* **2019**, *5*, 25. [CrossRef]

118. Schmitt, M.J.; Breinig, F. The viral killer system in yeast: From molecular biology to application. *FEMS Microbiol. Rev.* **2002**, *26*, 257–276. [CrossRef]

119. Mehlomakulu, N.N.; Setati, M.E.; Divol, B. Characterization of novel killer toxins secreted by wine-related non-*Saccharomyces* yeasts and their action on *Brettanomyces* spp. *Int. J. Food Microbiol.* **2014**, *188*, 83–91. [CrossRef]

120. Villalba, M.L.; Susana Sáez, J.; del Monaco, S.; Lopes, C.A.; Sangorrín, M.P. TdKT, a new killer toxin produced by *Torulaspora delbrueckii* effective against wine spoilage yeasts. *Int. J. Food Microbiol.* **2016**, *217*, 94–100. [CrossRef]

121. Comitini, F.; Ingeniis De, J.; Pepe, L.; Mannazzu, I.; Ciani, M. *Pichia anomala* and *Kluyveromyces wickerhamii* killer toxins as new tools against *Dekkera/Brettanomyces* spoilage yeasts. *FEMS Microbiol. Lett.* **2004**, *238*, 235–240. [CrossRef]

122. Santos, A.; San Mauro, M.; Bravo, E.; Marquina, D. PMKT2, a new killer toxin from *Pichia membranifaciens*, and its promising biotechnological properties for control of the spoilage yeast *Brettanomyces bruxellensis*. *Microbiology* **2009**, *155*, 624–634. [CrossRef] [PubMed]

123. Oro, L.; Ciani, M.; Bizzaro, D.; Comitini, F. Evaluation of damage induced by Kwkt and Pikt zymocins against *Brettanomyces/Dekkera* spoilage yeast, as compared to sulphur dioxide. *J. Appl. Microbiol.* **2016**, *121*, 207–214. [CrossRef] [PubMed]

124. De Ingeniis, J.; Raffaelli, N.; Ciani, M.; Mannazzu, I. *Pichia anomala* DBVPG 3003 secretes a ubiquitin-like protein that has antimicrobial activity. *Appl. Environ. Microbiol.* **2009**, *75*, 1129–1134. [CrossRef] [PubMed]

125. Fernández de Ullivarri, M.; Mendoza, L.M.; Raya, R.R. Characterization of the killer toxin KTCf20 from *Wickerhamomyces anomalus*, a potential biocontrol agent against wine spoilage yeasts. *Biol. Control* **2018**, *121*, 223–228. [CrossRef]

126. Lopes, C.A.; Sáez, J.S.; Sangorrín, M.P. Differential response of *Pichia guilliermondii* spoilage isolates to biological and physicochemical factors prevailing in Patagonian wine fermentations. *Can. J. Microbiol.* **2009**, *55*, 801–809. [CrossRef]

127. Csutak, O.; Vassu, T.; Corbu, V.; Cîrpici, I.; Ionescu, R. Killer activity of *Pichia anomala* CMGB 88. *Biointerface Res. Appl. Chem.* **2017**, *7*, 2085–2089.

128. Villalba, M.L.; Mazzucco, M.B.; Lopes, C.A.; Ganga, M.A.; Sangorrín, M.P. Purification and characterization of *Saccharomyces eubayanus* killer toxin: Biocontrol effectiveness against wine spoilage yeasts. *Int. J. Food Microbiol.* **2020**, *331*, 108714. [CrossRef]

129. Türkel, S.; Ener, B. Isolation and characterization of new *Metschnikowia pulcherrima* strains as producers of the antimicrobial pigment pulcherrimin. *Z. Naturforsch. C* **2009**, *64*, 405–410.

130. Oro, L.; Ciani, M.; Comitini, F. Antimicrobial activity of *Metschnikowia pulcherrima* on wine yeasts. *J. Appl. Microbiol.* **2014**, *116*, 1209–1217. [CrossRef]

131. Charpentier, C.; Escot, S.; Gonzalez, E.; Dulau, L.; Feuillat, M. The influence of yeast glycosylated proteins on tannins aggregation in model solution. *J. Int. Sci. Vigne Vin* **2004**, *38*, 209–218. [CrossRef]

132. Buerth, C.; Heilmann, C.J.; Klis, F.M.; de Koster, C.G.; Ernst, J.F.; Tielker, D. Growth-dependent secretome of *Candida utilis*. *Microbiology* **2011**, *157*, 2493–2503. [CrossRef] [PubMed]

MDPI
St. Alban-Anlage 66
4052 Basel
Switzerland
Tel. +41 61 683 77 34
Fax +41 61 302 89 18
www.mdpi.com

Fermentation Editorial Office
E-mail: fermentation@mdpi.com
www.mdpi.com/journal/fermentation